IRC-SET 2018

Huaqun Guo · Hongliang Ren ·
Aishwarya Bandla
Editors

IRC-SET 2018

Proceedings of the 4th IRC Conference
on Science, Engineering and Technology

 Springer

Editors
Huaqun Guo
Institute for Infocomm Research (I^2R)
Agency for Science, Technology
and Research (A*STAR)
Singapore, Singapore

Hongliang Ren
National University of Singapore
Singapore, Singapore

Aishwarya Bandla
Singapore Institute for Neurotechnology
National University of Singapore
Singapore, Singapore

ISBN 978-981-32-9827-9   ISBN 978-981-32-9828-6   (eBook)
https://doi.org/10.1007/978-981-32-9828-6

This Springer imprint is published by the registered company Springer Nature Singapore Pte Ltd.
The registered company address is: 152 Beach Road, #21-01/04 Gateway East, Singapore 189721, Singapore

Organizing Committee

General Chair

Dr. Huaqun Guo, Institute for Infocomm Research (I^2R), A*STAR, Singapore

General Co-Chair

Dr. Bhojan Anand, National University of Singapore (NUS), Singapore

Technical Program Chair

Prof. Chau Yuen, Singapore University of Technology and Design (SUTD), Singapore

Technical Program Co-Chair

Dr. Jizhong Luo, Institute of Chemical and Engineering Sciences, A*STAR, Singapore
Dr. Umayal Lakshmanan, ETC, A*STAR, Singapore
Prof. Bharadwaj Veeravalli, NUS, Singapore
Prof. Hwee-Pink Tan, Singapore Management University (SMU), Singapore
Dr. Victor Peng Cheng Wang, Singapore Institute of Technology (SIT), Singapore
Dr. Aishwarya Bandla, NUS, Singapore
Dr. Aldy Gunawan, SMU, Singapore
Dr. Boon Seng Chew, Singapore Polytechnic, Singapore

Dr. Mandar Godge, Temasek Polytechnic, Singapore
Dr. Guang Chen, Singapore General Hospital, Singapore
Prof. G. Roshan Deen, National Institute of Education, Singapore

Finance Chair

Dr. Yongqing Zhu, Data Storage Institute, A*STAR, Singapore
Mr. Dong Li, I^2R, A*STAR, Singapore

Publicity Chair

Dr. Aishwarya Bandla, NUS, Singapore
Ms. Yee Lin Tan, National Junior College, Singapore

Local Arrangements Chair

Dr. Rong Li, Experimental Therapeutics Centre (ETC), A*STAR, Singapore
Mr. Ivan Yu Hao Tan, Dunman High School, Singapore

Sponsorship Chair

Dr. Victor Peng Cheng Wang, SIT, Singapore
Mr. Andrew Qi Jun Lim, Hwa Chong Institution (HCI), Singapore

Publication Chair

Dr. Hongliang Ren, NUS, Singapore
Mr. Bo Xuan Tang, HCI, Singapore

International Advisory Panel Chair

Prof. Lawrence Wai Choong Wong, NUS, Singapore

International Advisory Panel

Prof. Janina Mazierska, James Cook University, Australia
Prof. Yong-Jin Park, Universiti Malaysia Sabah, Malaysia
Prof. Kristin L. Wood, SUTD, Singapore
Prof. Wee Sun Lee, NUS, Singapore
Prof. Gary Tan, NUS, Singapore
Prof. Chip Hong Chang, NTU, Singapore
Prof. Maode Ma, NTU, Singapore
Prof. Lun-De Liao, National Health Research Institute, Taiwan

TPC Members

Mr Krishnamoorthy Baskaran, Energy research Institute (ERIAN), NTU
Dr. Sher-Yi Chiam, NUS High School of Mathematics and Science
Dr. Chris Choy, StarHub
Dr. Keyu Gu, Temasek Life Sciences Laboratory, NUS
Dr. Claus Muschallik, Coffee Electronics Pte. Ltd.
Mr Bhaskaran David Prakash, ETPL, A*STAR
Dr. Muthu Sebastian, Environment Water Technology Center of Innovation
Dr. Guoxian Tan, Raffles Institution
Dr. Junxia Wang, Bioinformatics Institute
Ms Wai Ee Wong, WD Media (S) Pte Ltd
Mr Dajun Wu, Institute for Infocomm Research
Dr. Jien Wu, Institute of Molecular and Cell Biology
Dr. Dexin Xiong, NauticAWT Limited
Dr. Quanqing Xu, Data Storage Institute
Dr. Guisheng Zeng, Institute of Molecular and Cell Biology
Dr. Yi Zhou, Singapore Institute of Technology

Preface

International Researchers Club (IRC) (www.irc.org.sg) was set up in 2001 with the endorsement of Agency for Science, Technology and Research (A*STAR). The vision of IRC is to create a vibrant and innovative research community for Singapore, through the contributions of technical specialities and occupational experiences from its members, and fostering strong networking and social interactions of expatriates and new citizens with the local community.

With the vision of IRC, it is our great pleasure to organize an IRC conference on Science, Engineering and Technology (IRC-SET, www.ircset.org) for the younger talents and researchers. IRC-SET 2015 is the inaugural conference of IRC, and IRC-SET 2018 is the fourth conference. IRC-SET conference aims to provide a platform for young researchers to share fresh results, obtain comments, and exchange innovative ideas of the leading edge research in the multi-discipline areas. The students from universities, junior colleges, polytechnics and top secondary schools are invited to participate in this conference to showcase and present their research projects, results and findings. Unlike other academic conferences, this conference focuses specifically on Education and Youth development and has officially been given technically sponsorship from five Singapore universities, namely National University of Singapore (NUS), Nanyang Technological University (NTU), Singapore University of Technology and Design (SUTD), Singapore Management University (SMU) and Singapore Institute of Technology (SIT). IRC-SET 2018 conference is also supported by Science Centre Singapore, IT Wonders Web, IEEE Education Society Singapore Chapter, IEEE Intelligent Transportation Systems Society (ITSS) Singapore Chapter, IEEE Broadcast Technology Society (BTS) Singapore Chapter and IEEE Singapore Section Women in Engineering (WIE) Affinity Group.

The program of IRC-SET 2018 advocates the importance of innovative technology backed by the strong foundation of science and engineering education. By exposing students from universities, junior colleges, polytechnics and top secondary schools to the key technology enablers will encourage more interest into the fields of science, engineering and technology. To select students, the IRC-SET 2018 Call for Papers is broadcast to all universities, junior colleges, polytechnics and top secondary schools according to our plan. The students then submit their technical papers to the conference online system. For the criteria that papers have to meet, firstly, the submitted papers should follow the standard template. Secondly, the conference technical programme committee has allocated each paper to several reviewers from IRC researchers, professors, lecturers and teachers to review the paper and give the comments and recommendations based on novelty of the work, scientific, engineering and technology relevance, technical treatment plausible, and clarity in writing, tables, graphs and illustrations. Based on the review results, the technical programme committee has selected number of papers to present in the IRC-SET 2018 conference and publish in this proceeding.

IRC-SET 2018 Conference consists of the Opening Speech by Professor Tit Meng LIM (Chief Executive, Science Centre Singapore), Introduction of International Researchers Club by Dr. Huaqun GUO (President, IRC), a poster session and nine presentation sessions. The nine presentation sessions include Physics, Chemical Engineering I, Life Sciences, ITSS Session—Machine learning and IoT (Internet-of-Things), Mechanical Engineering, Education Session—Biomedical Engineering, Material Sciences, WIE session—Chemical Engineering II, and BTS Session—Electrical Engineering. In the closing ceremony, Prof. Lawrence Wai Choong WONG (International Advisory Panel Chair), Dr. Huaqun GUO (General Chair) and Dr. Bhojan Anand (General Co-Chair) present the best paper awards, best poster awards and best presenter award to the winners, and publication certificates to the authors.

Finally, this proceeding is dedicated to Agency for Science, Technology and Research (A*STAR), Singapore.

Singapore   Huaqun Guo
June 2019   Hongliang Ren
                                                                                   Aishwarya Bandla

Acknowledgements

We would like to extend our utmost gratitude to the people who have, in one way or other, inspired, aided and contributed to the successful completion of this book.

First of all, we would like to express our sincere gratitude to A*STAR and A*STAR Graduate Academy (A*GA) for the support toward IRC-SET conference.

We would also like to thank all members of conference organizing committee to contribute their time and professional knowledge to make the conference a big success.

Special thanks to all reviewers for their expertise, time, effort and timely response throughout the peer evaluation process.

We would also like to take this opportunity to thank Singapore Institute of Technology to sponsor the gifts for the conference participants and Science Centre Singapore to provide the free admission tickets to the conference participants to visit the Science Centre as a post conference event.

Last but not least, the heartiest gratitude is given to IRC members for their unity.

Contents

Prognostic Biomarkers for Hepatocellular Carcinoma

Koh Rui Qi, Pek Mi Xue Michelle and Tan Min-Han

1 Background and Purpose of Research

Hepatocellular carcinoma (HCC) affects approximately half a million patients world-wide and is the most rapidly increasing cause of cancer death in the United States owing to the lack of effective treatment options for advanced disease [1]. Numerous lines of clinical and histopathologic evidence suggest that HCC is a heterogeneous disease, but a coherent molecular explanation for this heterogeneity has yet to be reported [2]. Due to the phenotypic and molecular diversity of HCC, it is a challenge to determine a patient's prognosis [3]. It would be ideal to increase monitoring of patients with poor prognosis. Thus the inability to accurately predict prognosis leads to excessive or insufficient time spent following patients, resulting in unnecessary anxiety and cost for patients, and inefficient allocation of resources for hospitals.

In clinical settings, prognostic assessment and decision of surgical treatment are based on one of the tumour staging systems (i.e. Barcelona Clinic Liver Cancer [BCLC], cancer of the liver Italian program, Japan Integrated Staging, and TNM) [4, 5] These different staging systems are based mainly on the tumor size, number of nodules, and severity of the liver disease [5]. Some authors have proposed to improve the staging system by introducing tumor biomarkers, such as the level of α-fetoprotein in serum and pathological features, like microvascular invasion and tumour differentiation [4, 6]. To refine prognosis scoring, the search of molecular biomarkers is an expanding field [7, 8]. More than 18 different molecular signatures have been published but few have been externally validated [7–11]. One of these validated molecular prognostic classifications was the G3 signature, which has been shown to be associated with tumour recurrence in both fresh-frozen and paraffin-fixed HCC [12, 13]. Interestingly, the G3 subgroup of HCC also showed the strongest association with tumor recurrence among 18 different molecular signatures [13].

K. R. Qi · P. M. X. Michelle · T. Min-Han (✉)
Institute of Bioengineering and Nanotechnology, 31 Biopolis Way, the Nanos #07-01, Singapore 138669, Singapore
e-mail: mhtan@ibn.a-star.edu.sg

© Springer Nature Singapore Pte Ltd. 2019
H. Guo et al. (eds.), *IRC-SET 2018*,
https://doi.org/10.1007/978-981-32-9828-6_1

Concerning the cancer field effect in cirrhosis, a 186-gene signature derived from non-tumour liver sample was also able to predict late recurrence and survival by capturing biological signals of aggressive phenotype from the underlying cirrhosis [7, 14].

A technical challenge facing the use of gene-expression profiling to predict the outcome of hepatocellular carcinoma has been the lack of suitable specimens from patients. Current methods of genome wide expression profiling require frozen tissue for analysis, whereas tissue banks with clinical outcome data generally have formalin-fixed, paraffin-embedded (FFPE) specimens. Even today, the vast majority of specimens are formalin-fixed; the collection of frozen tissues has yet to become routine clinical practice [7].

Therefore, a simple, easy to use test remains to be identified and endorsed in HCC clinical guidelines. We aimed to identify a molecular signature able to accurately predict prognosis of patients with HCC using FFPE samples, to enhance clinical decision making. Our study comprised of two parts: (1) identification of a 9 key gene markers in a training set of patients; (2) validation of our gene markers in an independent cohort.

2 Hypothesis

Potential HCC prognostic biomarkers can be validated through RNA extraction and cDNA conversion from FFPE samples to develop a multi-gene qPCR assay.

3 Materials and Methods

3.1 Study Population

The retrospective study was conducted with a cohort of 82 first-time HCC patients treated at Singapore General Hospital (SGH) between 2011 and 2012. All patients had histologically confirmed HCCs for which FFPE primary tumor blocks were available. In the initial stage of study, prognostic genes were identified by my mentor based on prior microarray studies on frozen tissue samples of 23 patients (Fig. 1). The coefficients of variance (standard deviation/mean) of gene expression were calculated from microarray data. Three genes (PSMB2, RPS18, MRPL30) whose expression was the least variable were identified to serve as normalization genes for qPCR. Primers for the 9 potential prognostic transcripts selected for assay development and the 3 normalization genes were then provided by my mentor and used as received. In the second stage of study, an independent set of 82 FFPE samples were used for

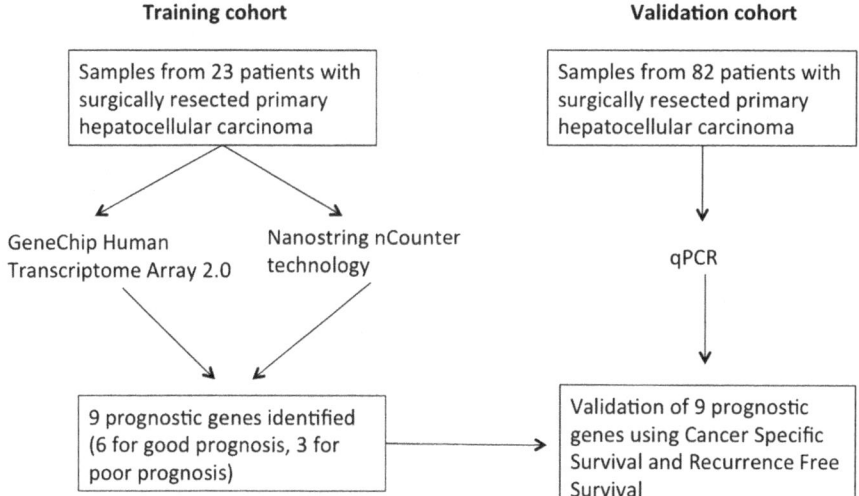

Fig. 1 Study design. In the training cohort, 2 microarray studies were used to select 9 potential prognostic genes. Filtering was done to keep the exons that displayed similar correlation to survival in the 2 assays. These 9 genes were then validated for their accuracy in predicting survival

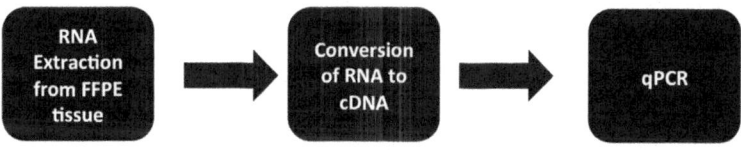

Fig. 2 Study design for validation cohort

validating the performance of these prognostic genes and developing a multi-gene qPCR assay (Fig. 2). Clinical and pathological data were obtained from ongoing chart review of medical records and electronic databases.

3.2 RNA Extraction from FFPE Tissue

FFPE blocks were sectioned in 5-μm sections and stained with hematoxylin-eosin for confirmation of histological diagnosis and tumour tissue content. For each sample, 1–3 FFPE sections were deparaffinated and microdissected with a sterile single-use scalpel to obtain tumour-specific parts. RNA was then extracted using RNeasy FFPE Kit (Qiagen, Hilden, Germany).

3.3   Design of qPCR Assay for FFPE Tissue RNA

1 μg of extracted RNA was reverse transcribed with random hexamer primers using High Capacity cDNA Reverse Transcription Kit (Life Technologies). Relative expression of each target gene was measured by real-time qPCR with Power SYBR Green Master Mix (Life Technologies) on a CFX96 machine (Bio-Rad Laboratories, Hercules, CA, USA). 12.5 ng of the four-fold diluted cDNA was used as template in a 10 μl reaction with primers at a final concentration of 200 nM. PCR amplicons were checked for specificity of amplification with melt curve. Negative controls were run for each plate.

3.4   Processing of qPCR Expression Data

We designed qPCR assays for a set of 12 genes (3 reference, 9 prognostic genes) identified from prior microarray studies. qPCR expression data collected as cycle threshold (Ct) expression was normalized by subtracting Ct values from the geometric average of Ct values for three normalization genes. The delta Ct value was then converted to a linear scale by the function $2^{-\text{delta Ct value}}$ to obtain the gene profile of the 9 genes: PGK1, CAD, ATF5, APOC1, IL32, HULC, CXCL16, CTSS and ALAS1.

3.5   Statistical Analysis

Using-delta Ct value for the 9 genes as input, K means analysis was performed using "fpc" package in R. Patients were clustered into 2 groups based on their levels of gene expression. Survival analysis was performed using "survival" package in R. The package was then used to evaluate the association of 9-gene prognostic signature with cancer-specific survival and relapse-free survival, and significance was determined by the log-rank test.

4   Results and Discussion

The 6 protective genes identified from the prior microarray studies were APOC1, IL32, HULC, CXCL16, CTSS and ALAS (blue). The 3 adverse genes identified were PGK1, CAD and ATF5 (red). Graphs 1 and 2 show samples with gene expression indicating good and poor prognosis respectively. A patient with good prognosis shows a relatively higher level of expression of protective genes compared to adverse genes

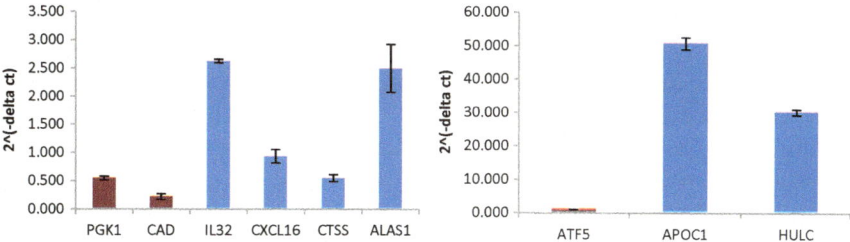

Graph 1 Example of patient with good prognosis

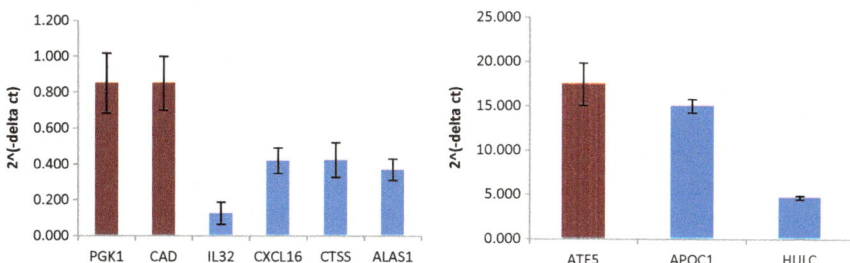

Graph 2 Example of patient with poor prognosis

(Graph 1). Conversely, a patient with poor prognosis shows a relatively higher level of expression of adverse genes compared to protective genes (Graph 2).

In our statistical analysis, the 82 patients were grouped into two clusters according to their levels of gene expression. We found our 9-gene assay to be significantly correlated with relapse-free survival (RFS) ($p = 0.0493$) (Fig. 3a), which includes patients who died of disease and those with disease recurrence. Patients were clus-

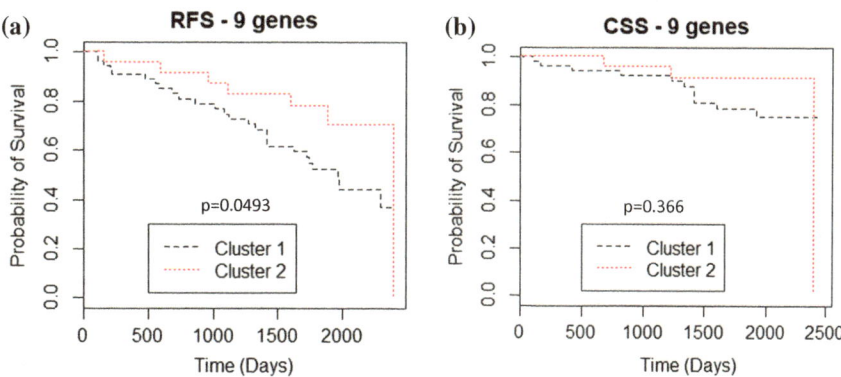

Fig. 3 Survival curves for relapse-free survival (**a**) and cancer specific survival (**b**) according to the level of expression of the 9 prognostic genes among the 82 patients

tered into 2 groups based on their distinct gene expression profiles. Clustering was found to be associated with RFS. However, when we tested the 9-gene assay for cancer specific survival (CSS), which includes only patients who died of disease, there was no statistical significance (p = 0.366) (Fig. 3b). The possible reason for the difference observed between CSS and RSS is the high probability of survival when patients with recurrence are identified and treated early. This further justifies the close monitoring of patients with poor prognosis. Our results support the validity of our assay in predicting the probability of relapse-free survival in HCC patients.

4.1   Clinical Application

We describe here a practical 9-gene assay capable of predicting the prognosis of HCC patients. The samples obtained from surgeries were FFPE tissue. Accordingly, our assay has been developed using RNA extracted from FFPE materials from surgeries and thus is expected to perform on such material in the clinical setting. The utility of the assay in abundantly available, routinely collected FFPE material greatly broadens the scope for rapid validation. Our assay can also be expanded to work on FFPE samples obtained from pre-operative core biopsies.

4.2   Cellular Functions of Prognostic Genes

The 9 genes in the prognostic assay—PGK1, CAD, ATF5, APOC1, IL32, HULC, CXCL16, CTSS and ALAS1—represent genes for angiogenesis, cell proliferation, transcription regulation, monocyte differentiation in the liver, chemokine signaling, MHC class II presentation and heme biosynthesis. The unbiased selection method in the 2 essays likely accounts for the wide variety of cellular functions encompassing in the prognostic gene set.

4.3   Limitations

The limitations of this study are its retrospective design, incomplete follow-up information of patients and the relatively limited number of subjects with poor prognosis for the validation cohort. Many patients with disease relapse survived for relatively longer than expected. This may have resulted in some difficulty in clustering data according to survival, resulting in the lack of statistical significance in the CSS value. External validation in prospective trials will be crucial to determine clinical value. Prognostic signatures ideally should be considered alongside optimal clinical predictors of outcome, such as the Barcelona Clinic Liver Cancer (BCLC), cancer of the liver Italian program, Japan Integrated Staging, and TNM. Future systematic studies will be important to address this.

Furthermore, the difference in level of protective genes and adverse gene expression in HCC is less distinct compared to that in other types of cancers. As it is more challenging to distinguish high gene expression from low gene expression, careful optimization and external validation in a larger cohort will be needed to further ensure the reliability of this assay. Additionally, predicting the survival for HCC patients is especially difficult because the liver plays crucial roles including detoxification, regulation of glycogen storage, plasma protein synthesis and hormone production, and is thus vital for the function of the human body. Many patients with HCC also have underlying fatal conditions such as cirrhosis and hepatitis B or C infection. In our study, it was difficult to ascertain whether the cause of death of certain patients was HCC or an underlying liver dysfunction. In a clinical setting, it may also prove challenging to predict survival of patients due to the plethora of other liver complications they may have in addition to HCC.

5 Conclusion and Recommendation for Future Work

In the future, the usefulness of our molecular 9-gene assay could be tested in clinical decision guidance. First, the 9-gene assay could be used to stratify the effectiveness of adjuvant therapy for various patients. This allows for more targeted and customized treatment that reduces cost and increases efficacy. Furthermore, the 9-gene assay could also modify transplantation indication, for example, by extending the Milan criteria to good molecular prognosis tumours even if it is >5 cm, whereas bad-prognostic molecular tumours within the Milan criteria could be excluded from liver transplantation or subjected to a more aggressive neoadjuvant strategy [15]. Despite the limited treatment options after liver resection in routine clinical practice, our 9-gene assay could also be tested to stratify the risk of relapse and death after liver resection in adjuvant randomized trial [16–18].

In conclusion we have designed a practical FFPE gene expression assay to predict the prognosis of HCC patients, with potential implications for therapeutic response. Incorporating the results from our assay allows an additional tool that can be integrated into the decision-making process, enhancing precision especially when it affirms the pathological assessment on core biopsy. We envision the use of this test to identify the patients with poorest prognosis in HCC to target intensive clinical follow-up and for predicting outcome and response to treatment.

References

1. El-Serag, H. B., & Rudolph, K. L. (2007). Hepatocellular carcinoma: Epidemiology and molecular carcinogenesis. *Gastroenterology, 132,* 2557–2576.
2. Yujin, H., Sebastian, M. B. N., Masahiro, K., et al. (2009). Integrative transcriptone analysis reveals common molecular subclasses of human hepatocellular carcinoma. *Cancer Research, 69*(18), 7385–7392.
3. Jean-Charles, N., Aurelien, D. R., Augusto, V., et al. (2013). A hepatocellular carcinoma 5-gene score associated with survival of patients after liver resection. *Gastroenterology, 145,* 176–187.
4. Forner, A., Llovet, J. M., & Bruix, J. (2012). Hepatocellular carcinoma. *Lancet, 379*(1245–125), 5.
5. Marrero, J. A., Fontana, R. J., Barrat, A., et al. (2005). Prognosis of hepatocellular carcinoma: comparison of 7 staging systems in an American cohort. *Hepatology, 41,* 707–716.
6. Roayaie, S., Blume, I. N., Thung, S. N., et al. (2009). A system of classifying microvascular invasion to predict outcome after resection in patients with hepatocellular carcinoma. *Gastroenterology, 137,* 850–855.
7. Hoshida, Y., Villaneuva, A., Kobayashi, M., et al. (2008). Gene expression in fixed tissues and outcome in hepatocellular carcinoma. *New England Journal of Medicine, 359,* 1995–2004.
8. Villanueva, A., Hoshida, Y., Toffanin, S., et al. (2010). New strategies in hepatocellular carcinoma: Genomic prognostic markers. *Clinical Cancer Research, 16,* 2688–4694.
9. Lee, J. S., Chu, I. S., Heo, J., et al. (2004). Classification and prediction of survival in hepatocellular carcinoma by gene expression profiling. *Hepatology, 40,* 667–676.
10. Roessler, S., Jia, H. L., Budhu, A., et al. (2010). A unique metastasis gene signature enables prediction of tumour relapse in early-stage hepatocellular carcinoma patients. *Cancer Research, 70,* 10202–10212.
11. Lee, J. S., Heo, J., Libbrecht, L., et al. (2006). A novel prognostic subtype of human hepatocellular carcinoma derived from hepatic progenitor cells. *Nature Medicine, 12,* 410–416.
12. Boyault, S., Rickman, D. S., de Reynies, A., et al. (2007). Transcriptome classification of HCC is related to gene alterations and to new therapeutic targets. *Hepatology, 45,* 42–52.
13. Villanueva, A., Hoshida, Y., Battison, C., et al. (2011). Combining clinical, pathology, and gene expression data to predict recurrence of hepatocellular carcinoma. *Gastroenterology, 140*(1501–1512), e1502.
14. Hoshida, Y., Villanueva, A., Sangiovanni, A., et al. Prognostic gene-expression signature for patients with hepatitis C-Related early stage cirrhosis. Gastroentology.
15. Clavien, P. A., Lesurtel, M., Bossuyt, P. M., et al. (2012). Recommendations for liver transplantation for hepatocellular carcinoma: An international consensus conference report. Lancet Oncology *13,* e11–e22.
16. Kudo, M. (2011). Adjuvant therapy after curative treatment for hepatocellular carcinoma. *Oncology, 81*(Suppl 1), 50–55.
17. Albain, K. S., Barlow, W. E., Shak, S., et al. (2010). Prognostic and predictive value of the 21-gene recurrence score assay in postmenopausal women with node-positive, oestrogen-receptor-positive breast cancer on chemotherapy: A retrospective analysis of a randomized trial. *Lancet Oncology, 11,* 55–65.
18. Cardoso, F., Van't Veer, L., Rutgers, E., et al. (2008). Clinical application of the 70-gene profile: The MINDACT trial. *Journal of Clinical Oncology 26,* 729–735.

Green Synthesis of Nanoparticles Using Dried Fruit Peel Extract

Clyve Yu Leon Yaow, Ian Ee En Sim, Feldman Kuan Ming Lee, Doreen Wei Ying Yong and Wee Shong Chin

Abstract Nanoparticles have been one of the leading topics in research due to their excellent catalytic properties and various applications. In this project, green synthesis of Manganese (III) Oxide (Mn_2O_3) and Silver (Ag) nanoparticles using beetroot peel extracts has been successful in achieving a more environmentally friendly method of production. The synthesized Ag and Mn_2O_3 nanoparticles were 17.0 ± 4.3 nm and 203.8 ± 61.1 nm respectively. UV-vis and X-ray Diffraction have confirmed the identities of the Ag and Mn_2O_3 nanoparticles respectively. Fourier Transform Infrared spectroscopy and a series of phytochemical tests were also conducted to determine the classes of chemicals involved in the synthesis method. Compared with traditional chemical methods, the synthesized Ag and Mn_2O_3 nanoparticles showed comparable characteristics and catalytic properties in the reduction of 4-nitrophenol and degradation of methylene blue solution respectively.

Keywords Nanoparticles · Green synthesis · Beetroot peel · Silver · Manganese (III) oxide

1 Introduction

Metal nanoparticles (NPs) are of increasing interest due to their remarkable physical and chemical properties for applications in photovoltaic cells, optical and biological sensors, catalysts, conductive materials, coating formulations and many more [1–4]. Among the nanoparticles, silver nanoparticles (Ag NPs) are relatively low cost, scalable and useful as catalysts in the reduction of 4-nitrophenol. 4-nitrophenol is used commonly in the manufacturing of pigments, dyes and plastics and is highly hazardous on release in environment [5]. Hence, the reduction of such a chemical is

C. Y. L. Yaow · I. E. E. Sim · F. K. M. Lee
NUS High School of Mathematics and Science, 20 Clementi Avenue 1, Singapore 129957, Singapore

D. W. Y. Yong · W. S. Chin (✉)
Department of Chemistry, National University of Singapore, 3 Science Drive 3, Singapore 117543, Singapore
e-mail: chmcws@nus.edu.sg

© Springer Nature Singapore Pte Ltd. 2019
H. Guo et al. (eds.), *IRC-SET 2018*,
https://doi.org/10.1007/978-981-32-9828-6_2

9

of importance to reduce pollution and health risk. Manganese (III) oxide (Mn_2O_3) is not only an inexpensive transition metal oxide but like Ag, they have good redox ability, and is suitable as a catalyst to degrade methylene blue dye, which can cause increased heart rate, cyanosis, and other health problems [6]. Unlike conventional chemical synthesis methods which include the use of hazardous chemicals, green syntheses using plant and fruit extracts have shown to be more environmental-friendly and cost effective [1, 2, 7, 8]. Furthermore, we observed the increasing problem of non-recycled food waste in Singapore. Accounting for about 10% of the total waste generated in Singapore, with only 14% of it recycled, it certainly is a problem worth tackling. The amount of food waste generated in Singapore has increased by about 40% over the past 10 years and is expected to increase with our growing population and economic activity.

In this project, we have successfully synthesized Ag and Mn_2O_3 NPs using beet-root peel extracts. The synthesized NPs have been characterised using various techniques and tested for their catalytic activity in reduction reactions. The methodology will be covered in the next section, and subsequently the characterisation and catalytic properties before some concluding remarks are made.

2   Materials and Methods

A. *Preparation of Fruit Peel Extract*

Fruit peels were screened (see annex Fig. 1). Fruit peels were first dried in a 70 °C oven for 3 days. 5 g of grounded fruit peel material was steeped in 100 mL of DI water at 80 °C for 30 min with constant stirring. The mixture was vacuum filtered to

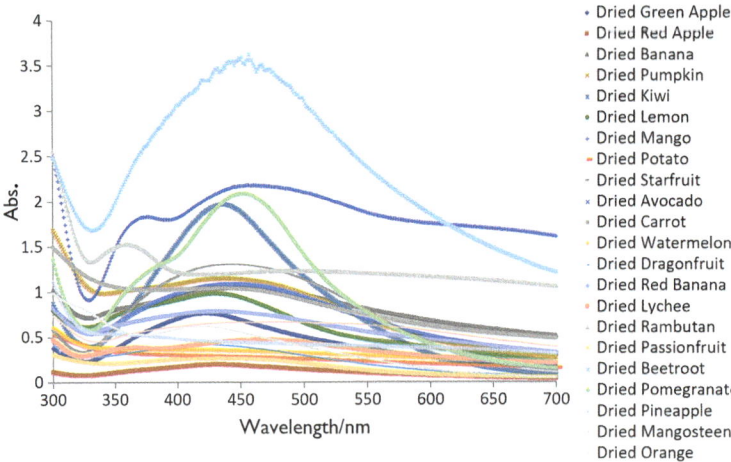

Fig. 1 UV-vis spectra of Ag NPs using different fruit extracts

make 5% w/v fruit peel extracts and the filtrate collected was stored in a refrigerator for further use [1].

B. *Synthesis of Silver Nanoparticles*

1 ml of each fruit skin extract was added to a 15 ml centrifuge tube and placed in a water bath at 100 °C for 5 min. 0.5 ml of 0.1 M silver nitrate ($AgNO_3$) solution was then added, the mixture was vortexed and placed in the 100 °C water bath again for 15 min [2, 9]. The mixture in the centrifuge tube was then tested for the presence of Ag NPs using UV-visible spectroscopy.

C. *Synthesis of Mn_2O_3 Nanoparticles*

10 ml of aqueous solution of 0.1 M Potassium Permanganate ($KMnO_4$) was added dropwise to 40 ml of the fruit peel extract at 25 °C. The solution was then left under continuous magnetic stirring for 3 days. After 3 days, the content is washed with DI water thrice by repeated centrifugation at 6000 rpm for 15 min. The precipitate is then dried in a Nabertherm muffle furnace at 80 °C for 24 h followed by calcinations at 600 °C for 4 h using a ramp rate of 30 °C/min [6, 10–12].

D. *Characterisation Methods*

UV-visible spectra were recorded using Shimadzu-UV 3600 spectrophotometer at a resolution of 5.0 nm. Powder X-Ray Diffraction (XRD) patterns were recorded on a Siemens D5005 diffractometer with Cu Kα radiation ($\lambda = 0.15406$ nm) in the 2θ range from 20° to 80°. Morphologies of the samples were characterised using field emission scanning electron microscope (FESEM) JEOL JSM-6701F and transmission electron microscope (TEM) microscope JEOL JEM 3011 operated at 300 kV accelerating voltage. SEM substrates were prepared by dropping NP sample solution on clean silicon substrates and dried under vacuum. TEM samples were prepared by placing a drop of NP sample solution on a copper grid and dried under vacuum. Particle sizes reported are an average and standard deviation of 50 measurements. Fourier transform infrared spectroscopy (FTIR) and phytochemical screenings were performed on both fresh and spent extracts. Spent extract was obtained by reacting multiple times until no AgNPs can be formed. FT-IR spectroscopy was carried out using Varian 3100 FT-IR Spectrometer (scan range 400 to 4000 cm^{-1}, resolution 4 cm^{-1}) on vacuum dried extracts pressed into Potassium Bromide (KBr) pellets.

Phytochemical tests were conducted. The type and methodology of the tests are as follows:

Test for Reducing Sugars

Benedict's reagent was prepared by dissolving 1 g of sodium carbonate and 1.73 g of sodium citrate dihydride in a final volume of 10 ml of water. Then 1.73 g of copper sulphite was added slowly with stirring. 2 ml of Benedict's reagent was then added to 1 ml of 2.5% w/v aqueous extract. The resulting solution was heated in a hot water bath for 5 min. A formation of red-orange precipitate would indicate the presence of reducing sugars.

Lead Acetate Test for Flavonoids

A 10% w/v aqueous lead acetate solution was prepared by dissolving 0.1 g of lead acetate in 1 ml of DI-water. 3 drops of this solution was added to 2 ml of the beetroot extract. The formation of a yellow precipitate indicated the presence of flavonoids.

Mayer's Test for Alkaloids

1 ml of the beetroot extract was combined with 3 ml of 10% ammonia solution and allowed to mix for 7 min. 10 ml of chloroform was then added to the mixture and the mixture was filtered via gravity filtration. Mayer's reagent was then prepared by adding 0.136 g of Mercuric Chloride in 6 ml of DI-water to 0.5 g of Potassium Iodide in 2 ml of DI-water and then adding another 2 ml of DI-water after. The filtered liquid was then left to evaporate before treating the residue with 3 ml of Mayer's Reagent. The formation of a cream coloured precipitate indicated the presence of alkaloids.

Ferric Chloride Test for Tannins

Alcoholic ferric chloride solution was prepared by dissolving 1 g of ferric chloride in 10 ml of ethanol. 4 drops of 10% w/v alcoholic ferric chlorides solution were added to 2.5 ml of extract and 2.5 ml of DI-water. The presence of blue-black colour indicated the presence of hydrolysable tannins and the presence of a green colour would indicate the presence of condensed tannins.

Test for Phlobatannins

19 ml of the beetroot extract was oiled in 1% HCl solution for 5 min. This procedure was then repeated with 2% HCl solution. The formation of a red precipitate indicated the presence of phlobatannins.

Chlorogenic Acid Test

1 ml of 10% ammonia solution was added to 0.5 ml of the beetroot extract along with 0.5 ml of DI-water. The solution was then exposed to air for 20 min. Chlorogenic acid is present if the solution turned green.

Coumarins Test

1 g of sodium hydroxide was added to 10 ml of DI-Water. 3 drops of the 10% w/v alcoholic sodium hydroxide solution was added to 1 ml of the beetroot extract and 2 ml of DI-water. A yellow colour change indicated the presence of coumarins.

Froth Test

20 ml of the beetroot extract was shaken vigorously for 20 s and allowed to sit for 20 min. The presence of a froth after 20 min was an indication of saponins.

Cardiac Glycosides Test

5 ml of the beetroot extract was mixed with 2 ml of glacial acetic acid and 1 drop of 10% w/v alcoholic ferric chloride solution. 1 ml of concentrated sulfuric acid was then added. The formation of a brown ring between the layers and a blue colour in the top layer indicated the presence of cardiac glycosides.

Quinones Test

3 drops of concentrated sulfuric acid were added to 1 ml of the beetroot extract and 2 ml of DI-water. A colour change indicated the presence of quinones.

Terpenoids Test

5 ml of the beetroot extract was mixed with 2 ml of chloroform and 3 ml of concentrated sulfuric acid. The presence of terpenoids was indicated by a reddish-brown ring at the interface of the layers.

E. Catalysis Test for Ag NPs

To test for the catalytic ability of beetroot synthesized Ag NPs, we used the reduction reaction of 4-nitrophenol (4-NP). This reaction is known to be thermodynamically favourable [2, 5]. However, the reaction is very slow as 4-NP is deprotonated in basic conditions due to the addition of sodium borohydride ($NaBH_4$). Ag NPs act as catalysts by facilitating the transfer of electrons from BH_4^- to the nitro group [2, 5]. 1.75 ml of DI-water, 150 μl of 1.5 mM 4-nitrolphenol solution and 1 ml of 30 mM $NaBH_4$ solution were mixed together in a 3 ml quartz cuvette [2, 5].

100 μl of Ag nanoparticle solution was then added into the solution and the reaction was then monitored using UV-vis over 300–500 nm every 2 min. A control was also monitored in the same manner. The control was created by adding 100 μl of water instead of Ag NPs. [2, 5] A total of 7 runs were performed, including one control.

F. Catalysis Test for Mn_2O_3 NPs

To test the catalytic ability of Mn_2O_3 NPs, we used the degradation of methylene blue with hydrogen peroxide [6, 11–13]. First, a beaker and a petri dish were wrapped in aluminum foil. 2 mg of Mn_2O_3 NPs were added to 2.6 ml of DI-water in the beaker. Then, 12.5 ml of 10 mg/L methylene blue solution and 400 μl of H_2O_2 (used as oxidiser) were added. The beaker was then covered with the petri dish and left to stir in the dark for 70 min to equilibrate the adsorption of methylene blue on catalyst surface [6, 11, 12].

3 ml of the solution from the beaker was drawn out and centrifuged with a HERMLE Z200A to remove the solid catalyst. The supernatant was tested for UV-vis at every 30 min interval.

A control was also performed using the same method but 2.6 ml of DI-water was added without the addition of 2 mg of Mn_2O_3 NPs. A total of 5 runs were studied, including 1 control.

3 Results and Discussion

A. *UV-vis Spectra of Silver Nanoparticles*

Figure 1 shows the representative UV-vis spectra of the aqueous nanoparticle mixtures synthesized by the different fruit peel extracts. Like previous work [2, 9], appearance of an absorbance peak at around 400 nm, which is attributed to the Surface Plasmon Resonance (SPR) absorption of AgNPs, suggests the successful formation of AgNPs.

The maximum absorbance (A_{max}) gives an estimation of the amount of AgNPs, with a larger concentration of AgNPs being reflected as a larger A_{max}. This indicates the yield of AgNPs given a fixed concentration of $AgNO_3$ precursor. The wavelength with maximum absorbance (λ_{max}) is an estimation of the modal size of AgNPs. Finally, Full Width at Half Maximum (FWHM), indicates the particle size distribution with small FWHM peak indicating more monodisperse AgNPs. From the UV spectra, we selected NPs synthesized using beetroot peel extract for further characterization as the AgNPs displayed the highest A_{max} and lowest FWHM at λ_{max} of 400 nm.

B. *Ag NPs Shape and Size Distribution*

Figure 2 shows a typical cluster of beetroot peel synthesised AgNPs. To determine the size distribution, we measured the size of 115 AgNPs using image editing software GIMP. The nanoparticles had a mean diameter of 17.0 nm and a standard deviation of 4.3 nm. As seen from Fig. 2, the AgNPs are generally spherical in shape with a small number of irregular shapes. The results are similar to other papers [9].

C. *Ag NP as Catalyst for Reduction of 4-Nitrophenol*

The UV-vis spectra in Fig. 3a shows a great decrease in the peak intensity during the reaction duration of 20 min. This corresponds to decolourising of the initial yellow

Fig. 2 TEM of Ag nanoparticles

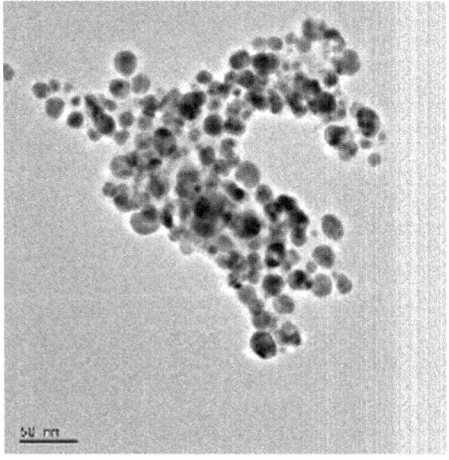

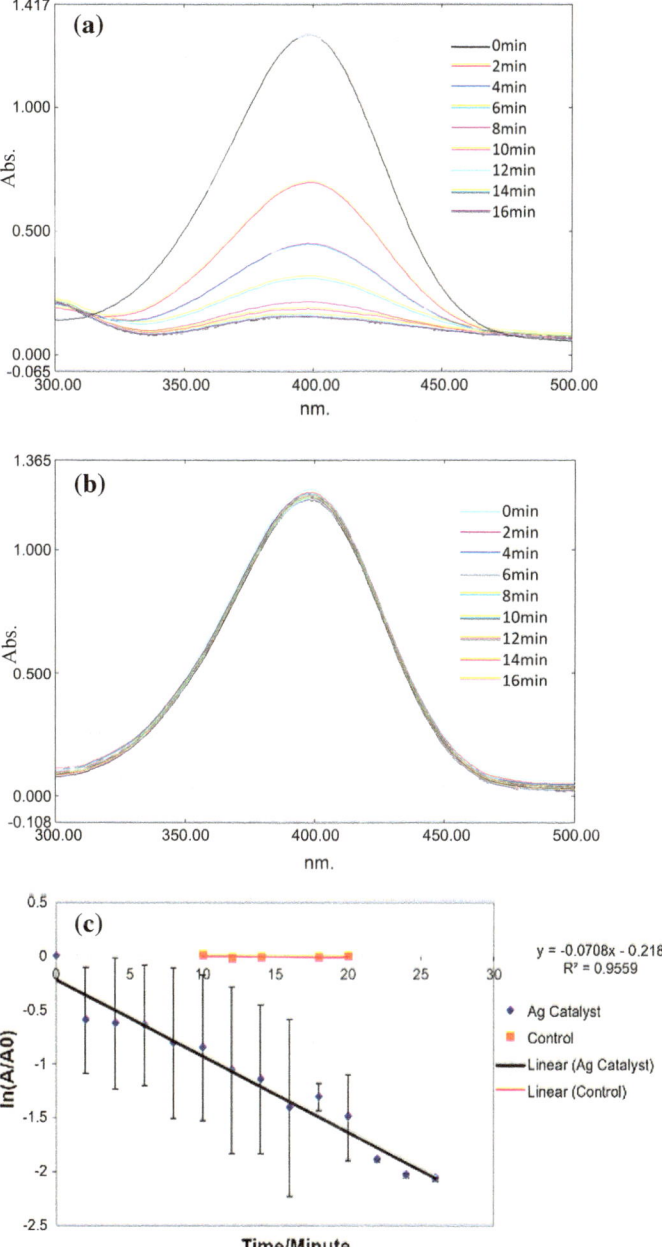

Fig. 3 **a** UV-vis spectra of Ag NPs catalysed reduction reaction of 4-nitrophenol. **b** UV-vis spectra of reduction reaction of 4-nitrophenol without Ag NPs. **c** Corresponding plots of ln (A/A_0) versus time for reduction reaction with Ag NPs and control

reaction mixture [2, 5, 14]. In the control, as shown in Fig. 3b, the peak did not decrease considerably for 20 min. This confirms that Ag NPs play a catalytic role in the reduction reaction.

In Fig. 3c, we have analyzed the reaction with and without the catalyst at its peak at 398 nm [2, 14] throughout the duration of the reaction and have plotted the corresponding graph of $\ln(A/A_0)$ against time [2, 14]. This resulted in a linear relationship which suggests a pseudo-first order kinetic with respect to the 4-nitrophenol concentration [2, 5]. Since the concentration of $NaBH_4$ is much larger than 4-nitrophenol, the reduction rate can be assumed to be independent of the $NaBH_4$ concentration. The rate constant, k_{exp} can be determined by taking the negative of the gradient, in this case 0.0708 min^{-1}, compared to the rate constant of the control which is 0.0005 min^{-1}.

D. *XRD Pattern of Synthesized Mn_2O_3 NPs*

XRD confirms the crystalline nature of green synthesized Mn_2O_3 NPs. The prominent peaks in the XRD pattern in Fig. 4, corresponded to pure Mn_2O_3 (JCPD 10-0069). Peaks other than those indicating presence of Mn_2O_3 were negligible and suggests that using dried beetroot peels to synthesize Mn_2O_3 NPs did not result in the formation of undesired nanoparticles [6, 11, 12].

E. *Mn_2O_3 NP Shape and Size Distribution*

Figure 5a shows a typical cluster of beetroot peel synthesized Mn_2O_3 NPs. To determine the size distribution, we measured the size of 115 Mn_2O_3 NPs using image editing software GIMP. The nanoparticles had a mean diameter of 203.8 nm and a standard deviation of 61.1 nm which is similar to other papers [6]. As seen from Fig. 5a Mn_2O_3 NPs are generally irregular in shape with a small number of spherical or spheroid shape.

Fig. 4 XRD of Mn_2O_3 NPs

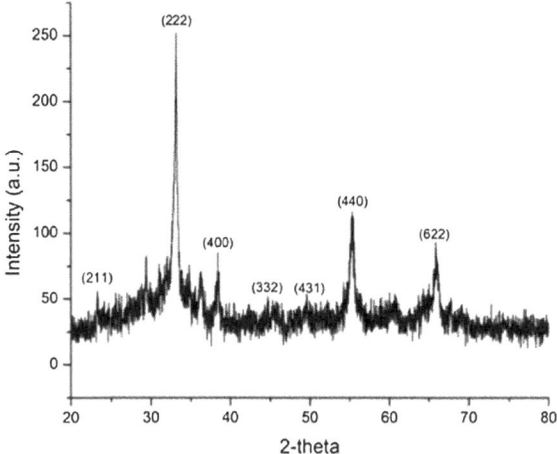

Fig. 5 SEM of Mn₂O₃ NPs

F. *Mn₂O₃ NPs as Catalyst for Oxidative Degradation of Methylene Blue*

The UV-vis spectra in Fig. 6a shows a great decrease in the peak intensity near 650 nm during the reaction duration of 190 min. This corresponds to decolourising of the initial blue reaction mixture [6, 11, 12]. In the control, as shown in Fig. 6b, the peak did not decrease considerably for 190 min. This confirms that Mn₂O₃ NPs play a catalytic role in the degradation reaction. In Fig. 6c, we have analyzed the reaction with and without the catalyst at its peak at 660 nm throughout the duration of the reaction and have plotted the corresponding graph of $\ln(A/A_0)$ against time [6, 11, 12]. This resulted in a linear relationship which suggests a pseudo-first order kinetic with respect to the methylene blue concentration. [6] Since the concentration of H_2O_2 is much larger than the methylene blue solution, the reduction rate can be assumed to be independent of the H_2O_2 concentration. The rate constant, k_{exp} can be determined by taking the negative of the gradient, in this case 0.0024 min^{-1}, compared to the rate constant of the control which is 0.0009 min^{-1}.

G. *FTIR Pattern of Synthesised Ag and Mn₂O₃ NPs*

In both FTIRs of AgNPs (Fig. 7a) and Mn₂O₃ NPs (Fig. 7b), a broad band appears at 3436 and 3443 cm^{-1} respectively. This is likely representative of the O–H stretching bond which is present in water, polyphenols and sugar found in the beetroot extract. Two and three less intense and narrower bands appear in Fig. 7a and 7b respectively between 2856 and 2959 cm^{-1}. These bands indicate a variety of N- and O-containing functional groups as well as the possible C–C and C–H. Both spectra also show a weak band at 1631 cm^{-1} which is indicative of a stretching alkene bond. The bands appearing between 1381 and 1358 cm^{-1} are C–H stretching modes from hydrocarbon chains. These features could result from polyphenols in the dried beetroot peel extract.

H. *FTIR Pattern on Beetroot Extract*

In both FTIR of fresh (Fig. 7c) and spent (Fig. 7d) beetroot extract, a broad band appears at 3389 and 3416 cm^{-1} respectively, which is representative of the O–H stretching bond present in sugars and water in the beetroot extract. One and two weak bands respectively appear between 2850 and 2939 cm^{-1}. These bands indicate a

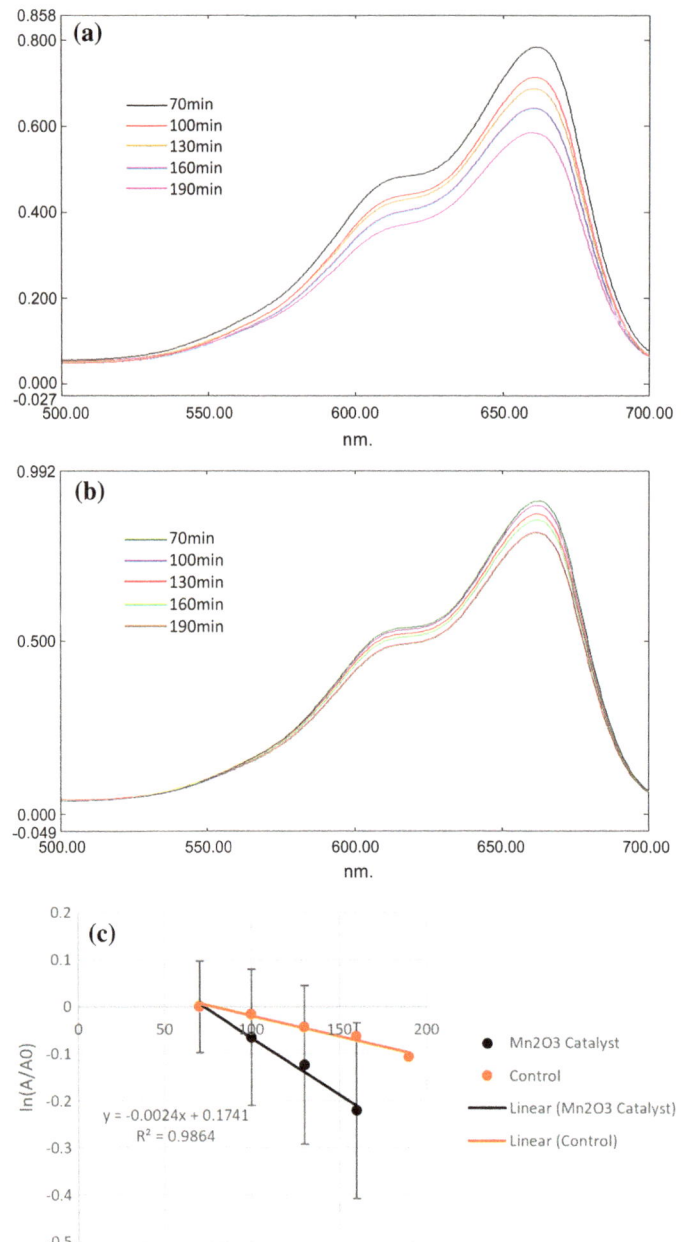

Fig. 6 **a** UV-vis spectra of Mn_2O_3 NPs catalysed degradation reaction of methylene blue. **b** UV-vis spectra of degradation reaction of methylene blue without Mn_2O_3 NPs. **c** Corresponding plots of $\ln(A/A_0)$ versus time for degradation reaction with Mn_2O_3 NPs and control

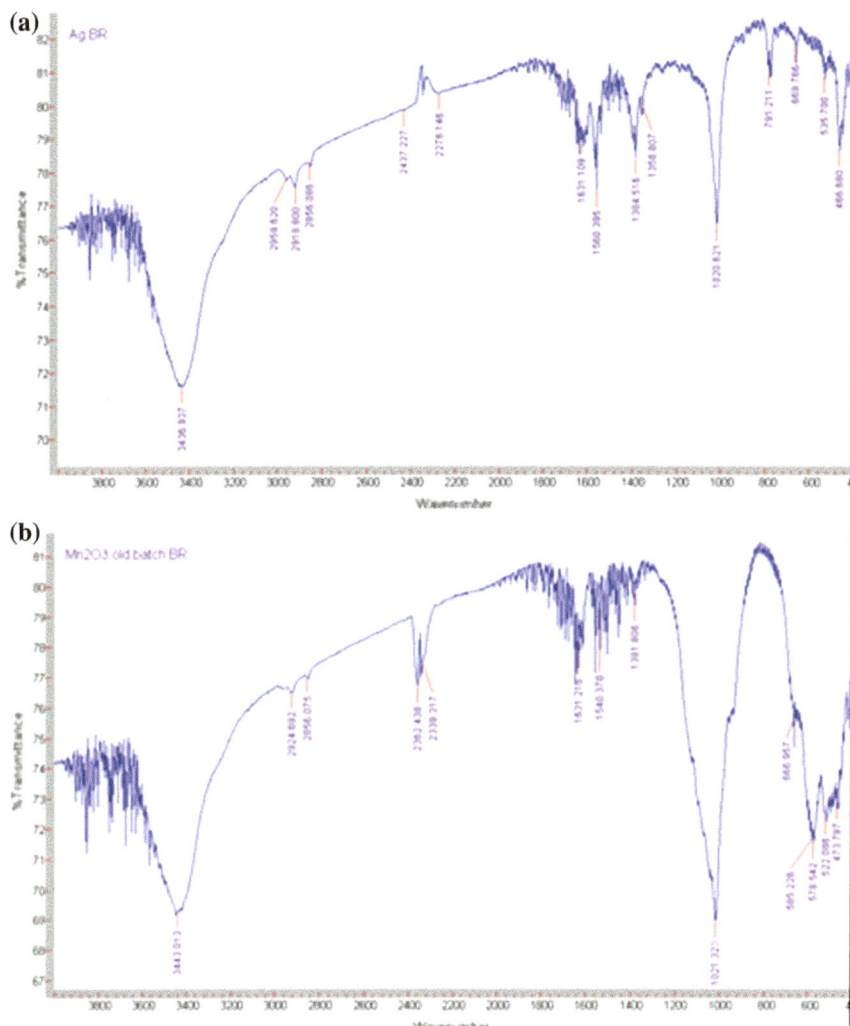

Fig. 7 **a** FTIR-spectra of Ag NPs. **b** FTIR-spectra of Mn$_2$O$_3$ NPs. **c** FTIR-spectra of fresh dried beetroot peel extract. **d** FTIR-spectra of spent dried-beetroot peel extract. **e** Table of FTIR frequencies and possible assignments

variety of N–H and O–H bonds as well as the possible hydrocarbons. Both spectra also show a weak band at 1638 and 1636 cm^{-1} respectively which indicates a stretching alkene bond. There was no observable change in peaks between the FTIR for the fresh and spent extract [15]. We would also like to note that there is a slight weakening in the peak heights, allowing us to deduce that the functional groups in the dried beetroot peel extract has been used during the reducing and capping process. All the

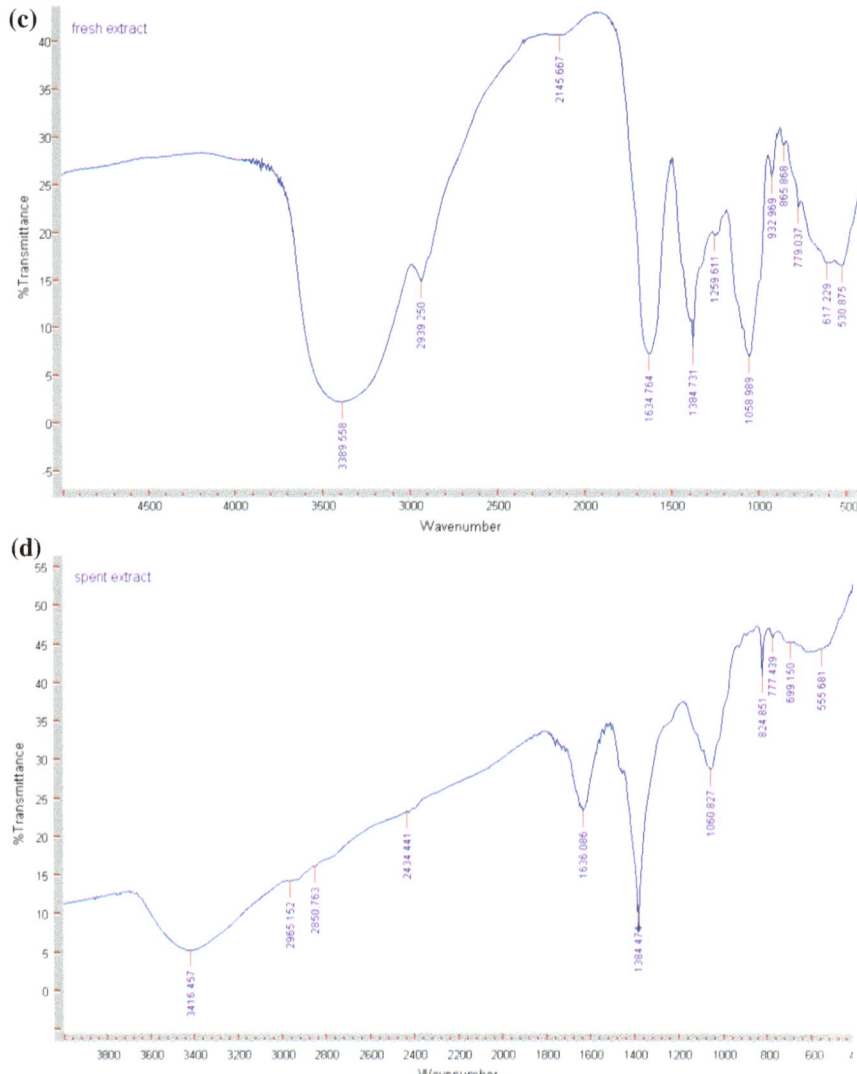

Fig. 7 (continued)

noted features were with reference to Fig. 7e which contains a table showing the possible assignment of vibrations corresponding to the frequency.

I. *Phytochemical Tests*

The phytochemical tests [2] on both the spent and fresh extracts have shown that both extracts do not contain reducing sugars [15], phlobatannins, coumarins, cardiac glycosides, or quinones. However, both extracts do contain alkaloids, and saponins,

(e)

Frequency/cm^{-1}	Possible Bonds
3550-3200	O-H Stretching
3400-3300 3330-3250	N-H Stretching
3000-2800	N-H Stretching
3000-2840	C-H Stretching
2275-2250	N=C=O Stretching
1648-1638	C=C Stretching
1550-1500	N-O Stretching
1390-1380	C-H Bending
1420-1330	O-H Bending
1410-1380	S=O Stretching
1250-1020	C-N Stretching
1075-1020	C-O Stretching
900-860	C-H Bending
690-515	C-Br Stretching

Fig. 7 (continued)

which indicated that these compounds were not used in the formation of nanoparticles.

In the fresh extract, phenolic compounds [15], chlorogenic acids [15], condensed tannins and terpenoids were present, but they were absent in the spent extract. These compounds were likely used during the reduction and capping process for the formation of nanoparticles. We believe that these compounds are all able to act as reducing agents due to their easily oxidisable O–H groups. Condensed tannins as well as Chlorogenic acids can act as good capping agents due to the high steric hindrance experienced between their compounds resulting from their rigid aromatic rings and the fact that these compounds can chelate the metal centre easily. These chemicals also correspond to the functional groups we deduced from our FTIR scans.

4   Conclusion

We have experimented with a substantial number of fruit and vegetable peel extracts and successfully tested a new method for green synthesis of nanoparticles, such as Ag and Mn_2O_3, using beetroot peel extract. The nanoparticles synthesized using this method were shown to have similar effectiveness in the degradation of methylene blue and reduction of 4-nitrophenol as compared to those available commercially. This is a significant breakthrough in the synthesis of nanoparticles as this method is low cost, helps to reduce waste, is environmentally friendly and can be easily scaled up to meet the needs of the market. With further research, we hope to be able to develop methods to produce other metal nanoparticles, such as copper and zinc oxide, using the green synthesis.

Acknowledgements FKM Lee, IEE Sim and CYL Yaow would like to acknowledge Ms Doreen Yong for her guidance and support throughout the research journey.

References

1. Aaron, D., & Brumbaugh, K. A. (2014). Ultrasmall copper nanoparticles synthesized with a plant tea reducing agent. *ACS Sustainable Chemistry & Engineering, 1933–1939.*
2. Abilash Gangula, R. P. (2011). Catalytic reduction of 4-nitrophenol using biogenic gold and silver nanoparticles. Derived from Breynia rhamnoides. *Langmuir,* 15268–15274.
3. Annavaram Viswadevarayalu, P. V. (2016). Fine ultrasmall copper nanoparticle (UCuNPs) synthesis by using terminalia bellirica fruit extract and its antimicrobial activity. *CrossMark,* 155–168.
4. Nguyen, T. H. (2014). Copper oxide nanomaterials prepared by solution methods, some properties, and potential applications: A brief review. *Hindawi,* 1–14.
5. Lunhong Ai, J. J. (2013). Catalytic reduction of 4-nitrophenol by silver nanoparticles stabilizedon environmentally benign macroscopic biopolymer hydroge. *Elsevier,* 374–377.
6. Ping Tao, M. S. (2016). Morphologically controlled synthesis of porous Mn2O3 microspheres and their catalytic applications on the degradation of methylene blue. *Desalination and Water Treatment,* 7079–7084.
7. Amit Kumar Mittala, Y. C. (2013). Synthesis of metallic nanoparticles using plant extracts. *Elsevier,* 346–356.
8. Mohd Sayeed Akhtar, J. P.-S. (2013). Biogenic synthesis of metallic nanoparticles by plant extracts. *ACS Sustainable Chemistry Engineering,* 591–602.
9. Basavegowda, N., & Lee, Y. R. (2013). Synthesis of silver nanoparticles using satsuma mandarin (Citrus Unshiu) peel extract: A novel approach towards waste utilization. *Materials Letters, 109,* 31–33.
10. Lina Sanchez-Botero, A. P. (2017). Oriented growth of α-MnO2 nanorods using natural extracts from grape stems and apple peels. *Nanomaterials,* 1–15.
11. Zeheng Yang, Y. Z. (2006). Nanorods of manganese oxides: Synthesis, characterization and catalytic application. *Journal of Solid State Chemistry,* 679–684.
12. Zhongchao Bai, B. S. (2012). Branched mesoporous Mn3O4 nanorods: Facile synthesis and catalysis in the degradation of methylene blue. *Full Paper,* 5319–5324.
13. Peng Su, D. C. (2010). Studies on catalytic activity of nanostructure Mn2O3 prepared by solvent-thermal method on degrading crystal violet. *Modern Applied Science,* 125–129.

14. Deka, P., et al. (2016). Hetero-nanostructured Ni/α-Mn2O3as highly active catalyst for aqueous phase reduction reactions. *Chemistry Select, 1*(15), 4726–4735.
15. Wruss, J., et al. (2015). Compositional characteristics of commercial beetroot products and beetroot juice prepared from seven beetroot varieties grown in upper Austria. *Journal of Food Composition and Analysis, 42*, 46–55.

Optimization of the Electrospinning Process to Create Pure Gelatin Methacrylate Microstructures for Tissue Engineering Applications

Srushti Sakhardande and Zhang Yilei

Abstract Synthetic scaffolds made from electrospun fibres have showed to be a novel method to solve issues related to shortage of donors and other complications. Pure gelatin methacrylate (GelMA) has showed to have excellent compatibility in tissue engineering applications as it closely resembles properties of native extracellular matrix. However, studies of electrospinning of GelMA largely use GelMA blends. In this study, pure GelMA fibres were successfully produced using electrospinning method. To facilitate development of pure GelMA fibers using electrospinning method as well as optimize the process, alterations to significant parameters of electrospinning, including concentration of solution, voltage of power supply, flow rate and temperature was done. It was observed that temperature is a significant parameter in the electrospinning of pure GelMA. A correlation between concentration and fibre diameter was observed that further emphasises the wide usage of electrospun GelMA fibers in tissue engineering application as fibers of differing diameters and mechanical strengths can be produced. This study sets the foundation for the usage of electrospun pure GelMA scaffolds in tissue engineering applications.

Keywords Electrospinning · Tissue engineering · Gelatin methacrylate

1 Introduction

As Singapore and the world faces societal problems like ageing population leading to increasing healthcare demands, there is a need to find effective and low-cost solutions to such problems. With problems associated with the shortage of donor organs, donor site morbidity and complications, risk of disease transmission and immuno-rejection problems, tissue engineered scaffolds have been developed as a novel perspective for organ repair.

Electrospinning is a fiber production method which uses electric force to draw charged threads of polymer solutions. It has emerged as a new scaffold fabrication

S. Sakhardande (✉)
Victoria Junior College, Singapore, Singapore
e-mail: srushti.sakhardande@gmail.com

Z. Yilei
Nanyang Technological University, Singapore, Singapore

© Springer Nature Singapore Pte Ltd. 2019
H. Guo et al. (eds.), *IRC-SET 2018*,
https://doi.org/10.1007/978-981-32-9828-6_3

method. The underlying rationale of using microfibres for scaffolding is based on the principle that electrospun fibers can mimic the physical structure of the native extracellular matrix (ECM). From the biological viewpoint, almost all of the tissues and organs, such as bone, nerve, blood vessel, ligament, tendon, and cartilage, are synthesized and hierarchically organized into fibrous structure [1–5]. The scaffolds made by electrospinning exhibit better cellular attachment, growth and differentiation compared with those made by other techniques. The large specific surface area provided by a low-dimension fibrous structure, which facilitates cell adhesivity to the electrospun fibers.

Gelatin methacrylate (GelMA) hydrogel, chemically modified gelatin, has been widely used for various biomedical applications due to its suitable biological properties and tunable physical characteristics. GelMA hydrogels closely resemble the aforementioned properties of native extracellular matrix (ECM) because of the presence of cell-attaching and matrix metalloproteinase responsive peptide motifs, which allow cells to proliferate and spread in GelMA-based scaffolds. GelMA is also versatile from a processing perspective. It crosslinks when exposed to light irradiation to form hydrogels with tunable mechanical properties [6].

Hydrogel electrospinning is recent invention. Hydrogels are difficult to spin as they have limited solubility in organic solvents, thus aqueous solvents are used. However, the combination of water's low volatility and hydrogel material's affinity for water slows the drying process for the polymer jet, resulting in deposition of wet material rather than dry polymer fibers. GelMA based electrospinning has largely been using GelMA blends with other polymers such as polyvinyl alcohol (PVA) or nanoparticles, and not using pure GelMA. This is due to pure GelMA's high affinity to water, making electrospinning challenging. However, there is a need to produce pure GelMA electrospun fibers due the excellent compatibility of pure GelMA in tissue engineering applications coupled with the effectiveness of electrospinning in producing synthetic scaffolds.

2   Aims and Objectives

We aim to create a pure GelMA fibrous structure that can be cell laden as electrospun fibers have shown to have excellent cellular attachment, growth and differentiation. Due to the lack of studies on the electrospinning of pure GelMA despite its excellent compatibility in producing tissue engineered scaffolds, this study aims to optimize parameters of electrospinning in order to produce pure GelMA fibers.

3   Methods and Materials

A. *Characterisation of GELATIN METHACRYLATE (GelMA)*

In order to make the GelMA solution for electrospinning, solutions were prepared by weighing GelMA and adding deionized water. The concentration of the solution was determined by the mass of GelMA in 0.5 ml of deionized water. The solutions was stirred at 50 °C using a heated magnetic stirrer, in order to ensure dissolution of GelMA and a homogenous mixture.

B. *Electrospinning Process*

The electrospinning method is a process that creates nanofibres through an electrically charged jet of polymer solution. A basic electrospinning setup (Fig. 1 [7]) consists of a syringe to hold polymer solution (often connected to a syringe pump), two electrodes and a DC voltage supply in the kV range.

Before electrospinning the needle was kept in a hot water bath of 50 °C, to ensure that the GelMA solution does not dry up when flowing through the needle.

Parameters of electrospinning are of paramount importance as they affect production of fibres and fibre morphology. Significant parameters include, viscosity and surface tension, voltage of power supply, distance between syringe and grounded surface, temperature, and flow rate of solution from syringe.

The following parameters are of significance when electrospinning polymers.

(1) *Suitable Solvent to dissolve polymer*
(2) *Vapour Pressure of the solvent so that the solvent*

    (i) Evaporates quickly enough for fibre to maintain its integrity
    (ii) Evaporates not too quickly to allow the fibre to harden before it reaches nanometre range

(3) *Viscosity and Surface Tension of the solvent*

    (i) Not too large to prevent the jet from forming
    (ii) Not be too small to allow the polymer solution to flow freely from the syringe

Fig. 1 Basic electrospinning setup

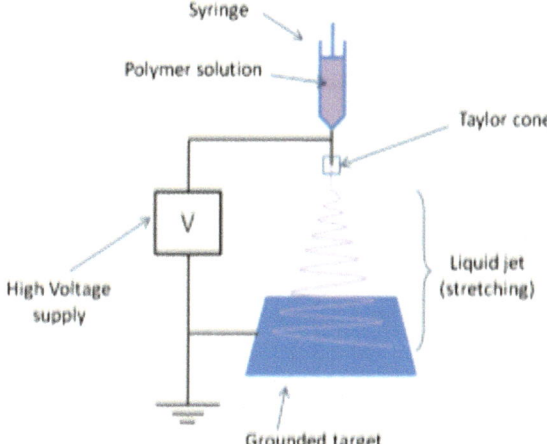

(4) *Power supply* should be adequate to overcome the viscosity and surface tension of polymer solution to form and sustain the jet of polymer from syringe

(5) *Distance between syringe and grounded surface* should not be too small to create sparks between the electrodes but should be large enough for the solvent to evaporate in time for fibres to form

(6) *Ambient Parameters* Temperature fluctuations can cause fluctuations in viscosity and surface tension of solution. Low humidity may dry the solvent and increase the rate of solvent evaporation. On the contrary, high humidity will lead to the thick fiber diameter.

4   Results and Discussion

A. *Optimising Parameters for Electrospinning of GelMA*

(1) *Temperature of GelMA Solution*

Since hydrogels like GelMA have a high affinity for water, the solution exist as gels at room temperature. This high viscosity solution does not produce fibres when electrospun as it results in the hard ejection of jets from solution. Generally, there is a linear relationship between concentration of solution and viscosity. However, in the case of GelMA solutions, even low concentration such as 10% solutions exist as a gel at room temperature. In order for electrospinning to occur, the mixture must exist as a solution to allow the evaporation of water and also the ejection of jets from the solution.

We found that heating the solution to an optimal temperature led to a significant increase in the success rate of fibre production.

As seen by Figs. 2, 3 and 4, the fibre morphology has drastically changed as temperature was increased. At room temperature wet gel was deposited, while at the optimal temperature (56 °C) smooth fibre were produced. At a high temperature

Fig. 2 Fiber morphology of 15% GelMA solution at room temperature (25 °C)

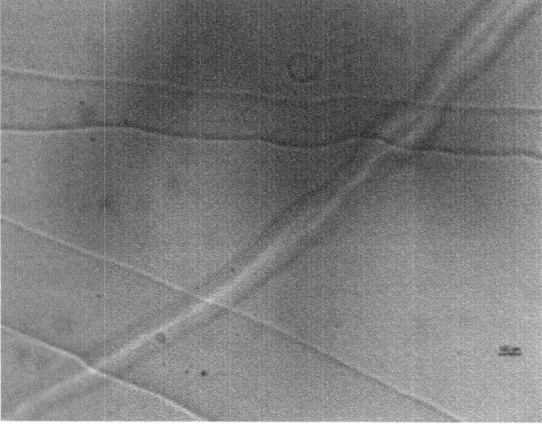

Fig. 3 Fiber morphology of 15% GelMA solution at 56 °C

Fig. 4 Fiber morphology of 15% GelMA solution at 70 °C

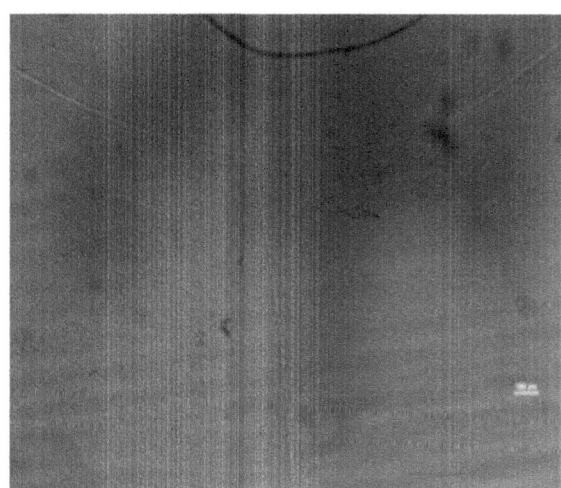

(70 °C), beaded fibre were produced which is not desirable. At a very high temperature, viscosity was very low and surface temperature of solvent was very high which leads to low extensional viscosity, thus leading to bead formation. Thus, temperature is a hugely important controllable variable to note in the electrospinning of GelMA.

(2) *Concentration and Flow Rate of GelMA Solution*

Concentration is an important parameter to consider in the electrospinning of all polymers. When the concentration is very low, polymeric micro (nano)-particles will be obtained and electrospray occurs instead of electrospinning because of low viscosity and high surface tensions of the solution [8]. As the concentration is little higher, a mixture of beads and fibers will be obtained.

Fig. 5 Electrospinning
setup for GelMA

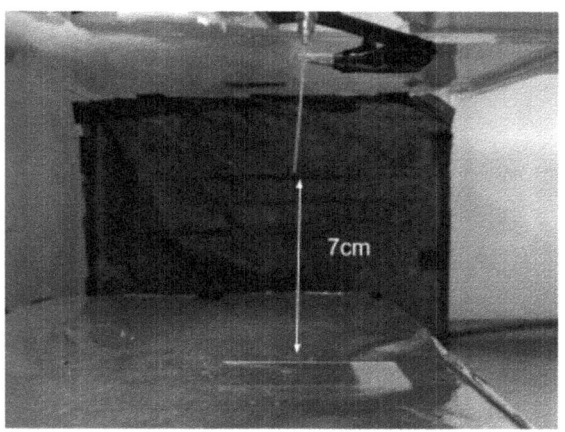

Generally, lower flow rate is recommended as the polymer solution will get enough time for evaporation. If the flow rate is very high, bead fibres with thick diameter will form rather than the smooth fibres with thin diameter owing to the short drying time prior to reaching the collector and low stretching forces. Flow rate is very important in the electrospinning of GelMA due to the low volatility of water, thus a lower flow rate ensures enough time for the evaporation of water and formation of polymer fibres. Generally, fibre production occurs when flowrate is between 0.5 and 1.0 ml/hr. In this study, flow rates of 0.6 and 0.8 ml/hr were tested.

For a certain concentration of GelMA solution and a certain flow rate, the temperature was increased until the mixture existed as a liquid solution rather than a gel. Moreover, in the electrospinning setup, the voltage was increased until formation of fibre was observed.

For all the experiments, the diameter of the needle was constant and the distance between the needle and grounded surface was maintained at 7 cm (Fig. 5).

In order to verify reproducibility, it was ensured that fiber collection with the parameters was repeatable at least twice.

Table 1 in Appendix shows the optimal conditions for electrospinning GelMA solutions of various concentrations as well as the respective fibre morphologies. As seen by the results in Table 1, fibre morphology significantly changed with change in concentration and change in feed rate. Generally, a lower feed rate produced more desirable fibres which had less beads and were smoother. Generally, an increase in concentration led to larger and less smooth fibres however, fibres produced were long and continuous as compared to the short and detached fibres produced using 10% and 12% GelMA solutions.

The temperature and voltage was optimized to the respective concentrations. Increase in concentration led to a more viscous solution thus a higher temperature was needed to ensure that the GelMA remained in solution form rather than in gel form. However, in the case of 20% GelMA solution, a high temperature of 70 °C was needed to maintain the solution in liquid form, causing the anomalous beading and discontinuous wet fibres. A very high concentration of 20% failed to produce dry fibres.

B. *Fiber Morphology*

The diameters significantly vary with concentration of solution, however, there was no significant change in diameter due to change in flow rate. Figures 6, 7, 8 and 9

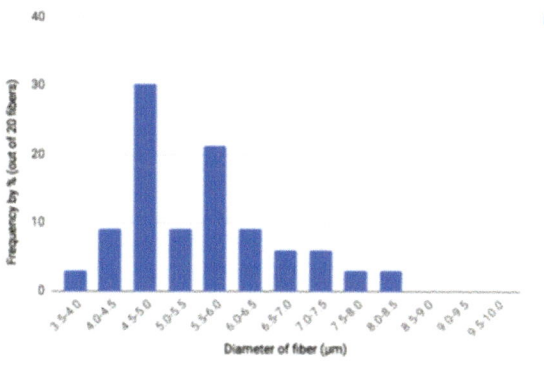

Fig. 6 Diameter of fibers produced from 10% GelMA solution

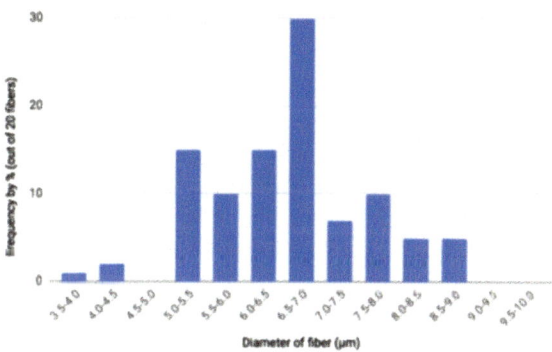

Fig. 7 Diameter of fibers produced from 12% GelMA solution

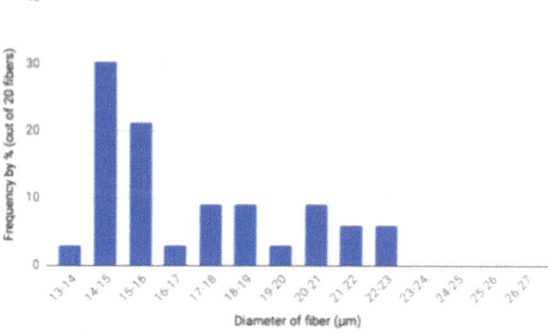

Fig. 8 Diameter of fibers produced from 15% GelMA solution

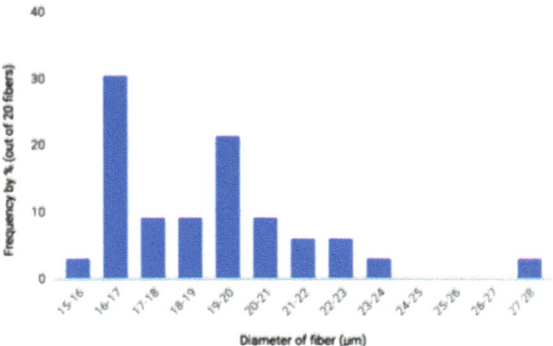

Fig. 9 Diameter of fibers produced from 18% GelMA solution

show the distribution of fibre diameters for 10, 12, 15 and 18% solution respectively. Generally, an increase in concentration led to a increase in average diameter.

This is of utmost importance for the application of GelMA electrospun structures in tissue engineering as engineering of different types of tissues require fibres different morphologies, for example, skeletal muscles require scaffolds to have high mechanical strength and larger fibre diameter.

5   Conclusion and Recommendations for Future Work

Electrospinning of pure GelMA under optimized conditions proved to be a novel method for tissue engineering applications due to the cost effectiveness of electrospinning, tunable fiber morphologies as well as the biocompatibility and mechanical properties of GelMA. Compared to other biocompatible polymers such as Polyvinyl alcohol (PVA), GelMA has a promising remarkable compatibility for a wide spectrum of applications. This study of optimization of noteworthy parameters specifically for the electrospinning of pure GelMA builds the foundation for the development of pure GelMA structures for various tissue engineering applications, tissue engineered scaffolds in particular.

In the future, studies should test the cell attachment, growth and differentiation in the pure GelMA fibrous structures formed using the different concentrations and flow rates, to further determine the optimal concentration for various tissue engineering applications.

Acknowledgements Srushti Sakhardande would like to express our sincere her sincere thanks to Assistant Professor Zhang Yilei for introducing and supervising her on this project as well as allow her to conduct my experiments in the lab at the School of Mechanical and Aerospace Engineering at NTU. She would also like to thank Vivek Damodar for his guidance and support in supervising this research and helping her with his expertise in electrospinning.

Appendix

See Table 1.

Table 1 Optimal conditions for electrospinning GelMA solution of various concentrations

Concentration (%)	Temperature (°C)	Voltage (kV)	Feed rate (ml/hr)	Fiber morphology as seen under light microscope under ×5 magnification
10	50	14.0	0.6	
10	50	14.0	0.8	
12	50	14.6	0.6	

(continued)

Table 1 (continued)

Concentration (%)	Temperature (°C)	Voltage (kV)	Feed rate (ml/hr)	Fiber morphology as seen under light microscope under ×5 magnification
12	50	14.6	0.8	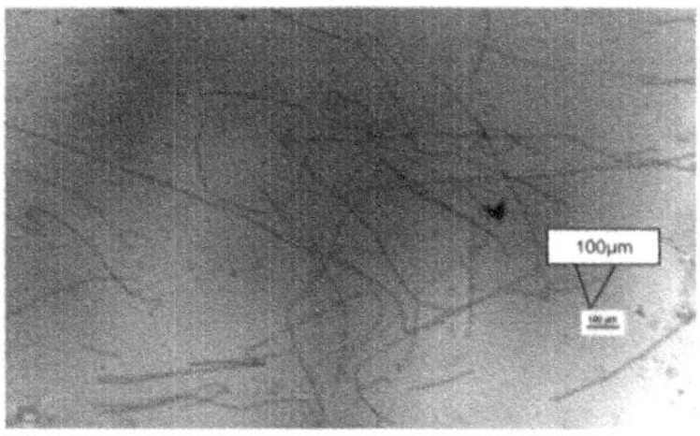
15	56	16.7	0.6	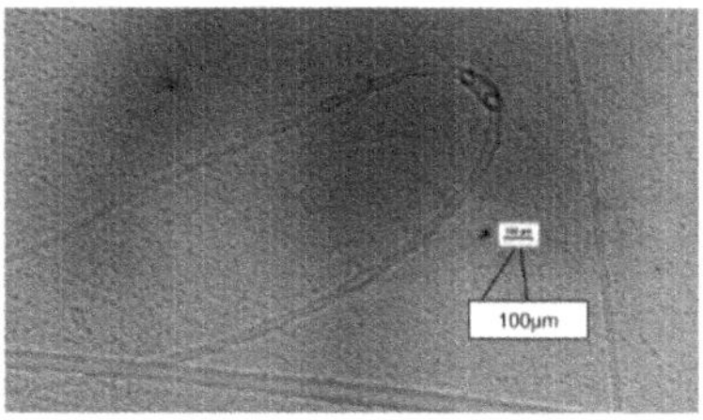
15	56	16.7	0.8	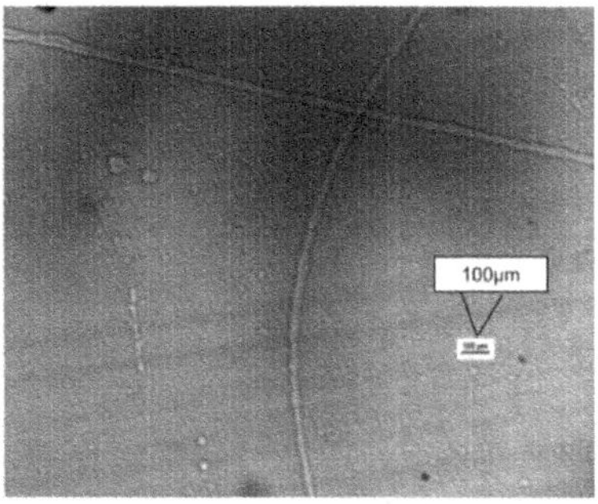
18	62	16.7	0.6	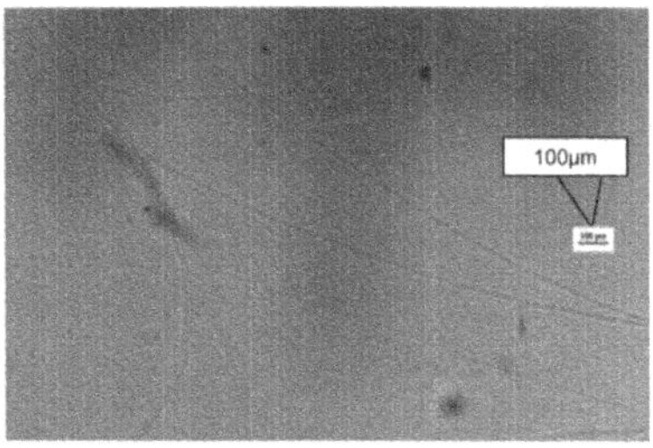

(continued)

Table 1 (continued)

Concentration (%)	Temperature (°C)	Voltage (kV)	Feed rate (ml/hr)	Fiber morphology as seen under light microscope under ×5 magnification
18	62	16.7	0.8	
20	68	16.7	0.6	
20	68	16.7	0.8	

References

1. Nishida, T., Yasumoto, K., Otori, T., & Desaki, J. (1988). The network structure of corneal fibroblasts in the rat as revealed by scanning electron microscopy. *Investigative Ophthalmology & Visual Science, 29*(12), 1887–1890.
2. Li, X., Gao, H., Uo, M., et al. (2009). Effect of carbon nanotubes on cellular functions in vitro. *Journal of Biomedical Materials Research A, 91*(1), 132–139.
3. Li, X., Liu, H., Niu, X., et al. (2012). The use of carbon nanotubes to induce osteogenic differentiation of human adipose-derived MSCs in vitro and ectopic bone formation in vivo. *Biomaterials, 33*(19), 4818–4827.
4. Li, X. M., Wang, L., Fan, Y. B., Feng, Q. L., Cui, F. Z., & Watari, F. (2013). Nanostructured scaffolds for bone tissue engineering. *Journal of Biomedical Materials Research A, 101*(8), 2424–2435.
5. Kadler, K. E., Holmes, D. F., Trotter, J. A., & Chapman, J. A. (1996). Collagen fibril formation. *Biochemical Journal, 316*(1), 1–11.

6. Kan, Y., Trujillo-de Santiago, G., Alvarez, M. M., Tamayol, A., Annabi, N., & Khademhosseini, A. (2015). Synthesis, properties, and biomedical applications of gelatin methacryloyl (GelMA) hydrogels. Retrieved from https://www.sciencedirect.com/science/article/pii/S014296121500719X.
7. Salles, V., Seveyrat, L., Fiorido, T., Hu, L., Galineau, J., Eid, C. … Guyomar, D. Synthesis and characterization of advanced carbon-based nanowires—Study of composites actuation capabilities containing these nanowires as fillers. Retrieved from https://www.intechopen.com/books/nanowires-recent-advances/synthesis-and-characterization-of-advanced-carbon-based-nanowires-study-of-composites-actuation-capa.
8. Deitzel, J. M., Kleinmeyer, J., Harris, D., & Beck Tan, N. C. (2011). The effect of processing variables on the Morphology of electrospun snanofibers and textiles. *Polymer, 42*(1), 261–272.

Study of Bird Feathers to Improve Design of Absorbent Pads for Greater Efficiency of Oil Spill Removal

Alicia Chua Hao Shan and Daphne Chu Zhiying

Abstract Oil spills in the ocean cause both short and long-term environmental damage and can pose threats to wildlife and ecosystems. The use of sorbents such as absorbent pads is one method which can remedy oil spills, because of their hydrophobicity and oleophilic nature. However, many of such absorbent pads are non-biodegradable and costly. Bird feathers are able to adsorb oil well due to their structure and are cheap organic materials. This research aims at studying the features of bird feathers which contribute to their oil adsorbing ability, so as to adapt these features to current absorbent pad models to determine an ideal structure that can increase efficiency of oil removal. In this research, pheasant and duck feathers were compared with four types of oil absorbent pads by determining which material absorbs the most oil relative to its initial mass. The materials were further observed using microscopes and a SEM machine. It was determined that pheasant feathers are better at oil adsorbing due to the size of their apparent contact angles, which affect the amount of oil which can enter the pores between the feather fibres (Cassie and Baxter in Trans Faraday Soc 40:546, [1]). It was also found that absorbent pads with larger fibre widths were able to adsorb and retain more oil, which may be due to a higher surface area or a lower contact angle of the oil on the surface of the pad.

Keywords Oil spills · Sorbents · Oil absorbent pad · Hydrophobic · Oleophilic · Bird feather · Adsorption

1 Introduction

Oil spills in the ocean are common occurrences which have great impact on the pollution and destruction of the environment. Being a major concern to the industry, certain methods such as the use of sorbents (absorbent pads) and biological agents have been used to tackle oil spills.

Pertaining to the use of biological agents, microbes are used to promote degradation of oil. However, due to the vast number of components in the oil, large numbers

A. Chua Hao Shan · D. Chu Zhiying (✉)
National Junior College, Singapore, Singapore
e-mail: daphnechu2001@gmail.com

© Springer Nature Singapore Pte Ltd. 2019
H. Guo et al. (eds.), *IRC-SET 2018*,
https://doi.org/10.1007/978-981-32-9828-6_4

of multiple types of bacteria are required to make changes to oil spill conditions. This can result in oxygen depletion in ocean waters in the areas where the microbes are active. Furthermore, a large amount of bacteria would require large amounts of nutrients from the ocean to work efficiently, which can lead to insufficient nutrients in the ocean and eventually result in inefficient bacteria activity, leading to the formation of sedimentation due to incomplete breakdown of hydrocarbons. Many hydrocarbons are also too big for microbes to clean up, resulting in ineffective oil spill cleanup.

Thus, the use of sorbents is ultimately the preferred method of cleaning up oil spills, because they are both highly hydrophobic and oleophilic and are able to retain oil due to adsorption and, though less commonly, absorption [2].

This research therefore focuses on sorbents rather than biological agents. Conventional sorbents such as oil absorbent pads are largely non-biodegradable and costly. Bird feathers are known for their hydrophobic as well as oleophilic properties. They are also low-cost, organic materials [3]. As such, this research aims to study the structure of feathers of both land and sea birds, as well as compare these structures to current absorbent pads used. This research uses two types of bird feathers, pheasant and duck, to represent the differences between land and sea birds. Four oil absorbent pads with different structural features were also compared and numbered for easier reference [Pad 1 (Cloversoft), Pad 2 (Joylife), Pad 3 (3M), Pad 4 (티|투컴)].

By observing and relating the oil adsorbing abilities of each of these materials to their structural properties, this research hopes to determine an ideal structure and feature for absorbent pads used to tackle oil spills, such that the efficiency of oil removal can be increased. In order to explain the reasons behind observations made during experimentations to allow the research to advance further, this research tapped on the previous research on the structural features of bird feathers. This research strictly focuses on the physical properties and features of the materials, which allow feathers and oil absorbent pads to adsorb and retain oil and repel water. The oils used in this research were gasoline and lubricants to mimic different types of oil spill conditions.

2 Hypothesis

With respect to the physical structural properties of feathers, feathers are able to adsorb oil more effectively than absorbent pads. In addition, feathers of sea birds are able to adsorb oil more effectively than that of land birds due to their water repellent properties.

3 Methodology

For each of these experiments, each material (absorbent pad 1, 2, 3 and 4, pheasant and duck feather) was dipped into their respective petri dishes for 10 s, then removed to be placed on a paper towel for 5 s, in order to drain excess oil. The materials' initial and final mass were recorded, in order to obtain the change in mass. The percentage of change in mass due to adsorbed oil (percentage of change) was derived by taking the change in mass over the initial mass.

3.1 Comparison of Oil Adsorption Ability of Feathers and Regular Paper Towels

The aim of this section was to determine the amount of oil that should be used in order to be able to observe how the structure of the bird feather changes with the addition of oil, as well as to compare the oil adsorbing ability of pheasant feathers, duck feathers and regular paper towels. The two types of feathers and the regular paper towel were cut to roughly the same length (between 3.80 and 4.50 cm). A total of 15 petri dishes were set up, including three negative controls, prepared with just tap water to prove that both types of feathers are hydrophobic, while the regular paper towel is not. The other 12 setups were prepared with either sunflower cooking oil or gasoline, with two different amounts of the oils, 0.5 and 10.0 mL. The recordings of the various setups and data are listed in the table of experimental data (found in Appendix). The materials were observed under a microscope before and after the experiments.

3.2 Effect of Surface Area to Mass Ratio on Oil Absorption Ability of Absorbent Pads

This experiment was conducted to test out the hypothesis that a stretched piece of absorbent pad is a more effective oil adsorber as compared to an untouched piece of absorbent pad, because the ratio of surface area to mass is larger for a stretched absorbent pad. Two pieces of the absorbent pad were cut out. One piece was stretched out until light was able to penetrate the pad, while the other piece remained at its original length and thickness. This experiment was conducted with petri dishes containing grapeseed oil. The table of experimental data shows the data collected (details in Appendix).

3.3 Comparison of Oil Adsorbency of Materials

In this experiment, the oil adsorbency of all the materials used in this research were compared. The structures of these materials were observed using the Scanning Electron Microscope (SEM). From there, the absorbent pad and feather that could adsorb

(a)

(b)

(c)

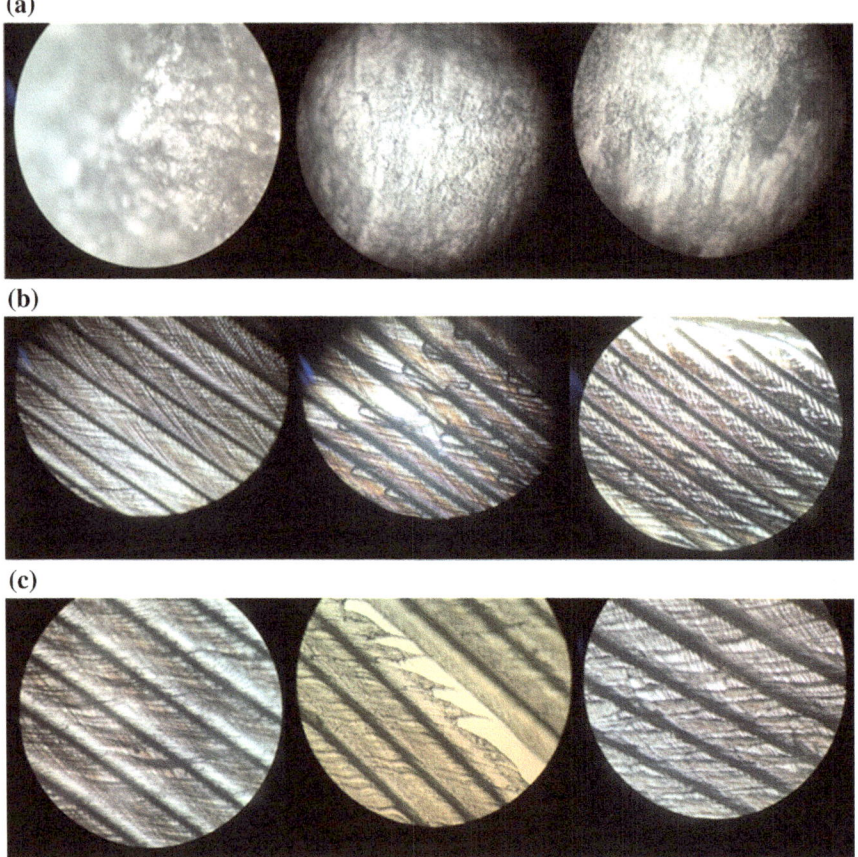

Fig. 1 **a** Duck feather—left to right: before oil, 0.5 mL cooking oil, 0.5 mL petroleum, **b** pheasant feather—left to right: before oil, 0.5 mL cooking oil, 0.5 mL petroleum, **c** paper towel—left to right: before oil, 0.5 mL cooking oil, 0.5 mL petroleum

oil the best were determined. The second part was done on the best absorbent pad and feather with motorcycle lubricant and salt water, which was made with sea salt to mimic sea water. The purpose of these experiments was to determine whether existing absorbent pads were better than bird feathers at adsorbing oil. Acetone was used to remove any traces of uropygial oil left on the feathers, to eradicate the possibility of the oil affecting the adsorbing ability of the feathers. The feathers were left in a beaker filled with acetone for roughly two minutes, then washed and dried thoroughly but carefully. The salt water was prepared by mixing 33.4 g of Red Sea Coral Pro Salt to 1 L of water. These feathers and absorbent pads were trimmed to similar lengths (~2.5 cm) and their initial masses were recorded.

(i) Comparing oil adsorbency of feathers and absorbent pads

A total of 18 petri dishes were set up in a way similar to the set-ups in Trial Run I. The oils used were 1.0 mL each of gasoline, motorcycle lubricant and automobile lubricant, and the materials being tested on were the four types of absorbent pads and the two types of feathers.

(ii) Comparing oil adsorbency of materials under conditions which mimic oil spills

The pheasant feather and absorbent pad 3 were tested on as they were determined to be the best out of the feathers and absorbent pads. Two set-ups were prepared with a mixture of saltwater and 1.0 mL of motorcycle lubricant (Fig. 1).

4   Results and Discussion

In the comparison of oil absorption ability of feathers and regular paper towels (Fig. 2a), when both petroleum and cooking oil were used, pheasant feathers had highest percentage of change, followed by duck feathers, then paper towels. In short, pheasant feathers were able to adsorb more petroleum and cooking oil as compared to duck feathers due to its structure, which was then observed using the SEM.

While comparing the oil adsorbency of absorbent pad 2 while varying the surface area, the percentage of change in the stretched absorbent pad was 6.530 (4 significant figures) times more than an untouched absorbent pad (Fig. 2b), confirming the hypothesis that an increased surface area contributes to a higher oil absorbent ability.

While comparing oil adsorbency of all the materials, for the absorbent pads that were dipped in motorcycle lubricant, the percentage of change was the highest for pad 3, followed by pad 1, then pad 4 and lastly pad 2 (Fig. 2c). Current sorbents rely on the use of mostly adsorption, which is determined by certain factors such as the difference in the critical surface tension of the material compared to the liquid and the surface area of the material [3]. It was observed that the thickness of absorbent pad 2 could not be reduced as it was not made up of layers like the other pads. This suggests that absorbent pad 2 might be able to adsorb more oil, which was proven when the stretched piece was able to adsorb more oil than an untouched piece (Fig. 2b). Therefore, the results that absorbent pad 2 adsorbed the least may be considered inaccurate. However, this may suggest that an absorbent pad that is unable to separate into layers is less efficient in oil adsorption, as a large part of the pad will be unable to make contact with the oil.

With reference to the SEM pictures (Fig. 3a–d), it can be observed that all the oil absorbent pads were made up of non-uniform network of fibres which have air spaces between them. These fibres differ in width (from ~10 to ~40 μm) as well as density. However, the fibres in absorbent pad 4 (Fig. 3d) are much thinner and inconsistent in terms of the width (differing from 2.988 to 12.02 μm) and more dense than the other structures. The fibres in absorbent pad 3 are mostly consistent in terms of width (from 22.38 to 23.66 μm) (Fig. 3c), while absorbent pad 1 consists of fibres that are less than 20 μm in width (Fig. 3a), and absorbent pad 2 consists of fibres of width from 31.65 to 31.65 μm (Fig. 3b).

Fig. 2 Comparison of oil absorption ability of feathers and regular paper towels

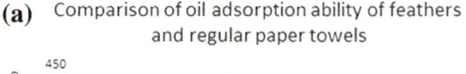

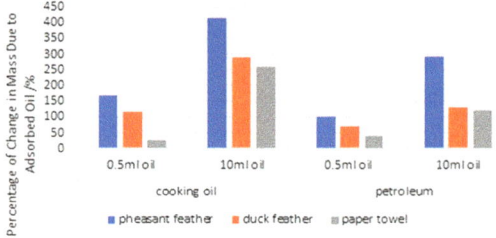

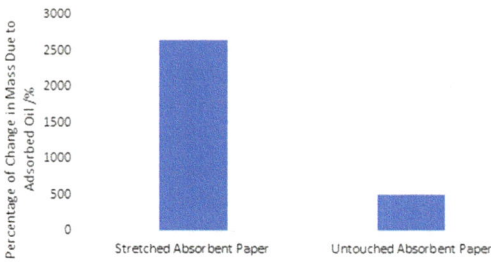

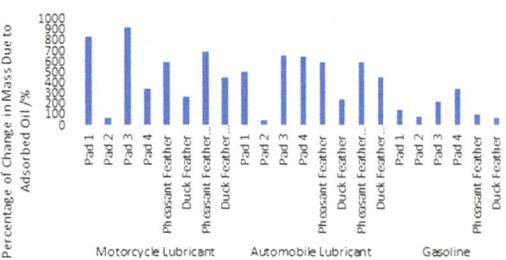

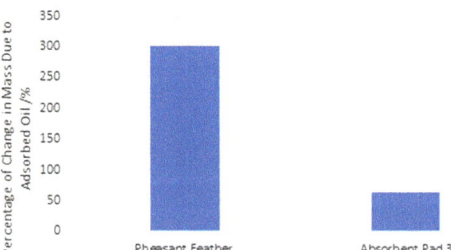

(a) (b) (c) (d)

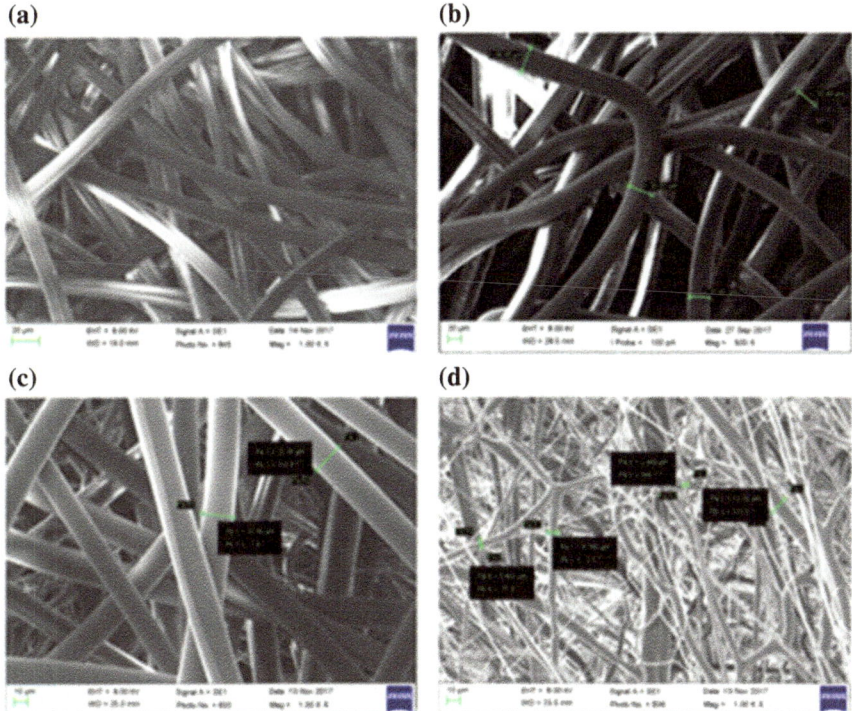

Fig. 3 **a** SEM of absorbent pad 1 at 1.0k×, **b** SEM of absorbent pad 2 at 500×, **c** SEM of absorbent pad 3 at 1.0k×, **d** SEM of absorbent pad 4 at 1.0k×

As such, an efficient absorbent pad should not have a network of thin fibres, but should contain fibres that are larger in width. This can be seen when absorbent pads 3 and 1 were able to adsorb more oil than 4. One theory as to why a larger fibre width allows for more oil to be adsorbed is that it lowers the contact angle of the oil on the surface of the pad and decreases the critical surface tension of the material. Alternatively, a larger fibre width may also enable a larger surface area for adsorption of oil to occur.

With regards to the feathers dipped into motorcycle lubricant, pheasant feathers had the higher percentage of change, as compared to duck feathers (Fig. 2c). Although the high water-repellency of duck feathers seem to indicate that they are better at adsorbing oil, when observed on a structural level, it can be seen that the features which contribute to the duck feather's high water-repellency also affect and decrease its ability to adsorb oil.

Large apparent advancing and receding contact angles of porous surfaces and mobile drops of liquids ensures that the substance can roll of the surface without wetting the material. Large apparent contact angles can be deduced when $(\gamma + d)/r$ is large, where $(\gamma + d)$ is the spacing of the fibres and r is the radius of the fibres [1].

Fig. 4 a SEM of pheasant feather at 300×, **b** SEM of duck feather at 300×, **c** SEM of pheasant feather at 800×, **d** SEM of duck feather at 800×

Table 1 Experimental data from comparing oil adsorbency of feathers and absorbent pads

	Initial mass (g)	Final mass (g)	Change in mass (g)	Percentage of change (%)
Pheasant feather	0.005	0.020	0.015	300
Absorbent pad 3	0.055	0.090	0.035	63.6

When any mobile drop of a liquid falls on the feather, due to the structure of the feather and depending on the surface tension of the liquid, the drop may or may not fall through the pores between the feather fibres. Water has higher surface tension than oil, due to the strong hydrogen bonds between water molecules, as compared to the weaker non-polar interactions between molecules in oil. Thus, the likelihood of oil droplets transiting from the Cassie state [1] to the Wenzel state [4, 5] by seeping into the pores between the fibres of bird feathers is higher than that of water, which is why feathers are able to adsorb oil, despite being water-repellent. Therefore, depending on their apparent contact angles, different types of bird feathers may or may not be more susceptible to adsorbing oil.

Based on the SEM pictures obtained in this research (Fig. 4a–d), although difficult to compare the spacing of the fibres due to the differences in angles, it can be seen that the radius of duck feather fibres is smaller than that of pheasant feather fibres, resulting in larger apparent contact angles for duck feathers. Thus, because of the structure of duck feathers, it is both highly water-repellent and more repellent to oil than pheasant feathers are. This explains why pheasant feathers have been shown to be better at adsorbing oil.

Therefore, it should be taken into account that the spacing and radius of the fibres must be such that $(\gamma + d)/r$ is substantially large, yet the spacing of the fibres must not be too big that droplets of liquid can pass into and out of the pores without difficulty. The data collected from this experiment is reflected in the table of experimental data (Table 1).

When the oil adsorbency of the pheasant feather and absorbent pad 3 were compared under conditions which mimic oil spills, the percentage of change for the pheasant feather was higher than that of absorbent pad 3 (Fig. 2d). This suggests that the structure of pheasant feathers enable it to adsorb more oil as compared to regular oil absorbent pads. The data collected is reflected in the table of experimental data (Table 2).

5 Conclusion and Future Works

Despite the fact that bird feathers are highly effective in adsorbing oil when observed on a volume per unit area level, the idea of directly using raw bird feathers to tackle oil spills is implausible and ineffective, because a large amount of bird feathers is required to make substantial impact on oil spill situations. On the other hand, if man-made materials were utilised, these materials can be used to produce absorbent pads in bulk to consistently tackle oil spills. As such, combining the features of bird feathers that aid them in their oil adsorbing abilities with current oil absorbent pad models would create an absorbent pad that is highly practical and effective in tackling oil spills.

Although this research managed to observe and determine the type and features of bird feathers that would be serve as an ideal template for absorbent pads, it was not able to explore the possibilities of physically creating an absorbent pad with said features due to the limitations of time and appropriate resources. If this research could be brought further, the following areas could be explored: varying the factors with which the experiments take place, such as temperature and humidity, to further mimic oil spill conditions. The type of oil used during experimentations could also become the dependent variable, to allow observation of how density and viscosity of different types of oils can affect the adsorbent properties of bird feathers and current absorbent pads. Once all thorough experimentations have been concluded, an ideal blueprint can be created and then 3D printed, such that the adsorbent properties of

Table 2 Experimental data from comparing oil adsorbency of materials under conditions which mimic oil spills

Material	Motorcycle lubricant				Automobile lubricant				Gasoline			
	Initial mass (g)	Filial mass (g)	Change in mass (g)	Percentage of change (%)	Initial mass (g)	Final mass (g)	Change in mass (g)	Percentage of change (%)	Initial mass (g)	Final mass (g)	Change in mass (g)	Percentage of change (%)
Pad 1	0.030	0.280	0.250	833	0.035	0.215	0.180	514	0.035	0.085	0.050	143
Pad 2	0.225	0.390	0.165	73.3	0.245	0.370	0.125	51.0	0.230	0.420	0.190	82.6
Pad 3	0.025	0.255	0.230	920	0.025	0.190	0.165	660	0.025	0.080	0.055	220
Pad 4	0.050	0.220	0.170	340	0.030	0.225	0.195	650	0.010	0.045	0.035	350
Pheasant feather	0.005	0.035	0.030	600	0.005	0.035	0.030	600	0.005	0.010	0.005	100
Duck feather	0.015	0.055	0.040	267	0.020	0.070	0.050	250	0.015	0.025	0.010	66.7
Pheasant feather (acetone)	0.005	0.040	0.035	700	0.005	0.035	0.030	600				
Duck feather (acetone)	0.010	0.055	0.045	450	0.010	0.055	0.045	450				

the new absorbent pad. Following that, different compositions of the absorbent pad can be explored, such as other forms of hydrophobic structures that may prove to be more effective and also biodegradable.

Appendix

See Tables 3 and 4.

Table 3 Experimental data from trial run I

Type of material used	Type of oil used (mL)	Volume of oil used (mL)	Length of material (cm)	Initial mass of material (Ml) (g)	Final mass of material (M2) (g)	Change in mass of material (M2 − M1) (g)	Percentage of change in mass due to adsorbed oil [(M2 − M1)/Ml * 100] (%)
Pheasant feather	–	0	4.50	0.015	0.016	0.001	
Duck feather	–	0	4.20	0.029	0.029	0	
Paper towel	–	0	4.40	0.040	0.106	0.066	
Pheasant feather	Cooking	0.5	4.20	0.013	0.035	0.022	169
Duck feather	Cooking	0.5	4.00	0.037	0.079	0.042	114
Paper towel	Cooking	0.5	4.20	0.028	0.036	0.008	28.60
Pheasant feather	Petroleum	0.5	3.90	0.013	0.026	0.013	100
Duck feather	Petroleum	0.5	4.20	0.041	0.070	0.029	70.70
Paper towel	Petroleum	0.5	4.10	0.030	0.042	0.012	40
Pheasant feather	Cooking	10.0	4.00	0.014	0.072	0.058	414
Duck feather	Cooking	10.0	4.10	0.033	0.128	0.095	288
Paper towel	Cooking	10.0	4.30	0.036	0.130	0.094	261
Pheasant feather	Petroleum	10.0	3.80	0.010	0.039	0.029	290
Duck feather	Petroleum	10.0	3.90	0.040	0.091	0.051	128
Paper towel	Petroleum	10.0	3.90	0.035	0.078	0.043	123

Table 4 Experimental data from trial run II

Type of material used	Initial mass of material (Ml) (g)	Final mass of material (M2) (g)	Change in mass of material (M2 − M1) (g)	Percentage of change in mass due to adsorbed oil [(M2 − M1)/Ml * 100] (%)
Stretched absorbent pad 2	0.080	2.115	2.035	2540
Untouched absorbent pad 2	0.700	3.425	2.725	389

References

1. Cassie, A. B., & Baxter, S. (1944). Wettability of porous surfaces. *Transactions of the Faraday Society* 546–551.
2. ITOPF, I. T. (2014, May 19). *Use of sorbent materials in oil spill response.* Retrieved from ITOPF Technical Information Paper: https://www.itopf.org/knowledgeresources/documents-guides/document/tip-08-use-ofsorbent-materials-in-oil-spill-response/.
3. Ifelebuegu, A. O., & Chinonyere, P. (2016). *Oil spill clean-up from sea water using waste chicken feathers.* The fourth international conference on advances in applied science and environmental technology (ASET) (pp. 61–64).
4. Wenzel, R. N. (1936). Resistance of solid surfaces to wetting by water. *Industrial & Engineering Chemistry* 988–994.
5. Dai, X., Stogin, B. B., Yang, S., & Wong, T.-S. (2015). *Slippery Wenzel state.* Pennsylvania.
6. Limited, I. T. (2012). *Use of sorbent materials for oil spill response.*

Design of Smart Antimicrobial Materials Based on Silver-Silica Nanocapsules

Zi Chyng Elizabeth Tan, Chenxin Zhang and You Wei Hsu Benedict

Abstract Silver nanoparticles have been commonly used as an antibacterial agent and are often delivered in a burst release manner to the site of infection. However, a drawback of this release mode is the limited lasting duration of the antibacterial properties of the particles. Hence, in order to achieve a more effective and sustained protection against bacteria growth, this project aims to design and create smart antimicrobial materials based on silver-silica nanocapsules that can respond to an acidic environment to release Ag^+ ions in a targeted, slow and sustained manner. In this project, the as-synthesized silica-silver nanocapsules were found to exhibit excellent colloidal stability, thus allowing for a homogenous distribution within different polymer matrix materials. Explored applications include the incorporation of the silver-silica nanocapsules into F127 hydrogel and poly(vinyl alcohol) (PVA) film so as to develop antibacterial biomaterials that can effectively prevent bacteria growth for a sustained period of time. In subsequent proof-of-concept studies, both the F127 hydrogel and PVA film were able to respond to acidic conditions for a gradual release of Ag^+ ions. Interestingly, the as-released Ag^+ ions from the PVA film were effectively entrapped within the polymer matrix, thereby demonstrating their promising potential to sterilize absorbed fluid from wound sites when applied as a wound dressing. On the other hand, the F127 hydrogel exhibited a slow and sustained release of Ag^+ ions into the surrounding environment, hence affirming their capacity for topical administration in the form of lotions or creams for antibacterial purposes.

Keywords Silver nanocapsules · Silver-silica nanocapsules · Antimicrobial · Sustained release · Acid responsive · Stimulated response

Z. C. E. Tan · C. Zhang
Hwa Chong Institution (College), Singapore, Singapore

Y. W. H. Benedict (✉)
Institute of Materials Research and Engineering, Agency for Science, Technology and Research (A*STAR), Singapore, Singapore
e-mail: benedict-hsu@imre.a-star.edu.sg

© Springer Nature Singapore Pte Ltd. 2019
H. Guo et al. (eds.), *IRC-SET 2018*,
https://doi.org/10.1007/978-981-32-9828-6_5

49

1 Background and Purpose of Research Area

Silver ions have been established to have antibacterial properties as they are able to destroy bacteria cells by damaging the cell envelope of bacteria and intracellular content [1]. So far, the antibacterial properties of silver has been utilized in many areas, such as in water purification [2] or in the treatment of burn victims [3]. Such silver ions are utilized in a burst release manner, where all the silver ions present are instantly exposed to their surroundings. However, high concentrations of either silver ions or metallic silver at once can be toxic to the human body [4]. On the other hand, a more controlled release will ensure that the amount of silver ions remain at a level below the toxic threshold of the body as the ions will be slowly exposed to their surroundings and used up accordingly.

Besides, a more limited exposure to silver will also allow for a larger reservoir of silver to be present without being toxic to the human body. Coupled with the design of a stimuli-responsive antibacterial material to allow for a more targeted and controlled release of silver ions, this can potentially increase the lasting duration of antibacterial properties. In this regard, it is noteworthy that certain types of bacteria can create an acidic medium by secreting acids [5], which can act as a stimuli for the release of silver ions. Therefore, this project aims to develop a smart antibacterial material that can respond to acidic conditions to release the silver ions in a sustained and controlled manner. This will hopefully improve current silver-based antibacterial materials by enabling the material to be more effective for a sustained protection against bacteria growth.

2 Engineering Goal of Project

To design and create smart antimicrobial materials based on silver-silica nanocapsules. The silver-silica nanocapsules (AgNCs) will act as a reservoir of silver ions stored in the form of metallic silver on the surface of the nanocapsules. Upon exposure to an acidic environment, the metallic silver will be converted into silver ions for enhanced antibacterial properties. Silver nanocapsules are chosen because they have been found to provide a sustained release of silver ions instead of a burst release when at sizes larger than 50 nm [6]. This aligns with the goal of the project for a sustained protection against bacteria. Their smaller size also allows for more silver to be stored due to a larger surface area to volume ratio.

3 Methods and Materials

3.1 Synthesis of Silica Nanocapsules

30 mg of Pluronic F127 (F127) was first dissolved in 900 μL of tetrahydrofuran (THF), followed by the addition of 40 μL of tetramethoxysilane (TMOS) to obtain a homogeneous mixture. The resulting solution mixture was then slowly injected into 10 mL of deionized water while stirred at 800 rpm, and left to stir for an additional 3 days at 300 rpm (For a schematic illustration of the synthesis protocol, refer to Appendix 1).

3.2 Synthesis of Silver Nanocapsules (AgNCs)

A templating strategy was explored in the synthesis of AgNCs, where the prepared hollow silica nanocapsules were encapsulated in metallic silver particles. Metallic silver particles were deposited onto the hollow silica nanocapsules templates via the Tollens reaction of silver mirroring. Herein, the silver complex $[Ag(NH_3)_2]^+$ stock solution was first prepared by dissolving silver nitrate (200 mg) in deionized water (10 mL), to which ammonium hydroxide (400 μL) was added dropwise until the precipitate that formed redissolved. Next, the silica nanocapsules, glucose and silver complex $[Ag(NH_3)_2]^+$ were mixed together and left to agitate. Finally, the conditions of the Tollens' reaction were then optimized by varying the concentration and volume of the glucose solution, concentration of silica nanocapsules and the concentration of the silver complex $[Ag(NH_3)_2]^+$ solution (Refer to Appendix 2 for details of the concentration variation).

3.3 Characterization Methods

Characterization of the AgNCs by UV-Vis spectroscopy was carried out to determine if silver has indeed formed on the surface of the silica nanocapsules. The AgNCs were then analyzed under the absorbance spectra in 400–4000 nm range with a Fourier Transform Infrared (FTIR) spectrophotometer to identity organic species in the AgNCs to confirm the silica still remained in the AgNCs. Dynamic light scattering (DLS) was used to determine the size of the nanocapsules as well as the critical micelle concentration. A scanning electron microscope (SEM) was used to determine the morphology of the AgNCs and the contents of the poly(vinyl alcohol) (PVA) film.[1] To test for the antibacterial properties of AgNCs, a culture of *Pseudomonas aeruginosa* ATCC 9027 strain was incubated with varying AgNC concentrations of

---

[1]Done by supervisor.

5, 10, 20 and 40% in respect to the cell medium. The culture was then left at room temperature for two days before quantifying for the bacteria growth in comparison to a bacteria culture containing no AgNCs (see footnote 1).

To test for the acid-responsive properties of AgNCs, the AgNC solution was first freeze-dried to obtain its dry powder form. Next, the AgNC powder was added to different types of acids at various pH. It was noted that 1 M nitric acid (HNO_3) was able to completely breakdown AgNCs (dry powder form) into Ag^+ ions as shown by the decolorization of the brown mixture into a clear solution (Refer to Appendix 3 for overnight results).

3.4 Preliminary Processing of Composite Antimicrobial Biomaterial Using PVA Film

Thin films incorporating AgNCs were created using 10% weight per volume (w/v) PVA. PVA powder of 4 different molecular weights were experimented: (i) 80% hydrolyzed, MW 9000–10,000, (ii) 87–89% hydrolyzed, MW 30,000–50,000, (iii) 99+% hydrolyzed, MW 1–30,000, and (iv) 99+% hydrolyzed, MW 85,000–124,000. For each type of PVA, 1 g of PVA was dissolved in 5 mL of deionized water by heating in an oven for 3 h at 95 °C. After cooling to room temperature, 5 mL of AgNC solution was added and left on an orbital shaker to be homogenized overnight. After homogenization, the mixture was spread over a petri dish to form a thin layer, after which it was placed in a freezer at −20 °C to freeze overnight. The samples were then taken out to thaw for 30 min, before freezing overnight again. The freeze-thaw cycle was repeated thrice.

To test if AgNCs can still release silver ions in acidic conditions while being embedded in PVA polymer matrix, the as-synthesized PVA films were cut into squares of size 2 cm × 2 cm, and then immersed into 10 mL of HNO_3 with concentrations of 1, 0.5 and 0.1 M. The supernatants were then extracted and tested for silver ions using qualitative analysis. The composite antibacterial films were also subsequently observed for possible decolorization, which would indicate the conversion of metallic silver into silver ions, upon leaving to stand overnight.

3.5 Preliminary Processing of Composite Antimicrobial Biomaterial Using F127 Hydrogel

F127 is a FDA-approved thermoresponsive polymer with a sol-to-gel transition that can be triggered by the average body temperature of 37 °C. Hence it can be hopefully used as a gel to be applied on the body. 16%, 18%, 20% and 22 wt% F127 were prepared by weighing 0.8 g, 0.9 g, 1.0 g and 1.1 g of F127 respectively and topping up to 5 mL using deionized water, after which it was left in a fridge (4 °C) overnight for complete dissolution. To verify the sol-to-gel transition behavior of F127 hydrogel, the glass vial was inverted at both 25 and 37 °C. Visual inspection was then carried out to observe for any flow of liquid after being inverted.

Acid response to AgNCs-incorporated F127 hydrogel was tested by immersion in 1500 cm^3 of 0.1 M HNO$_3$ using a dialysis cassette of 10,000 MW cutoff, which would prevent the F127 hydrogel from passing through while allowing the Ag$^+$ ions to diffuse out into the HNO$_3$. Visual inspection of these hydrogels was then carried out to observe for color change to indicate the breakdown of AgNCs into Ag$^+$ ions. At 12-h intervals over 60 h, small aliquots of the dialysis fluid were extracted to determine the released Ag content using Inductively Coupled Plasma Mass Spectrometry (ICP-MS), and the surrounding medium was also replaced by fresh HNO$_3$. A control was also set up by replacing the HNO$_3$ with distilled water under the same experimental conditions to prove that Ag$^+$ ions were only released in acidic conditions.

4 Results and Discussion

4.1 Silica Nanocapsules Templates

The silica nanocapsule templates were monodisperse and on the nanoscale dimensions, with well-structured morphology (Fig. 1c). Its mechanical stability was attested by the DLS results, where the critical micelle concentration of the silica nanocapsules occurred at a higher dilution level than that of F127 (Fig. 1a, b). This showed that the silica nanocapsules were able to maintain their integrity and withstand high levels of dilution, thereby rendering it a good template to be adopted for the silver nanoparticle coating on it.

4.2 Optimization of the Synthesis Conditions of AgNCs

The optimum reaction conditions were determined from the brown color intensity of the supernatant obtained by the end of the reaction. The brown intensity gave an indication of the presence of metallic silver, which would be formed over the silica nanocapsule templates. Herein, the aim was to obtain a clear solution with a high brown color intensity, yet without the presence of any precipitation since it would otherwise indicate the severe aggregation of the silver nanoparticles.

Based on the solution color intensity obtained after varying the conditions of (i) glucose concentration (Fig. 2a); (ii) glucose amount (Fig. 2b); (iii) silica nanocapsule concentration (Fig. 2a); and (iv) silver complex concentration (Fig. 2c), the optimal combination was (i) 10 mg/mL glucose concentration, where (ii) 40 mg of glucose was dissolved in (iii) 3 mL silica nanocapsules, and (iv) 1 mL silver complex synthesized from 10 mg/mL silver nitrate.

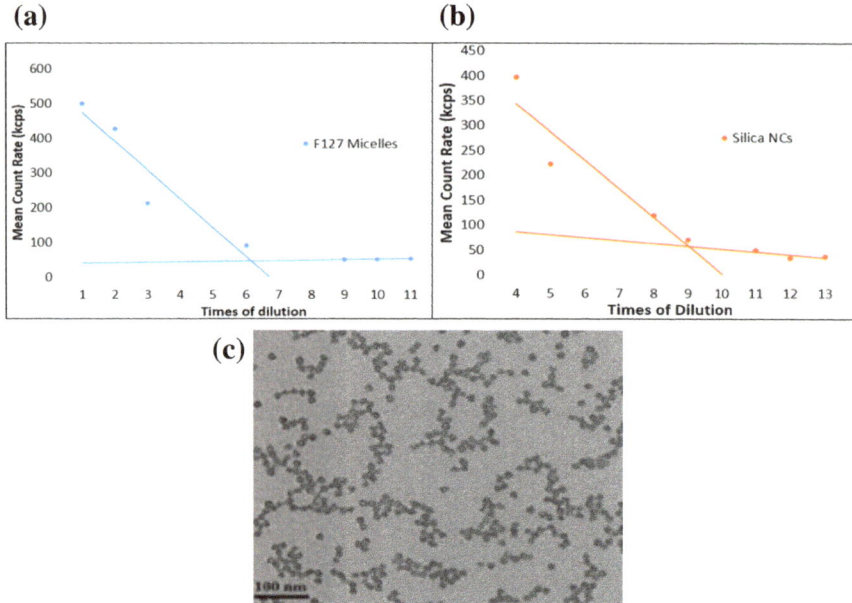

Fig. 1 Characterization of F127 Micelles and Silica NCs. **a** Stability measurement of F127 micelle. **b** Stability measurement of silica NCs. **c** Transmission electron microscopy (TEM) image of silica NCs synthesized [13]

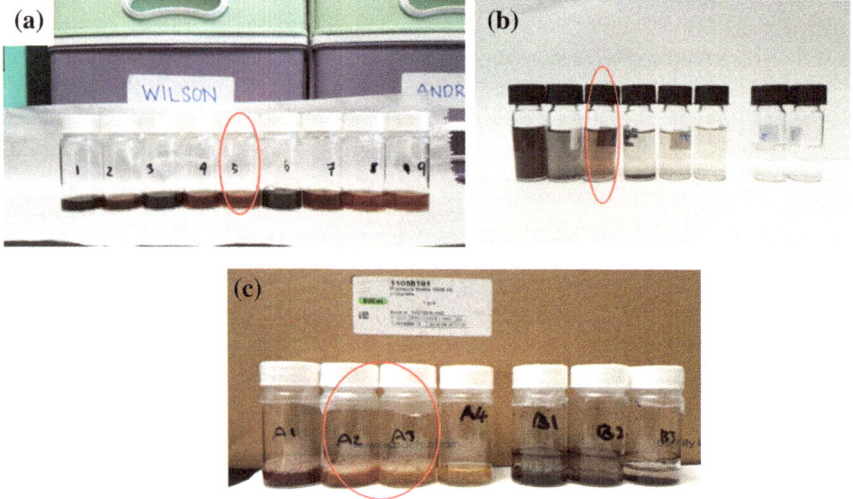

Fig. 2 Results of optimization of synthesis conditions of AgNCs after 2 days, best result(s) circled in red. **a** Variation of the concentration of silica nanocapsules and amount of glucose added, shown left to right, samples 1–9. **b** Variation in glucose concentration, shown left to right: g0–g5, last 2 bottles are controls with no glucose added and no silver complex added respectively. **c** Variation of silver complex solution shown from left to right, A1–A4, B1–B3

4.3 Properties of AgNCs

Formation of AgNCs was determined by UV-vis spectroscopy, where the surface plasmon absorbance maxima can be observed at around 420 nm from the UV-vis spectrum (Fig. 3a). This peak corresponds with what is expected of silver nanoparticles [7]. FTIR spectrum suggested that silica remained intact after its encapsulation with silver due to similar peaks (Fig. 3c). The AgNCs synthesized were uniform and spherical in shape, with a small particle size of around 100 nm (Fig. 3e). This was further supported by the DLS results, where the average hydrodynamic diameter of AgNCs was around 100 nm. DLS also showed a single peak, with little aggregation, which made the AgNCs ideal to be incorporated into polymers to form a homogeneous composite material (Fig. 3d). The as-synthesized AgNCs were also more stable than the silica nanocapsules as the former could withstand higher levels of dilution cycles before reaching its critical micelle concentration (Figs. 3b and 1b). This excellent colloidal stability could be attributed to the hydrophilic PEO chains of F127 in the AgNCs, which increased the water solubility and decreased non-specific protein adsorption.

The antibacterial property of AgNCs was measured by comparing bacteria growth in cell culture mediums with no AgNCs, as well as with AgNCs of various concentrations. Based on the results, it was evident that the AgNC had a dose-dependent antibacterial effect, albeit limited in its inhibition of the bacterial growth. Notably, it was only able to prevent 47.1% of bacteria growth after two day exposure to a 40% nanoparticle concentration (Fig. 3f). This was likely because the AgNCs had not been converted into free silver ions, which would otherwise have a stronger antibacterial effect [8]. Hence, this further justified the design a biomaterial that allowed a stimuli controlled release of Ag^+ ions.

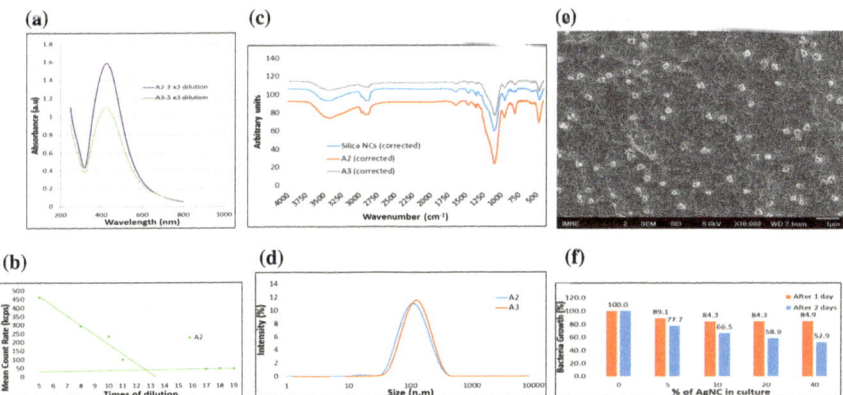

Fig. 3 Properties of AgNCs. **a** UV-Vis spectrum. **b** Stability results. **c** FTIR spectrum. **d** DLS results. **e** Morphology and size. **f** Antibacterial properties

4.4   Incorporation of AgNCs into PVA Film to Form Composite Biomaterial

PVA was chosen as it is biodegradable, cheap to make and has low cytotoxicity, thus making it suitable for medical usage [9, 10]. It has also been widely used as a wound dressing due to its ability to provide a suitable environment for quick healing [11]. After 3 freeze-thaw cycles during the processing of PVA films, only PVA films synthesized using 99+% hydrolyzed PVA formed the films successfully (Refer to Appendix 4 for images of the films). Thus for the subsequent test for their acidic response behavior, only PVA films synthesized using 99+% hydrolyzed PVA, of MW 1–30,000 and MW 85,000–124,000 were used for experimentation.

The PVA films were observed to decolorize from brown to near colorless after immersing overnight in HNO_3 for all 3 experimented concentrations (Fig. 4b). This indicated the successful conversion of AgNCs to Ag^+ ions, stimulated by the acidic medium. Hence, HNO_3 could be used as the basis for the design of antibacterial material that can respond to acidic conditions for a slow release of Ag^+ ions for sterilization purposes.

However, qualitative analysis of the supernatant with sodium chloride yielded negative results as no white precipitate was observed, suggesting that instead of being dissociated into the surrounding aqueous acidic medium, the Ag^+ ions were most probably trapped inside the PVA polymer matrix. Subsequent SEM analysis of the air-dried PVA films (Fig. 4a) confirmed that the AgNCs and any silver complex formed were indeed entrapped in the PVA polymer matrix.

This holds much potential in pharmaceutical applications. Since PVA films are known to have high water intake that allows them to absorb fluid from the wound site [11], the embedding of AgNCs in this polymer matrix kills bacteria and other microorganism that has been absorbed from the wound, preventing the growth of bacteria after being entrapped in the matrix. This keeps the wound sterile, thereby enhancing the property of PVA film as an ideal wound dressing.

(a)   **(b)**

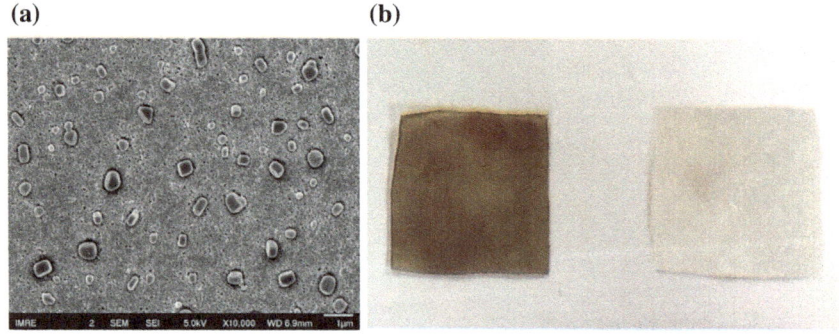

Fig. 4 Properties of PVA film. **a** SEM image of transverse section of PVA film. **b** Shown left to right, PVA film incorporating AgNC, before immersion in HNO_3 (1 M), after immersion in HNO_3 (1 M)

4.5 Incorporation of AgNCs into F127 Hydrogel to Form Composite Biomaterial

F127 was chosen as it is a Food and Drug Administration (FDA) approved thermoresponsive polymer, which can be explored for various unique applications while being safe for medical use [11]. It is also regarded as an effective drug delivery material, thus making it ideal for medical usage [12]. As such, the incorporation of AgNCs into the F127 hydrogel would allow for a safe and effective delivery of the silver ions.

It was observed that the hydrogels formed using 18%, 20%, 22% and 24 wt% F127 could undergo the transition to solid hydrogel at room temperature (25 °C) (Refer to Appendix 5 for images of the hydrogels). Hence, these concentrations of F127 used were less ideal for bioapplication, as the intended sol-to-gel transition was to be induced at around 37 °C, the average body temperature. On the other hand, the hydrogel synthesized using F127 of 16 wt% turned to solid gel at 37 °C within 1–2 min, while transitioning reversibly back to liquid solution at 25 °C within 10 min (Refer to Appendix 5 for image of the hydrogel). This ensured that the F127 hydrogel acted as an effective reservoir of AgNCs that could be both stored and used conveniently.

However, after immersion into HNO_3 maintained at 37 °C, the 16 wt% F127 gel reverted back to solution form, suggesting that F127 was unable to remain in gel form in extreme pH. In this regard, the F127 hydrogel could be treated as a lotion when exposed to extreme pH instead of a gel. From visual inspection (Fig. 5a), and analyzing the aliquots taken at regular 12-h intervals using ICP-MS, the F127 hydrogel was seen to be able to release silver ions at a slow, sustained rate (Fig. 5b), aligning with the goals of the project.

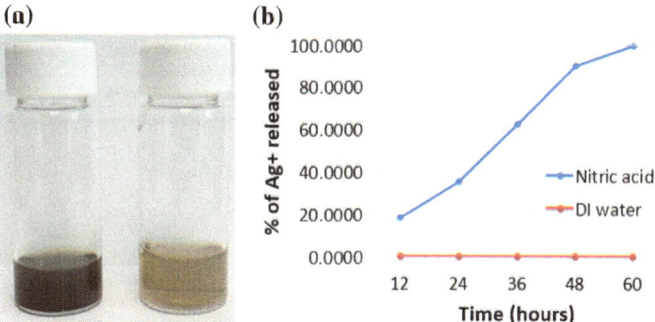

Fig. 5 Release of Ag^+ ions from F127. **a** Left to right, picture of gel before and after immersion in HNO_3. **b** Release profile of Ag^+ in HNO_3 and DI water, adjusted by treating the amount released in HNO_3 at 60-h as 100% released

5   Results and Discussion

From the experiments conducted, it can be seen that the silver-silica nanocapsules
were a stable antibacterial agent that could be easily incorporated into common
polymers, to be further used in various applications. Its colloidal stability arose from
the presence of hydrophilic PEO chains of F127 which enabled the AgNCs to have
excellent antifouling behavior, and also avoided aggregation for a homogenous distribution within different materials. Explored applications include a thermoresponsive
hydrogel or a bandage-like film, both of which were helpful in sterilization and
to prevent bacteria growth. The incorporated AgNCs were also able to respond to
an acidic stimuli within the materials chosen, thus allowing for a more controlled
and sustained release of Ag^+ ions for bactericidal purposes, greatly enhancing its
antibacterial effects.

Due to time constraints, the PVA film and hydrogel were not tested with bacteria
cultures under acidic conditions. Future work can involve testing for the effectiveness
of the materials in preventing bacterial growth in actual bacterial cultures in an acidic
medium.

Appendices

Appendix 1: Synthesis of Silica Nanocapsules

See Fig. 6.

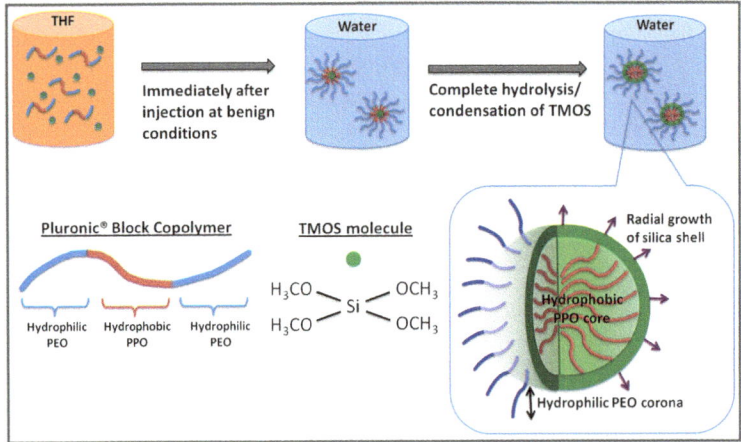

Fig. 6 Diagram showing the formation of micelle and subsequently the silica nanocapsule [13]

Table 1 Variations of glucose concentration for samples g1 to g5 to find the optimal glucose concentration

Sample	Concentration of glucose (mg/mL) in total reaction volume
g0	40.00
g1	20.00
g2	10.00
g3	5.00
g4	2.50
g5	1.25

The glucose solutions (1 mL) were added to the silica nanocapsules (1 mL) and silver complex $[Ag(NH_3)_2]^+$ (1 mL), and left to agitate for 2 days

Table 2 Variations of reagents amount for reactions A1 to A7 to find the optimal glucose amount, based on the optimum glucose concentration of 10 mg/mL, as determined in Table 1

Sample	Mass of glucose (mg)	Volume of deionized water (mL)	Volume of silica nanocapsules (mL)	Volume of silver complex $[Ag(NH_3)_2]^+$ (mL)	Total reaction volume (mL)
A1.	25.0	0.5	1.0	1.0	2.5
A2.	30.0	1.0	1.0	1.0	3.0
A3.	40.0	2.0	1.0	1.0	4.0
A4.	60.0	4.0	1.0	1.0	6.0
A5.	100.0	8.0	1.0	1.0	10.0
A6.	150.0	13.0	1.0	1.0	15.0
A7.	200.0	18.0	1.0	1.0	20.0

Appendix 2: Variation of Reaction Conditions to Synthesize AgNCs

See Tables 1, 2, 3 and 4.

Appendix 3: Response of AgNCs to Acidic Conditions

See Fig. 7.

Table 3 Variations of the reagents amount for samples 1 to 9 to find the optimal silica nanocapsule concentration

Sample	Mass of glucose (mg)	Volume of deionized water (mL)	Volume of silica nanocapsules (mL)	Volume of silver complex [Ag(NH$_3$)$_2$]$^+$ (mL)	Total reaction volume (mL)
1.	30.0	1.0	1.0	1.0	3.0
2.	30.0	0.0	2.0	1.0	3.0
3.	40.0	2.0	1.0	1.0	4.0
4.	40.0	1.0	2.0	1.0	4.0
5.	40.0	0.0	3.0	1.0	4.0
6.	50.0	3.0	1.0	1.0	5.0
7.	50.0	2.0	2.0	1.0	5.0
8.	50.0	1.0	3.0	1.0	5.0
9.	50.0	0.0	4.0	1.0	5.0

The mass of glucose added was also varied in accordance to the total reaction volume so as to fix at the optimal glucose concentration of 10 mg/mL

Table 4 Variations of the silver complex concentration for samples A1 to A4

Sample	Concentration of silver complex (mg/mL)
A1	20.0
A2	10.0
A3	5.0
A4	2.5
B1	80.0
B2	40.0
B3	20.0

The silver complex [Ag(NH$_3$)$_2$]$^+$ (1 mL) was added to the silica nanocapsules (1 mL) and glucose solutions (1 mL) at their respective optimum concentrations, then left to agitate for 2 days

Fig. 7 Shown left to right, image of AgNCs (s) immersed overnight in 1 M HNO$_3$, 1 M acetic acid and deionized water respectively

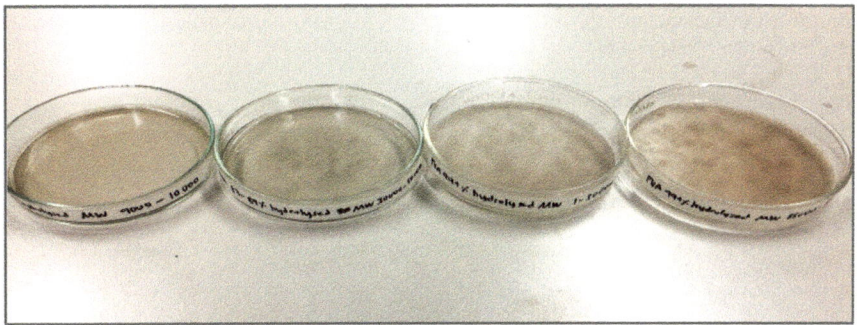

Fig. 8 Results of the composite biomaterials formed from different types of PVA (shown left to right: (i) 80% hydrolyzed, MW 9000–10,000, (ii) 87–89% hydrolyzed, MW 30,000–50,000, (iii) 99+% hydrolyzed, MW 1–30,000, and (iv) 99+% hydrolyzed, MW 85,000–124,000), where only the latter two were formed successfully after 3 freeze-thaw cycles

Appendix 4: Incorporation of AgNCs into PVA Film to Form Composite Material

See Fig. 8.

Appendix 5: Sol-to-Gel Transition of F127 Hydrogel at Different wt%

See Figs. 9 and 10.

Fig. 9 F127 hydrogels of 18 and 20% F127 at 25 °C, observed to be in gel state

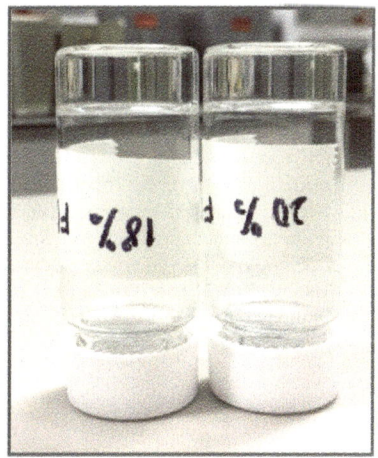

Fig. 10 F127 hydrogel of
16% F127 after heating up to
37 °C in a warm water bath
for 1–2 min, observed to be
in gel state

References

1. Jung, W. K., et al. (2008). Antibacterial activity and mechanism of action of the silver ion in *Staphylococcus aureus* and *Escherichia coli*. *Applied and Environment Microbiology, 74*, 2171–2178.
2. The National Aeronautics and Space Administration (NASA). (2017). *Water purification*. https://ntrs.nasa.gov/archive/nasa/casi.ntrs.nasa.gov/20020083175.pdf. Accessed November 24, 2017.
3. Clement, J. L., & Jarret, P. S. (1994). Antibacterial silver. *Metal-Based Drugs, 1,* 467–482.
4. Drake, P. L., & Hazelwood, K. J. (2005). Exposure-related health effects of silver and silver compounds: A review. *Annals of Occupational Hygiene, 49,* 575–585.
5. Special Pathogens Laboratory. (2017). *Acid producing bacteria*. http://www.specialpathogenslab.com/acid-producing-bacteria.php. Accessed November 7, 2017.
6. Dobias, J., & Bernier-Latmani, R. (2013). Silver release from silver nanoparticles in natural waters. *Environmental Science and Technology, 47,* 4140–4146.
7. Saware, K., Sawle, B., Salimath, B., Jayanthi, K., & Abbaraju, V. (2014). Biosynthesis and characterisation of silver nanoparticles using *Ficus benghalensis* leaf extract. *International Journal of Research in Engineering and Technology, 5,* 867–874.
8. Jiraroj, D., Tungasmita, S., & Tungasmita, D. N. (2014). Silver ions and silver nanoparticles in zeolite A composites for antibacterial activity. *Powder Technology, 264,* 418–422.
9. Wang, L.-C., Chen, X.-G., Yu, L.-J., & Li, P.-W. (2007). Controlled drug release through carboxymethyl-chitosan/poly (vinyl alcohol) blend films. *Polymer Engineering & Science, 47,* 1373–1379.
10. Sharonova, A., et al. (2016) Synthesis of positively and negatively charged silver nanoparticles and their deposition on the surface of titanium. *IOP Conference Series: Materials Science and Engineering, 116,* 012009.
11. Oliveira, R. N., et al. (2014). Mechanical properties and in vitro characterization of polyvinyl alcohol-nano-silver hydrogel wound dressings. *Interface Focus, 4,* 20130049.
12. Nie, S., Hsiao, W. W., Pan, W., & Yang, Z. (2011). Thermoreversible Pluronic® F127-based hydrogel containing liposomes for the controlled delivery of paclitaxel: In vitro drug release, cell cytotoxicity, and uptake studies. *International Journal of Nanomedicine, 6,* 151–166.
13. Hsu, B. Y. W., et al. (2012). PEO surface-decorated silica nanocapsules and their application in in vivo imaging of zebrafish. *RSC Advances, 2,* 12392–12399.

Investigation of Low Cost Eye Tracker and EEG for Objectively Assessing Vigilance Level

Yuqing Xue, Wenjing Tan, Jia Ning Shermaine Ang
and Aung Phyo Wai Aung

Abstract Vigilance is important for jobs requiring sustained attention for prolonged times, especially in this century where terrorism poses a challenge to the world. This research investigates a sensorised approach to objectively evaluate vigilance levels to maintain optimal work performance. We conducted two experiments on 25 subjects using low cost EEG and eye tracker. The first experiment identified the optimal working conditions for the selected eye tracker. By applying these conditions in vigilance testing, both accuracy and precision of the eye tracker achieved above 90%. Vigilance level was observed to have decreased over time by analysing eye gaze and using reaction time in quantifying vigilance level. These findings were supported by EEG band power features; showing a decrease in frontal asymmetry index corresponding to vigilance level. The increased theta and decreased beta powers in temporal lobes were identified; where of high beta reflects alertness while high theta results drowsiness. From our investigation, the decline in vigilance level corresponds to an increase in reaction time, blink rates, decrease in frontal alpha asymmetry index and change in beta and theta bands. With the sensorised objective measures of vigilance levels, appropriate countermeasures can be taken when respondents' vigilance level is low to alleviate these undesirable and unproductive states.

Keywords Vigilance · Eye tracker · EEG

Y. Xue · W. Tan · J. N. S. Ang
Nanyang Girls' High School, Singapore, Singapore
e-mail: 7114170385@nygh.edu.sg

W. Tan
e-mail: 7114150025@nygh.edu.sg

J. N. S. Ang
e-mail: 7114150098@nygh.edu.sg

A. P. W. Aung (✉)
Nanyang Technological University, Singapore, Singapore
e-mail: apwaung@ntu.edu.sg

© Springer Nature Singapore Pte Ltd. 2019
H. Guo et al. (eds.), *IRC-SET 2018*,
https://doi.org/10.1007/978-981-32-9828-6_6

1   Background and Purpose of Research

Terrorism is prevalent around the world and has been a rising threat to humanity. However, it can be minimised significantly if the country has good security measures, vigilance is thus vital in this case when combating terrorism. In Singapore, SGSecure movement was launched in September 2016 to sensitise, train and mobilise the community in fight against terror, where every member of the community must do his part by staying alert to ever-present security threats [1]. For instance, security personnel on patrol who are viewing CCTV monitors should be increasingly vigilant of unusual or out-of-place behaviour in these perilous times [2]. Low vigilance level will cause the security personnel to spot suspicious people or items at a lower frequency—taking a longer time to react. Thus, it is important to have a reliable method to facilitate assessment of the symptoms of decline in vigilance level, so as to increase the security level in a country and minimise the possibility of terrorist attacks due to low vigilance level.

While EEG is the most commonly studied physiologic measure of vigilance, various measures of eye movement have also been used [3]. There have been studies using high-end eye trackers (Eyelink 1000) to assess vigilance level [4], but there has not been any studies using low-end eye trackers, such as Tobii EyeX. In our study, we first evaluated the performance of low-cost eye trackers in tracking gaze. To do so, we evaluated the optimal conditions when using a low cost remote eye tracker, Tobii EyeX eye tracker. The optimal conditions required when using the eye tracker, which include extrinsic parameters that will ultimately affect the precision and accuracy of the eye tracker (e.g. presence of chin rest, light intensity, distance from eye tracker), can then be obtained.

We determined how the gaze can be used to quantify vigilance and lastly, how EEG and gaze features are related to assess vigilance. With these, we evaluated if the information collected from the users can be used to optimize devices, or software interfaces to maintain vigilance level. In other words, whenever vigilance levels fall below a certain value, respondents can be notified to initiate counteraction.

2   Objective and Hypothesis

Our goal is to evaluate how vigilance level can be assessed reliably using a low cost eye tracker. We hypothesised that the repetitive visual stimuli induce mental fatigue and cause the subjects to feel drowsy. This caused a decline in vigilance level which can be observed from results of subjective rating based on reaction time, the change in raw EEG data and frontal asymmetry. Blink rate (BR) can be compared with these results to find out how it can be used to evaluate user's vigilance level.

3 Experiment Design and Data Collection

Figure 1 shows the experimental setup with different sensors. Table 1 explains the experimental design and how data collection will be done in our experiment.

We sent out a pre-experiment survey to a high school student population who are females and received a total of 222 responses. Through this survey, we have gathered information that the average time majority can stay vigilant for is within 5–15 min and that loud music is the most common form of distraction.

3.1 Experimental Design for Sensor Evaluation

Sensor Evaluation will consist of two parts: Eye Calibration & Sensors Checking and Random Dot Experiment. The Eye Calibration & Sensors Checking would take a total of 10 min. The Random Dot Experiment will have 6 sessions with different variables and each session would last for 5 min, with a rest period of 1 min in between.

An experiment will first be designed to assess the hypotheses: What are the optimal conditions in which a low cost eye tracker can perform best in, and to evaluate the performance of a low cost eye tracker. This experiment will show a dot at different places around the screen. The subject would then be asked to look at the dot at the position it appears (Fig. 3), which the subject then have to press the spacebar, to indicate that he/she has seen the dot. In each session, we repeated the experiment and change some variables as seen in Table 2. This experiment also ensures that

Fig. 1 Experimental setup

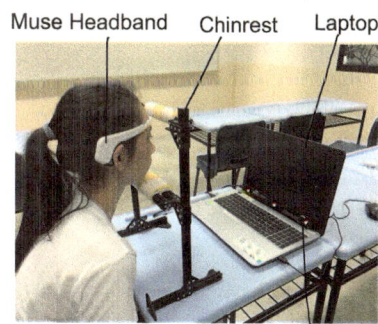

Muse Headband   Chinrest   Laptop

Tobii EyeX

Table 1 Devices and their measured parameters in experiment

Modality	Device	Measured parameters
Brain signals	Muse 2014	5 EEG band powers (delta, theta, alpha, beta, gamma)
Eye gazes	Tobii EyeX	Eye gaze coordinates on screen (x, y)

Table 2 Variables in each experiment session

Experiment session	Chin rest (yes/no)	Light intensity (lx)	Distance (cm)
1	No	250	45
2	Yes	250	45
3	Yes	0	45
4	Yes	50	45
5	Yes	250	60
6	Yes	250	75

the data collected later in our actual controlled experiment is the most accurate and precise.

3.2   Experimental Design for Vigilance Testing

This controlled experiment is approximately 40 min in length, and consists of two tasks: Psychomotor Vigilance Task (PVT) and a Spot the hidden item (SPOT) test which is a simulation of security checks. This study is designed to assess the hypothesis that low cost remote eye trackers can objectively assess vigilance level.

At the start and during resting periods of experiment, subjects are to rate their sleepiness level on the 9-level Karolinska Sleepiness Scale (KSS) (Fig. 2) that is used for evaluating subjective sleepiness level.

As shown in Table 3, the sensor setup and calibration will be conducted to check the accuracy and performance of the eye tracker. After every task, rest periods are planned with KSS survey, so that the participants can start every task in the same relaxed condition (about 1 min for the subjects to close their eyes to relax), regardless of what the previous task is.

Subsequently, a special test will be used in our experiment to determine the relationship between the vigilance level and sleepiness of a person. In the experiment, the modified PVT is a simple task where the subject presses a button as soon as the

Fig. 2 Karolinska sleepiness scale

1. Extremely alert
2. Very alert
3. Alert
4. Rather alert   Active
5. Neither alert nor sleepy
6. Some signs of sleepiness

7. Sleepy, but no difficulty remaining awake
8. Sleepy, some effort to keep alert   Sleepy
9. Extremely sleepy, fighting sleep

Table 3 Experiment flow with different tasks

Sensor setup and calibration		10 min
Rest (KSS)		1 min
PVT 1.1	PVT 1.2	5 min each, with rest of 1 min in between
Rest (KSS)		1 min
Spot the hidden item		15 min
Rest (KSS)		1 min
PVT 2.0		5 min
End of experiment (rest) (KSS)		

light appears to assess the reaction time. The dot will turn on randomly every few seconds for 5 min. The test will last for 5 min with a dot appearing at the different spots, changing every 3 s. The purpose is to measure vigilance level, and give a numerical measure of sleepiness by analysing the lapses in attention of the tested subject (Fig. 3).

We designed 2 parts of this test, PVT and PVT (music) where the difference is the presence and absence of loud music that act as distractor. We have chosen to use loud music as based on our survey outcomes, 34% of the respondents have stated that out of the different sound distractions (soft music, loud music, whispers, conversations), they are most distracted by loud music. Other studies have also shown that sustained attention, or vigilance, can be related to many aspects of musical influence [5].

The SPOT test will be a simulation of security check where subjects will be required to spot certain shapes that are hidden between different shapes. Subjects are then tasked to click on the 10 hidden shapes (given to subjects in on-screen instructions) that they can find to successfully complete one round. The entire test will only end after 15 min.

There would be a total of 3 different levels of this test. Starting with the easiest level to find 10 hidden yellow circles filled with many different shapes of many

Fig. 3 Screenshot of PEBL's perceptual vigilance test

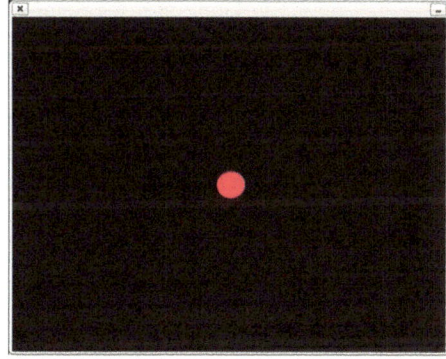

Fig. 4 Screenshot of easy
level of SPOT

different colours (seen in Fig. 4). Yellow circles have been used as yellow is the
most fatiguing color and causing more eye irritant. Subjects will only be given 3 min
for this difficulty level, and subjects are required to finish as many rounds of this
difficulty level as possible.

After 3 min, subjects enter the medium level where subjects will have to find 10
red squares, the shapes would be more hidden, and may even be hidden behind red
rectangles. This makes it more difficult for subjects to find the hidden red squares.
A longer period of time of 5 min will be given to complete the search task. Next
stage is hard level, where subjects would be required to find 10 dark green triangles.
Different shades of green triangles are around the screen to confuse the subjects, and
the dark green triangles will be placed behind rectangles and squares, making this
level the most difficult. This last difficulty level would take a total of 7 min. The
SPOT test would therefore take a total time of 15 min.

After 1 min rest period, the subject will have to complete another round of PVT.
The only difference would be that the subject will have a high possibility of losing
their original vigilance level, due to the long period of time spent completing the
entire controlled experiment. In order to ensure that there is a significant change in
vigilance level, we had gathered the results from our pre-experiment survey, regarding
the average time one can stay vigilant. From the survey of 220 respondents, more than
half (55.2%) can only stay vigilant within an average time of 10–20 min. Therefore,
the time period between the first and last PVT is more than 20 min. This will be
the end of the controlled experiment. Subjects will then be asked again about their
9-level KSS, to evaluate the subjects' subjective sleepiness.

From the sensor evaluation we determined the optimal conditions which a low
cost eye tracker can perform best in. Subjects have to stare at the dot when it appears,
showing randomly around the screen. During the 6 different sessions, variables will
be changed as shown in Table 2 and the presence of a chin rest. We used the optimal
conditions obtained to conduct the vigilance testing. The modified PVT assesses the
reaction time of subjects. For PVT (music), loud music was played as a form of
distraction, causing their vigilance level to drop. For the Spot Test, it was to simulate
security checks and assess the users' vigilance level with the sensors. Subjects were
asked to find 10 hidden shapes within a limited time. While resting, we instructed

subjects to complete 9-level Karolinska Sleepiness Scale (KSS) to evaluate their subjective vigilance rating. We used OpenSesame to implement data collection application and performed different statistical testing using Excel Analysis Toolpak. We then obtained reaction time for PVT and KSS rating which can be used as surrogate measures of vigilance levels to evaluate both the eye tracker and EEG.

4 Data Analysis and Evaluation

We collected data from 10 and 15 healthy students (aged 14–17, all female) in the sensor evaluation and vigilance testing respectively. Each experiment lasted approximately 1 h. Students' EEG signals and eye gaze coordinates were recorded.

Sensor Evaluation All eye tracker data were mapped to [0–1], making all the x and y values within the range of 0–1, which is expressed as:

$$v' = \frac{v - min\ A}{max\ A - min\ A} \times (new\_max\ A - new\_min\ A) + new\_min\ A \quad (1)$$

Following that, we had to calculate the accuracy of each point, by firstly calculating

$$Pixel\ Accuracy\ Both$$
$$= \sqrt{(target\ X - gaze\ point\ X) + (target\ Y - gaze\ point\ Y)} \quad (2)$$

The accuracy of each point is calculated by taking Eq. (1) and (2). To calculate the precision of each points, we first had to find the mean of the two eyes which is expressed by.

$$dX = diff\left(\frac{(Temp\ Gaze\ Point\ X\ Right + Temp\ Gaze\ Point\ X\ Left)}{2}\right) \quad (3)$$

$$dY = diff\left(\frac{(Temp\ Gaze\ Point\ Y\ Right + Temp\ Gaze\ Point\ Y\ Left)}{2}\right) \quad (4)$$

$$diffs = \sqrt{dX^2 + dY^2} \quad (5)$$

To calculate the precision of each points, we then used

$$Pixel\ Precision\ Both = \sqrt{mean(diffs)^2} \quad (6)$$

We conducted the *T*-test on the accuracy and precision values of each dot during the different sessions. The objective is to find out if the different variables significantly affect the precision and accuracy of the eye tracker.

The following features were considered to assess vigilance levels. **Reaction time**: where eye gaze coordinates were labelled with the times participants pressed the "space". Reaction time is used to find out the change in vigilance level, as reaction time indicates perceptual vigilance.

Blink Rate Eye position coordinates revealed that a decline in vigilance level, was characterised by a higher blink rate. To identify instances of a decline in vigilance level during the experiment, eye position coordinates that were at 0 for both left and right eyes that last for between 0.1 and 0.4 s were labelled as blinks. The number of blinks per minute is then calculated to find the blink rate.

Frontal Asymmetry From EEG data, we used the alpha relative band from Fp1 and Fp2 electrodes to compute the frontal asymmetry which can be expressed as:

$$Frontal\ Asymmetry = \log \frac{(fp1 - fp2)}{(fp1 + fp2)} \tag{7}$$

The calculation simply means that higher scores are indicative of approach behavior—a form of behaviour in which the person is moved towards, or in this case, more vigilant, and lower scores are indicative of avoidance motivation—a behavioral act that enables an individual to avoid a situation, or in this case, less vigilant [6].

Sensor Evaluation Using data from the random dot in the 5 sessions in sensor evaluation (the sixth session was excluded from the data analysis, as the eye tracker could not detect the subjects' eyes), we achieved the following accuracy and precision shown in Fig. 5.

From the accuracy and precision of 6 different sessions from the Sensor Evaluation, the optimal condition at which the eye tracker works best is with a distance of

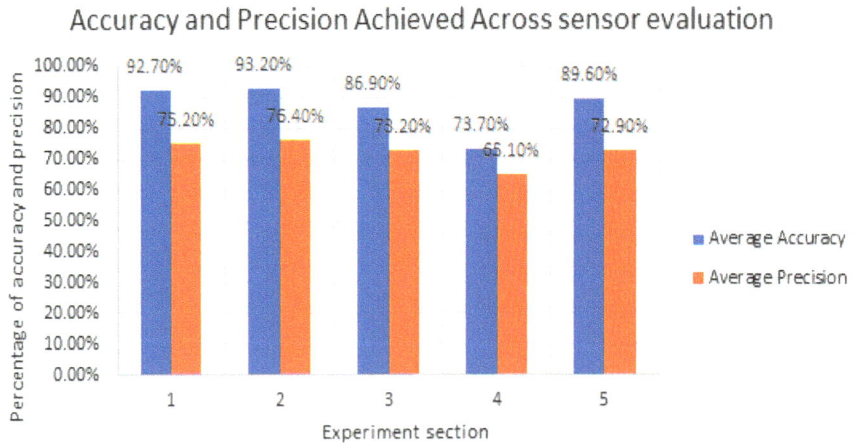

Fig. 5 Accuracy and precision achieved across sensor evaluation

45 cm with 250 lx light intensity. We conducted the *t*-test on the accuracy and preci- sion data and realised that the presence of chin rest does not affect both the accuracy and precision of the eye tracker significantly. Hence, to make the experiment more similar to a real-life simulation, we have decided to not use the chin rest. Therefore, we have used 45 cm distance, 250 lx light intensity and absence of chinrest in the vigilance testing to ensure that the eye tracker data we collect in the experiment is the most accurate and precise.

Vigilance Testing The KSS was analysed and we picked out the subjects who became sleepy over time. Using these subjects' data from the eye tracker, we achieved the following blink rates trend shown in Fig. 6a. Together with the timings when the subject pressed the spacebar, we obtained the reaction time of the subjects throughout the experiment shown in Fig. 6b. An increase in blink rate and reaction time shows a decline in vigilance level.

From above graphs, we concluded that the increase in blink rate and reaction time proves that the subject's vigilance level declined over time. The frontal asymmetry

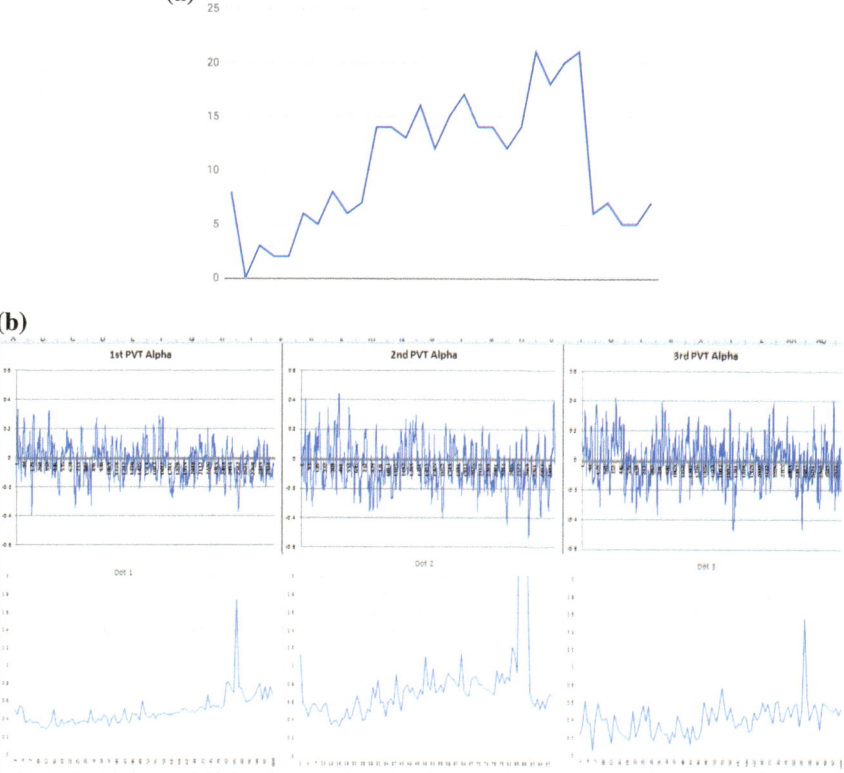

Fig. 6 a Blink rate during vigilance testing. **b** Reaction time and frontal asymmetry of alpha relative

of alpha relative correlates with the reaction time. When the reaction time increases, the frontal asymmetry becomes more negative. As mentioned previously, the higher scores are indicative of approach behavior which shows a higher vigilance level and lower scores are indicative of avoidance motivation which shows a lower vigilance level. From Fig. 6b, it can be seen that the top row of graphs became more negative over time, confirming that the vigilance level has declined over time. Moreover, the raw β and θ activities in both the temporal lobes were analysed and observed to have decreased and increased respectively as β waves associated with alertness and θ waves with drowsiness [7].

5 Discussion

While the participant's vigilance level was detected due to the symptoms of decline in vigilance level observed, we were unable to find out the exact characteristics of when one's vigilance level is low. True to the limitation of using a low cost eye tracker to measure vigilance level as mentioned in the introduction, vigilance level cannot be assessed using exact values, but instead trends over time have to be observed. Moreover, we designed the vigilance testing experiment in which the repetitive visual stimuli should be able to induce mental fatigue and cause subjects' vigilance level to decrease, as mentioned in our hypothesis. However, according to our KSS results, not all subjects became less alert as the experiment progresses.

To further improve analysis, we wish to have subjects of different genders and age to get the common regularity and the possibility to get values to differentiate high vigilance level from low vigilance level. It is more accurate to have more adults as our experiment subjects since we would be applying the results to security guards. Furthermore, other than blink rates, we can analyse more features from the eye tracker data (e.g. PERCLOS) to allow us to have a better evaluation on vigilance level.

With a good understanding of the trends and symptoms of decline in vigilance, our findings could be used to optimize devices or software interfaces in tracking one's vigilance level. This would allow jobs that require high vigilance levels to maintain a high vigilance level as if any decline in vigilance level is detected, they would be alerted to take counteractions. However, since we were only able to obtain trends, developing an algorithm to evaluate instantly one's vigilance level will be the next step to develop a software that can quantify a person's vigilance level which can be used in the selection of security guards or alert the supervisor if on-duty guard's vigilance becomes too low.

6 Conclusion

From the results of sensor evaluation, we found that a low cost eye tracker performs best with a distance of 45 cm, in a 250 lx light intensity place in the absence of chinrest. These results were useful in the experimental design of the vigilance testing.

We were able to evaluate the symptoms of decline in vigilance level from vigilance testing which are: increase in blink rate, decrease in β and increase in θ activities in the both temporal lobes and increase in reaction time. We are able to conclude that a low cost eye tracker can ultimately assess vigilance level. Overall, our project seeks to objectively evaluate vigilance states to maintain desirable vigilance level in an automated approach. In the long term, our project can function as a feasible method to assess and maintain people's vigilance level using a low cost eye tracker.

Acknowledgements Firstly, we are particularly grateful for the assistance given by 2 NTU undergraduate students, Nageshwari and Jamie Yap Yi Qi, for helping us with the data analysis. We would also like to express our appreciation to our teacher-mentor, Mr. Ang Joo Liak, for helping us with numerous administrative tasks during this entire journey. Last but not least, we would like to acknowledge with much appreciation to our school's (NYGH) ICT department, for loaning us required equipment for experiments.

References

1. Ministry of Home Affairs. (2017, June). *Singapore terrorism threat assessment report 2017.* Retrieved January 3, 2018, from https://www.mha.gov.sg/Newsroom/press-releases/Pages/Singapore-Terrorism-Threat-Assessment-Report-2017.aspx.
2. Securitas. (n.d.). *Vigilance key in combating terrorism* (online). Retrieved August 27, 2017, from http://www.securitasinc.com/globalassets/us/files/knowledge-center/spotlights/antiterrorism/vigilance-is-key-to-fighting-terrorism_aug-2005.pdf.
3. Oken, B. S., Salinsky, M. C., & Elsas, S. M. (2006). Vigilance, alertness, or sustained attention: Physiological basis and measurement. *Clinical Neurophysiology: Official Journal of the International Federation of Clinical Neurophysiology, 117*(9), 1885–1901. https://doi.org/10.1016/j.clinph.2006.01.017.
4. Bodala, I. P., Li, J., Thakor, N. V., & Al-Nashash, H. (2016). EEG and eye tracking demonstrate vigilance enhancement with challenge integration. *Frontiers in Human Neuroscience, 10,* 273, https://doi.org/10.3389/fnhum.2016.00273.
5. Wang, M. H. (2015). *The influence of preferred music on vigilance and mental workload (Honors thesis projects. 17).* http://digitalcommons.otterbein.edu/stu_honor/17.
6. Frontal asymmetry 101—Motivation and emotions from EEG. (2017, September). Retrieved January 06, 2018, from https://imotions.com/blog/frontal-asymmetry-101-get-insights-motivation-emotions-eeg/.
7. Zeid, S. K. (2017). Assessment of vigilance using EEG source localization. In *Conference Abstract: 2nd International Conference on Educational Neuroscience. Frontiers in Human Neuroscience.* https://doi.org/10.3389/conf.fnhum.2017.222.00025.
8. Allison, B. Z., Dunne, S., Leeb, R., del R. Millán, J., & Nijholt, A. (2012). Recent and upcoming BCI progress: Overview, analysis, and recommendations. In B. Allison, S. Dunne, R. Leeb, J. Del R. Millán, & A. Nijholt (Eds.), *Towards practical brain-computer interfaces. Biological and medical physics, biomedical engineering.* Berlin, Heidelberg: Springer.
9. Franzen, P. L., Siegle, G. J., & Buysse, D. J. (2008). Relationships between affect, vigilance, and sleepiness following sleep deprivation. *Journal of Sleep Research, 17*(1), 34–41. https://doi.org/10.1111/j.1365-2869.2008.00635.x.
10. Fukuda, K., Stern, J. A., Brown, T. B., & Russo, M. B. (2005). Cognition, blinks, eye-movements, and pupillary movements during performance of a running memory task. *Aviation, Space and Environmental Medicine, 76*(7 Suppl), C75–C85.

11. In pictures: Eight reasons why you can't pay attention. (2008, October). Retrieved October 15, 2017, from https://www.forbes.com/2008/10/15/short-attention-span-forbeslife-cx_avd_1015health_slide.html.

12. Millán, J. D., Rupp, R., Müller-Putz, G. R., Murray-Smith, R., Giugliemma, C., Tangermann, M., et al. (2010). *Combining brain–computer interfaces and assistive technologies: State-of-the-art and challenges*. Retrieved August 10, 2017.

13. Nicolas-Alonso, L. F., & Gomez-Gil, J. (2012). Brain computer interfaces, a review. *Sensors (Basel, Switzerland), 12*(2), 1211–1279. https://doi.org/10.3390/s120201211.

14. Nijboer, F. (2015). Technology transfer of brain-computer interfaces as assistive technology: Barriers and opportunities. *Annals of Physical and Rehabilitation Medicine, 58*(1), 35–38. https://doi.org/10.1016/j.rehab.2014.11.001. ISSN 1877-0657.

15. The science of attention: How to capture and hold the attention of easily distracted students. (2017, April). Retrieved October 15, 2017, from https://www.opencolleges.edu.au/informed/features/30-tricks-for-capturing-students-attention/.

16. Leeb, R., Lancelle, M., Kaiser, V., Fellner, D. W., & Pfurtscheller, G. (2013). Thinking penguin: Multimodal brain-computer interface control of a VR game. *IEEE Transactions on Computational Intelligence and AI in Games, 5*(2), 117–128. https://doi.org/10.1109/TCIAIG.2013.2242072.

17. Unknown. (n.d.). Retrieved October 16, 2017, from https://www.colormatters.com/color-and-vision/color-and-vision-matters.

18. Eye tracker prices—An overview of 15 eye trackers. (2017, August). Retrieved October 22, 2017, from https://imotions.com/blog/eye-tracker-prices/.

19. Irimia, R.-E., & Gottschling, M. (2016). Taxonomic revision of *Rochefortia* Sw. (Ehretiaceae, Boraginales). *Biodiversity Data Journal, 4*, e7720. Advance online publication. http://doi.org/10.3897/BDJ.4.e7720.

Human Attribute Classification for Re-identification Across Non-overlapping Cameras

Wen Jun Calvin Gao, Poh Say Keong and Bingquan Shen

Abstract This project makes use of a Convolutional Neural Network (CNN) to perform multi-class attribute recognition, in which this information is used to perform person re-identification (re-ID). From our research, we discovered that a trained CNN model performs better when given less attributes per image to focus on, as it decreases chances of error when making predictions of attributes of a person based on an image. Moreover, we found out that re-ID is done more effectively when a CNN is tasked to identify attributes that causes a person to stand out from others. Thus, salient attributes that can be clearly identified from cameras of different viewpoints are the most important attributes to focus on to perform re-ID effectively, while more common attributes can perform a filtering role in the re-ID problem. By modifying the Inception v3 model [1] for multi-label classification, the model is able to output probabilities for each attribute for every input image. Experiments on the PETA (PEdesTrian Attribute) dataset [2] has shown that the model performs better while recognising salient attributes only compared to recognising both common and salient attributes.

Keywords Convolutional neural networks (CNN) · Person re-identification (Person Re-ID) · Attribute · Salient · Inception v3

1 Introduction

This project aims to train a deep Convolutional Neural Network (CNN) to perform multi-label image classification, which can be applied to perform person re-identification (re-ID) tasks optimally. Person re-ID is the ability to spot a person of

W. J. C. Gao · P. S. Keong
Dunman High School, Singapore, Singapore
e-mail: wen.jungao.calvin@dhs.sg

P. S. Keong
e-mail: poh.saykeong@dhs.sg

W. J. C. Gao · P. S. Keong · B. Shen (✉)
DSO National Laboratories, Singapore, Singapore
e-mail: SBingqua@dso.org.sg

© Springer Nature Singapore Pte Ltd. 2019   75
H. Guo et al. (eds.), *IRC-SET 2018*,
https://doi.org/10.1007/978-981-32-9828-6_7

interest in different cameras. In real-life camera surveillance scenarios, person re-ID through multi-attribute learning is not widely researched on, but if done successfully, has the potential to reduce manpower, time required, and boost accuracy when performing laborious tasks involving looking through large volumes of image data, for tracking specific people from a large dataset of surveillance camera images. Multi-attribute learning algorithms using CNN has become more accurate and feasible in recent years. Using a CNN for the re-ID task is effective as the performance of a CNN on attribute recognition can increase much greater than that of older learning algorithms given a large dataset, such as the PEdesTrian Attribute (PETA) dataset-the largest and most diverse pedestrian attribute dataset—which we used for training the CNN. Therefore applying a robust CNN model on Singapore's surveillance camera images to perform person re-ID can increase efficiency in performing security tasks such as finding suspicious people using images from every surveillance camera in Singapore, contributing to the security of the nation.

2   Hypothesis

In this project, we hypothesize that making use of one round of filtering of images based on common attributes, followed by a 2nd round of filtering using salient attributes, as well as focusing on a small number of attributes per round, will increase re-ID performance of the CNN. Experiments conducted to validate our hypothesis are carried out and will be elaborated further in detail in this paper.

3   Methods and Materials

3.1   *Inception V3*

Inception v3's architecture [1, 3] is used for this project. Code to retrain Inception v3 in TensorFlow was obtained from the official GitHub repository. In order to perform multi-label image classification, certain adaptations were made:

(1)   *From Softmax to Sigmoid*

    (a)   *Changing the Last Layer*

Inception v3, a single-class classifier, makes use of the softmax function in its final layer to predict the probabilities of the existence of each attribute in the form of a vector, based on the assumption that attributes are mutually exclusive. However, in multi-class image classification, attributes may not be mutually exclusive and the output probabilities for each attribute do not sum up to 1, a property of the algorithm softmax uses. Thus the softmax function was replaced with the sigmoid function to generate the probability vector, which is more suitable for multi-class classification:

Each image is represented as x_i, $i \in 1, ..., N$

$$p_{il} = \frac{1}{1 + \exp(-x_{il})}, \qquad (1)$$

where p_{il} is the output probability for the lth attribute of example x_i.

The output probability vector is compared with the ground truth labels vector to evaluate the accuracy of the CNN model. The output vector for Inception v3 is a one-hot vector, as it takes the output probability vector and returns the index with the highest value as the predicted class, with a value of 1 in the resultant vector. For the multi-class classification task, the one-hot vector is replaced with a vector with multiple 1s by rounding the values of each index of the probability tensor (output tensor of the sigmoid function) to 0s and 1s.

(b) *Sigmoid Weighted Loss*

The loss function for Inception v3 made use of the softmax function as well. By the same reasoning in the previous section, the softmax loss had to be changed into a sigmoid loss. Furthermore, to counter the imbalanced distribution of attributes in the PETA dataset, we added weights to be multiplied to the final predictions:

$$w_l = \exp(1 - 2 * r_l), \qquad (2)$$

where w_l is the loss weight for the lth attribute and r_l is the positive ratio of the lth attribute in the training set.

With the weights, the loss function would be as follows:

$$Loss = -\frac{1}{N} \sum_{i=1}^{N} \sum_{i=1}^{L} w_l(t_{il} \log(p_{il}) + (1 - t_{il}) \log(1 - p_{il})), \qquad (3)$$

where N is the number of images used in total; L is the number of labels an image contains; t_{il} is the ground truth label which represents whether example x_i has the lth attribute or not.

3.2 *Person Re-ID Base-Net*

The base-net (Fig. 1) is used as the primary criterion to evaluate the performance of our adaptation of Inception v3 in the person re-ID task. The Viewpoint Invariant Pedestrian Recognition (VIPeR) dataset is split equally into a probe set and a gallery set, such that every image in each set is unique and one image of each image pair is in either the gallery or probe set. The gallery images and probe images are analysed by a trained Inception v3 model to generate a rank of all the attributes used in training for every image, with the attribute of highest probability holding top rank. The top

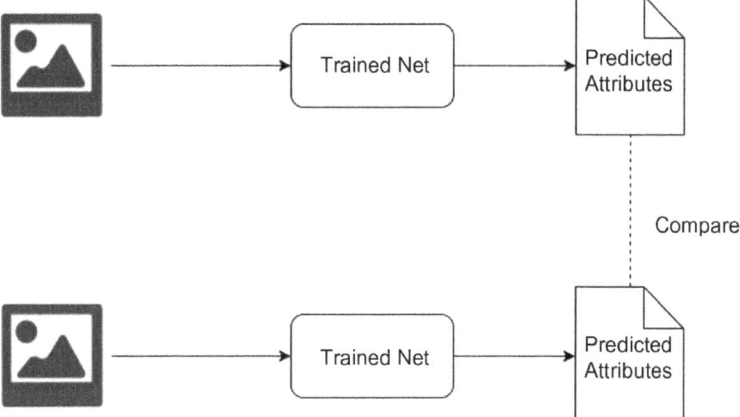

Fig. 1 Diagram illustrating the Re-ID base-net

3 attributes of each probe image are used to compare with the top 3 attributes of all the gallery images, and a custom distance algorithm, Eq. (4), is used to compare between the probe and gallery images. For every probe image, a ranking of all the gallery images are obtained based on the distance, with the image the model predicts as most similar to a specific probe image holding top rank.

These rankings are then used to evaluate the performance of the CNN in person re-ID model by calculating Cumulative Marching Characteristic (CMC) scores [4], as elaborated in Sect. 3.3.

3.3 Evaluation Methods

(1) Cumulative Matching Characteristic (CMC)

For every gallery image compared against the probe image, a score S is allocated to that particular gallery image:

$$S = \frac{\left(R_{nq} \cap R_{ng}\right)/R_n}{\left(\sum_{R_{nq} \cap R_{ng}} \left|S_q - S_g\right|\right)^{R_{nq} \cap R_{ng}}}, \tag{4}$$

where R_{nq} and R_{ng} are the attributes used for comparison in probe and gallery datasets respectively; R_n is the number of attributes used for comparison; S_q and S_g are the confidence scores of an attribute from the probe (or query) and gallery datasets respectively. The higher the score, the more similar the model thinks the gallery image is compared to the probe image.

Three models are being trained: 1 using 15 common attributes, chosen based on the attributes used to train DeepMAR [5], with the highest 15 positive ratios, 1 using 15 salient attributes determined based on attributes provided by the PETA dataset [2], with the lowest 15 positive ratios, and 35 attributes that consist of both common and salient attributes. The 3 models are trained with the same hyperparameters (learning rate at 0.001, 56,000 training steps, train batch size at 1000) which are optimized by us based on Figs. 2 and 3, thus acting as the controlled variables to ensure a fair test. Learning rate is optimized by attaining the ideal gradient of the graphs in Figs. 2 and 3. Train steps are optimized by determining the point where the validation curve (blue) stagnates while train curve (red) continues to rise. Train batch size is optimized by attaining minimal fluctuations in Figs. 2 and 3, while ensuring that the size is not too large that the model overfits while training.

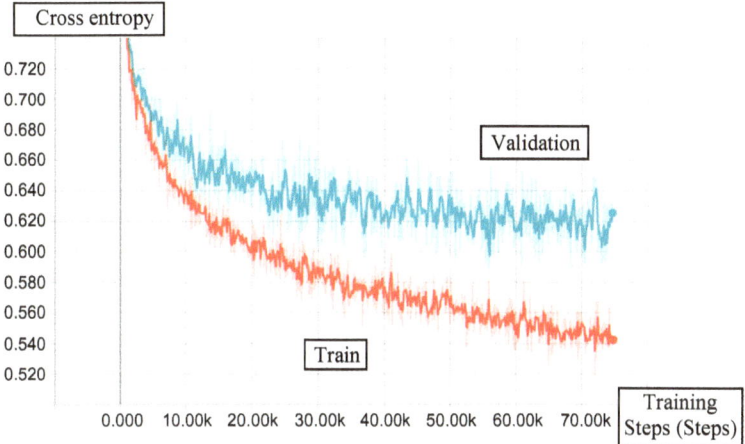

Fig. 2 Graph of cross entropy against number of training steps

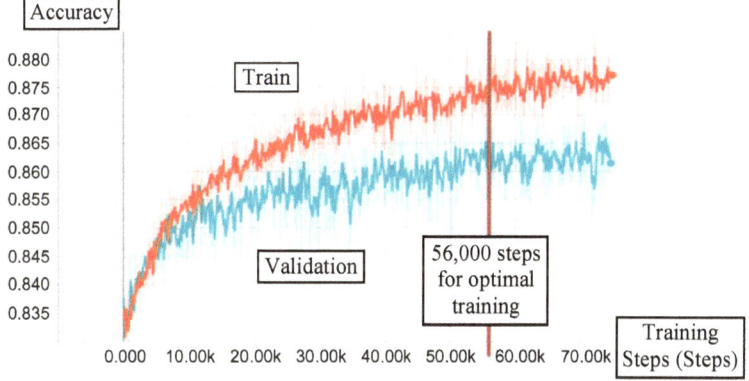

Fig. 3 Graph of accuracy against number of training steps

Three comparisons will be made: between the common and salient attributes, with the number of attributes constant, to find out the effect of saliency on CMC score; between the 15 salient attributes and 35 attributes, to find out the effect of number of attributes on CMC score; and between the 15 salient attributes and the 2-stage filtering method, whereby the 15 common attributes are used to eliminate images from the gallery set that are highly unlikely the probe image, and the remaining images are re-ranked based on salient attributes, to test the effectiveness of re-ranking in improving CMC score.

(2) *Precision and Recall*

Precision shows the frequency the model predicts correctly the presence of an attribute out of all its guesses that the attribute exists in the images in the test set. Recall shows the frequency the model predicts the existence of the attribute correctly out of all the instances the attributes really exist in the images in the test set:

$$Prec = \frac{1}{N} \sum_{i=1}^{N} \frac{|t_i \cap p_i|}{|p_i|}, \tag{5}$$

$$Rec = \frac{1}{N} \sum_{i=1}^{N} \frac{|t_i \cap p_i|}{|t_i|}, \tag{6}$$

where t_i is the ground truth positive labels of example x_i, p_i is the predicted positive labels of example x_i and $|\cdot|$ means the set cardinality.

Since, the re-ID task is solved by attribute recognition, precision and recall for the salient and common attributes, 15 each, used by the CNN model is calculated to show attribute recognition performance of the CNN. To obtain these results, we tested 2 trained models using the sets of attributes shown in Tables 1 and 2, with 1262 test set images, from PETA's various sub-datasets, that the model has not seen before.

4   Results and Discussion

In Tables 1 and 2, positive ratios shown are based on the entire PETA dataset. Attributes with 0% precision and recall indicate that there are no such attributes within the test set and are not considered in the average.

From Table 1, the model used to recognise salient attributes yields a low precision and recall. This is due to the lack of positive examples of salient attributes to train the model, resulting in relatively poorer results than training with common attributes. Salient attributes with a positive ratio of more than 0.15 have a higher recall than precision, indicating that the model is making many guesses that the attribute exists, but not very accurately. For attributes below 0.15, the precision and recall are both

Table 1 Precision and recall of 15 salient attributes

S/N	Category	Attributes	Positive ratio	Recall (%)	Precision (%)
	Salient attributes				
1	Accessories	Messenger bag	0.294	89.9	25.9
2		Backpack	0.191	13.8	70.5
3	Upper body	T-shirt	0.085	0	0
4		Grey	0.177	3.8	50
5		White	0.201	36.3	51.9
6		Blue	0.073	0	0
7		Brown	0.069	0	0
8		Brown hair	0.2	14.1	32.2
9		Grey hair	0.08	0	0
10	Lower body	Jeans	0.293	99.8	42.7
11		Blue	0.178	32.6	47.9
12		Grey	0.241	40.9	41.6
13	Footwear	sneakers	0.213	15.8	30.7
14		Brown	0.063	0	0
15		Grey	0.001	0.9	3.7
Average				34.8	39.7

Table 2 Precision and recall of 15 common attributes

S/N	Category	Attributes	Positive ratio	Recall (%)	Precision (%)
	Common attributes				
1	Accessories	Nothing	0.749	0	0
2	Upper body	Casual	0.846	82.8	97.2
3		Long sleeve	0.802	82.1	72.3
4		Short hair	0.742	36.2	79.1
5		Long hair	0.235	1.6	80
6	Lower body	Casual	0.854	75.5	97.9
7		Trousers	0.511	4.9	92.3
8		Jeans	0.293	0.9	100
9	Footwear	Sneaker	0.213	0	0
10		Leather shoes	0.307	1.4	75
11	Gender	Male	0.549	11.1	65.4
12	Age	<30	0.477	0.9	88.9
13		31–45	0.353	0	0
14	Carrying	Messenger bag	0.294	0.32	100
15		Nothing	0.274	0	0
Average				27.1	86.2

Table 3 CMC score of 15
salient versus 15 common
attributes

CMC (out of 631)	15 salient attributes	15 common attributes
Rank-1 accuracy (%)	0.00	0.00
Rank-5 accuracy (%)	1.11	0.01
Rank-10 accuracy (%)	2.54	0.03
Rank-20 accuracy (%)	5.39	0.06
Average rank	250	257
Variance	300	303

extremely low. These indicate that the model requires more positive examples and a dataset with better attribute distribution to learn salient attributes.

From Table 2, for the case of common attributes, the model is moderately well trained as the much higher precision and lower recall compared to salient attributes suggests that it is making fewer guesses, which are most often correct. Larger volumes of data used for training may help to increase recall.

Individual attributes with a positive ratio of more than 0.8 have very high precision and recall, indicating that the model is well trained in identifying these attributes. However, as positive ratio decreases, the recall becomes lower than the precision, and the difference between recall and precision increases. This issue arises from the unbalanced data distribution between attributes, thus the model is more confident in identifying certain attributes with higher positive ratios more than others. This results in the model making very reserved guesses for attributes that have lower positive ratios. This can be solved by using a dataset with better attribute data distribution, which is not publicly available currently.

From Table 3, although the CMC accuracies are similar for both salient and non-salient models, the average score for the salient model is much higher than that of the non-salient model. This shows that while both models are unable to pin-point the exact gallery image that corresponds to the probe image, but by using salient attributes, the model is able to eliminate more gallery images that are not similar to the probe images than by using non-salient attributes. From the higher recognition precision of common attributes based on Tables 1 and 2, even though the model is better at identifying common attributes, they can be found on many people, but the uniqueness of salient attributes sets people apart from others, thus identifying such attributes allows the model to perform person re-ID more effectively.

Based on Table 4, CMC scores for the model using 35 attributes to profile images were lower than that for the model using only 15 salient attributes. This is due to the fact that with more attributes, there will be a greater margin for error made by the model when doing attribute recognition. As a result, attribute profiles generated from the 35-attributes model will be inconsistent between images of the same identity.

Table 4 CMC score of 15 salient versus 35 mixture of salient and common attributes

CMC% (out of 631)	15 salient attributes	35 attributes
Rank-1 accuracy (%)	0.00	0.30
Rank-5 accuracy (%)	1.11	1.00
Rank-10 accuracy (%)	2.54	2.54
Rank-20 accuracy (%)	5.39	5.39
Average rank	250	262
Variance	300	283

Table 5 CMC score of 2-stage filtering

CMC% (out of 631)	2-stage filtering
Rank-1 accuracy (%)	1.74
Rank-5 accuracy (%)	5.86
Rank-10 accuracy (%)	10.62
Rank-20 accuracy (%)	16.8
Average rank	152
Variance	298

Moreover, with a difference in occlusion between the probe and gallery images, difference in confidence scores for the same attribute may be larger as well, which contributes to greater error in performing the person re-ID task, based on our CMC scoring Eq. (4). Thus, attributes used to train the CNN should be kept to a small number.

Comparing Table 5 with Tables 3 and 4, the 2-stage filtering method yields the best results as the CNN can eliminate the gallery images that are highly unlikely the probe image using common attributes. As a result during the 2nd round of image analysis using the salient attributes, there will be a smaller margin of error made by the trained CNN when working with a smaller data size, allowing it to perform re-ID the most accurately.

From Tables 3, 4 and 5, it was noticed that the variance, which indicates spread of corresponding images relative from the average, is quite large, indicating a certain degree of inconsistency of the CNN model in perform person re-ID, which is an area to improve on.

5 Conclusion and Future Work

This project uses the Inception v3 architecture that is adapted to do multi-attribute image classification, and makes use of this trained net to perform re-ID of the probe image amongst a set of gallery images. Making use of attributes is a viable way to perform person re-ID since a person's identity can be determined by garments worn and anatomical details. However, attribute recognition by the trained CNN requires a high accuracy and consistency for it to be able perform person re-ID effectively, which is difficult in real-life scenarios due to differences in background, occlusion and brightness. Due to these factors, much more image data is required for training a CNN in order to prevent confusing the CNN despite processing images of different brightness and backgrounds. Even under lack of comprehensive input data, our CNN model is still able to reduce the dataset size required for manual person re-ID to just 25% of the original, reducing the time required for man to look through the surveillance camera images. Therefore, we conclude that salient attributes are more significant features the CNN should focus on to achieve better person re-ID results.

Future works include applying Support Vector Machine (SVM) algorithms on top of the CNN model to improve the classification ability by identifying which attributes are actually present instead of ranking attributes according to how probable their existence are.

Image datasets other than the PETA dataset can be considered for training as well, for example [6]. Training and testing on more evenly distributed (in terms of attributes) datasets allows our CNN model to be more robust and gain the ability to detect attributes under different conditions.

Acknowledgements The authors, Poh Say Keong and Wen Jun Gao Calvin, would like to thank mentor Dr. Shen Bing Quan from DSO National Laboratories and teacher mentor Mr. Lee Wei Keong from Dunman High School for their support and guidance.

Appendix: Code to Calculate Precision and Recall of Each Attribute

```
1.  import tensorflow as tf
2.  import os, sys
3.  import math
4.  import operator
5.
6.  images_dir = sys.argv[1]              # Directory with images
7.  true_dir = sys.argv[2]                # Directory with true labels for images
8.  rank_number = int(sys.argv[3]) * 2    # Number of attributes used to predict identity
9.
10. # Unpersists graph from file, used as trained model for attribute recognition
11. with tf.gfile.FastGFile("trained_model.pb", 'rb') as f:
12.     graph_def = tf.GraphDef()
13.     graph_def.ParseFromString(f.read())
14.     _ = tf.import_graph_def(graph_def, name='')
15.
```

```
16. # Controls GPU usage
17. gpu_options = tf.GPUOptions(per_process_gpu_memory_fraction=0.333)
18. sess = tf.Session(config=tf.ConfigProto(gpu_options=gpu_options))
19. os.environ["CUDA_VISIBLE_DEVICES"]="0"
20.
21. # Generates attributes txt files for each image
22. with tf.Session() as sess:
23.     softmax_tensor = sess.graph.get_tensor_by_name('final_result:0')
24.
25. # Array of all attributes used for attribute recognition
26.     label_lines = [line.rstrip() for line
27.                     in tf.gfile.GFile("all_labels.txt")]
28.
29.     images_data = []
30.     filenames = []
31.     for filename in os.listdir(images_dir):
32.         dataset = tf.gfile.FastGFile(images_dir + "/" + filename, 'rb').read()
33.         predict = sess.run(softmax_tensor, \
34.             {'DecodeJpeg/contents:0': dataset})
35.         images_data.append(predict)
36.         filenames.append(filename)
37.
38. # Sorts attributes according to confidence score (how probable attribute exists)
39.     top_k = []
40.     for i in range (len(images_data)):
41.         labels = "predicted_labels_dir/" + filenames[i] + ".txt"
42.         rank = images_data[i][0].argsort()[-len(images_data[i][0]):][::-1]
43.         top_k.append(rank)
44.         top = top_k[i][:15]
45.         with open(labels, 'a+') as f:
46.             f.write('**%s**\n' % (filenames[i]))
47.             for node_id in top:
48.                 human_string = label_lines[node_id]
49.                 score = images_data[i][0][node_id]
50.                 f.write(human_string + '\n')
51.                 f.write('(score = %.5f)\n' % (score))
52.
53.     prediction_count = []   # Number of times a positive label is predicted correctly
54.     negative_count = []     # Number of times a negative label is predicted correctly
55.     for i in range (len(label_lines)):
56.         count = 0
57.         false = 0
58.         for filename in os.listdir('PR_labels'):
59.             with open('PR_labels/' + filename, 'r') as f1, open(true_dir + '/' + filename, 'r') as
    f2:
60.                 a = f1.readlines()
61.                 x = a[:rank_number]
62.                 b = f2.readlines()
63.                 label = label_lines[i] + '\n'
64.                 if label in x and label in b:
65.                     count += 1
66.                 if label not in x and label not in b:
67.                     false += 1
68.         prediction_count.append(count)
69.         negative_count.append(false)
70.
71.
72.     recall_count = []   # Number of times a positive label is true
73.     for i in range (len(label_lines)):
74.         count = 0
75.         for filename in os.listdir(true_dir):
76.             with open(true_dir + '/' + filename, 'r') as f:
77.                 a = f.readlines()
78.                 if (label_lines[i]+'\n') in a:
79.                     count += 1
80.         recall_count.append(count)
```

```
81.
82.         precision_count = []       # Number of times model predicts a label is true
83.         for i in range (len(label_lines)):
84.             count = 0
85.             for filename in os.listdir('PR_labels'):
86.                 with open('PR_labels/' + filename, 'r') as f:
87.                     a = f.readlines()
88.                     x = a[:rank_number]
89.                     if (label_lines[i]+'\n') in x:
90.                         count += 1
91.             precision_count.append(count)
92.         print(precision_count)
93.
94.   # Calculation of the precision and recall of every attribute used for prediction by model

95.         recall = []
96.         precision = []
97.         for i in range (len(label_lines)):
98.           if prediction_count[i] == 0 or precision_count[i] == 0 or recall_count[i] == 0:
99.             recall.append(0)
100.                 precision.append(0)
101.                 print(str(label_lines[i]) + ' not predicted')
102.                 continue
103.             else:
104.                 recall.append(prediction_count[i]/recall_count[i]*100)
105.                 precision.append(prediction_count[i]/precision_count[i]*100)
106.                 mean_accuracy = (1/2) * ((prediction_count[i]/recall_count[i]) + (negative_count[i
       ]/(len(images_data)-recall_count[i]))) *100
107.
108.                 print(str(label_lines[i]) + " recall:" + str(recall[i]))
109.                 print(str(label_lines[i]) + " precision:" + str(precision[i]))
110.                 print(str(label_lines[i]) + "mean average: " + str(mean_accuracy))
```

References

1. Szegedy, C., Vanhoucke, V., Ioffe, S., Shlens, J., & Wojna, Z. (2016). Rethinking the inception architecture for computer vision. In *Proceedings of the IEEE conference on computer vision and pattern recognition* (pp. 2818–2826).
2. Deng, Y., Luo, P., Loy, C. C., & Tang, X. (2014, November). Pedestrian attribute recognition at far distance. In *Proceedings of the 22nd ACM international conference on Multimedia* (pp. 789–792). ACM.
3. Szegedy, C., Liu, W., Jia, Y., Sermanet, P., Reed, S., Anguelov, D., et. al. (2015). Going deeper with convolutions. In *Proceedings of the IEEE conference on computer vision and pattern recognition* (pp. 1–9).
4. Xu, M., Tang, Z., Yao, Y., Yao, L., Liu, H., & Xu, J. (2017). Deep learning for person reidentification using support vector machines. *Advances in Multimedia, 2017*.
5. Li, D., Chen, X., & Huang, K. (2015, November). Multi-attribute learning for pedestrian attribute recognition in surveillance scenarios. In *2015 3rd IAPR Asian Conference on Pattern Recognition (ACPR)* (pp. 111–115). IEEE.
6. Li, D., Zhang, Z., Chen, X., Ling, H., & Huang, K. (2016). A richly annotated dataset for pedestrian attribute recognition. arXiv:1603.07054 (arXiv preprint).

The Power of the Micro-tide

Su Minn Jeilene Ho and Rei Ying Nadine Wang

Abstract Self-sustaining water induced energy generation is achieved when verti-cally aligned multi-walled carbon nanotubes (CNTs) are used as an active medium for water-induced separation of ions. The concept of using a stern double layer to induce voltage is optimised by varying the pattern on the CNTs array and varying the solution used. Using a focused laser beam setup, engineered patterning of microar-ray of CNTs with high precision is achieved. Further enhancement to such effect is achieved in the presence of Na^+ and Cl^- ions in water. In a place that is both humid and surrounded by the ocean, our findings opens the doorway to an alternative source of sustainable energy.

Keywords Water · Carbon Nanotubes · CNT · Power · Electrical energy ·
Streaming potential · Zeta potential · Sustainable energy

1 Literature Review and Background

Due to their physical, electronic, and mechanical properties, carbon nanotube (CNT) are promising materials for next generation electronic devices [1, 2]. In a recent report in Nature Nanotechnology [3], it was found that voltage can be generated when water evaporates off the surface of ethanol treated carbon black samples. No doubt, results from such evaporation-induced production of electrical energy is highly feasible in relatively dry and windy climates. However, in the context of Singapore and around the region where tropical equatorial climate is experienced, we believe evaporation would be less effective. With a drive towards developing materials for application in sustainable and renewable energy in Singapore, we propose to develop a means to generate potential by withdrawing water from the surface of our proposed sample. With Singapore being an island surrounded by seawater and having a massive network of drainage that runs throughout our nation, retracting water level by means of tidal waves and water flowing through drains would make our work highly applicable in Singapore's context.

S. M. J. Ho (✉) · R. Y. N. Wang
Fairfield Methodist School (Secondary), Singapore, Singapore
e-mail: lavenderedcorgi@gmail.com

© Springer Nature Singapore Pte Ltd. 2019
H. Guo et al. (eds.), *IRC-SET 2018*,
https://doi.org/10.1007/978-981-32-9828-6_8

In this project, instead of using carbon black which involves a very complex preparation process of heating and oxidation, CNTs were selected for the following reasons: (1) CNTs are one of the most conductive and versatile materials. Furthermore, its high porosity allows ions to flow through the CNTs; (2) CNTs can easily be synthesized on semi-conducting (SiO_2), conductive (Si) and transparent (Quartz) substrates and (3) CNTs can also be transferred onto flexible substrates such as polymers. Additionally, to improve its functionalities, a focused laser beam can also be used to assist in the optimization of the microarray to generate optimal voltages. At the same time, the process also allows tuning of the degree of transparencies of the sample. A customized setup was also designed and constructed to test both water evaporation and tidal effect induced generation of potential from the sample.

2 Hypothesis

It is hypothesized that as water retracts along the CNTs array, zeta potential will be generated. This process involved separation of ionic charges within the liquid medium due to the attraction between ions in the solution and that on the CNT surface. The distribution is affected by the net charge at the CNT surface, dissociation of ions in the liquid medium and decrease in the water level. The outcome is the formation of an electrical double layer at the particle-liquid interface, at the point where shear is experienced. Using this principle, it is believe that CNTs can generate potential as water moves along its surface.

3 Methodology

Carbon nanotubes (CNTs) with typical length of 80 μm are grown on clean silicon dioxide (~5 mm × 5 mm, SiO_2) or quartz (~7 mm × 7 mm) substrates. Before growth, a layer of iron film is coated on the substrates as catalyst using a magnetron sputtering system (model: RF Magnetron Denton Discovery 18). The coating rate is 4 nm min^{-1} lasting for 6 min. These CNTs are synthesized using a plasma enhanced chemical vapor deposition (PECVD) system, and details of the growth process are reported elsewhere [4]. Quality of the CNT used in the experiment is quantified by Raman spectroscopy. A typical Raman spectrum from the CNT sample is presented in Fig. 8. Further characterisation of the samples are conducted using Raman Spectroscopy, Scanning Electron Microscopy (SEM, JOEL 6700), Keithley 6430 and Keithley 2636 Sourcemeter.

4   Results and Discussion

For electrical measurements, CNT sample are designed to adhered 1 cm above the edge of the glass slide, and two supports were adhered lengthwise to either side of the CNT. The supports ensured the wires would not bend upwards at the edge of the CNT, consequently causing defects in the connection of the CNT and wire (Fig. 9). The wires are then taped using thermal tape to minimise contact with water. Silver paste is used to connect the wire to the CNT array (Fig. 1a). To determine the viability of generating a potential through a process of withdrawing water from the surface of CNTs, a customized setup was designed and fabricated by us (Fig. 1b). In our design, water is contained in a glass tank of $110.48 \times 140.53 \times 145.21$ mm. The sample is partially submerged in the water while water is withdrawn by a tubing fed from the base of a glass tank to the tip of the syringe. 100 and 250 mL beakers are arranged at the base of the tank to decrease the overall base area. In doing so, a smaller volume of water is needed to be drawn for the identical rate of the decrease of water level in the tank. Thus, counteracting the limited volume of the syringe (20 mL). Additionally, in the study of evaporation effect, the surfaces of the beakers allow water vapour to condense, avoiding condensation from disrupting the recording process. A CCD camera is used to record video of water receding from the surface of the sample. Figure 1d shows time lapsed images of water receding from the CNT surface (with

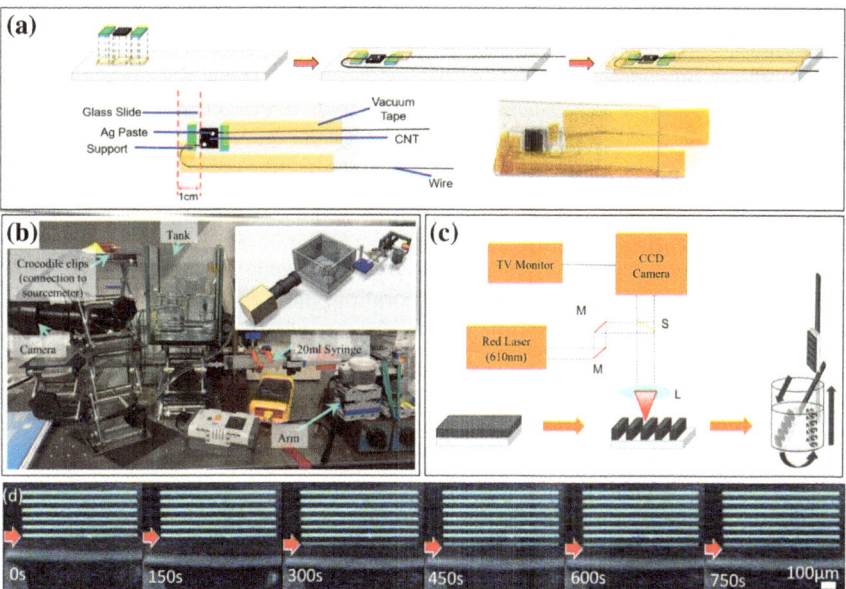

Fig. 1 **a** Schematic of preparation of electrodes. **b** Labeled experimental setup and digital schematic. **c** Schematic of the preparation of collapsed samples and the focused laser patterning system for the cutting of channels in CNTs array. **d** Time lapsed snapshots of water receding from the surface of the CNTs

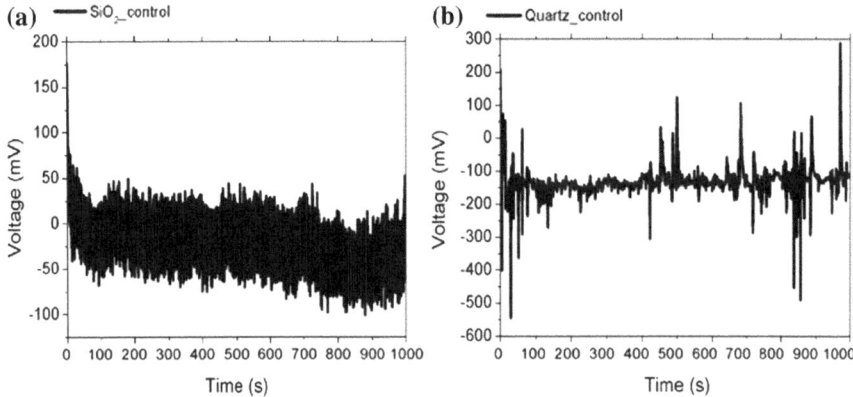

Fig. 2 Potential generated from **a** SiO$_2$ and **b** quartz substrate in water

horizontal channels, details will be discussed in later section). Images are captured at 150 s intervals and the red arrow indicates the water level at each time. Live demonstration of this tidal effect can be obtained from the link provided at the end of the report. As it is set to be a slow process, the uploaded video has been speed up by 8× and it only shows a portion of the experimental process.

Prof. Sow Chorng Haur, Dr. Sharon Lim X.D.

Physics Department

National University of Singapore

Singapore, Singapore

To verify that the base substrate did not contribute to the generated potential, two control electrodes using the base substrates, SiO$_2$ and quartz are made. The results (Fig. 2) show insignificant voltage detected from both types of base substrates. Thus, subsequent measurements presented are true values obtained from the different types of CNT structures.

Synthesized CNTs is used to test the difference between potential generated from evaporation and tidal method. During evaporation, the sample is heated to 50 °C at ambient condition and the directional flow of water was upwards, along the surface of the CNT. Measurements obtained from CNT arrays with respect to temperature during the evaporation process is presented in Fig. 10. Contrary to the former, the tidal method induces a laminar flow of water in the downwards direction (Fig. 3a). Figure 3b–c shows top and side view SEM images of the CNT sample used. Tidal method generated an average of 3.60E−4 V which was greater than using the evaporation average of 3.43E−04 V (Fig. 2d). Given the consistency and reliability, and relevant application to Singapore's context, subsequent experiments will be conducted using tidal method.

To recede water from the CNT surface, a syringe is connected to a custom water retracting system (Fig. 4). Using Lego Mindstorms, the rate at which the water is withdrawn can be varied through the program and maintained at a constant rate for a fixed duration. The motor turns a worm gear connected to a 40-tooth gear, the gear

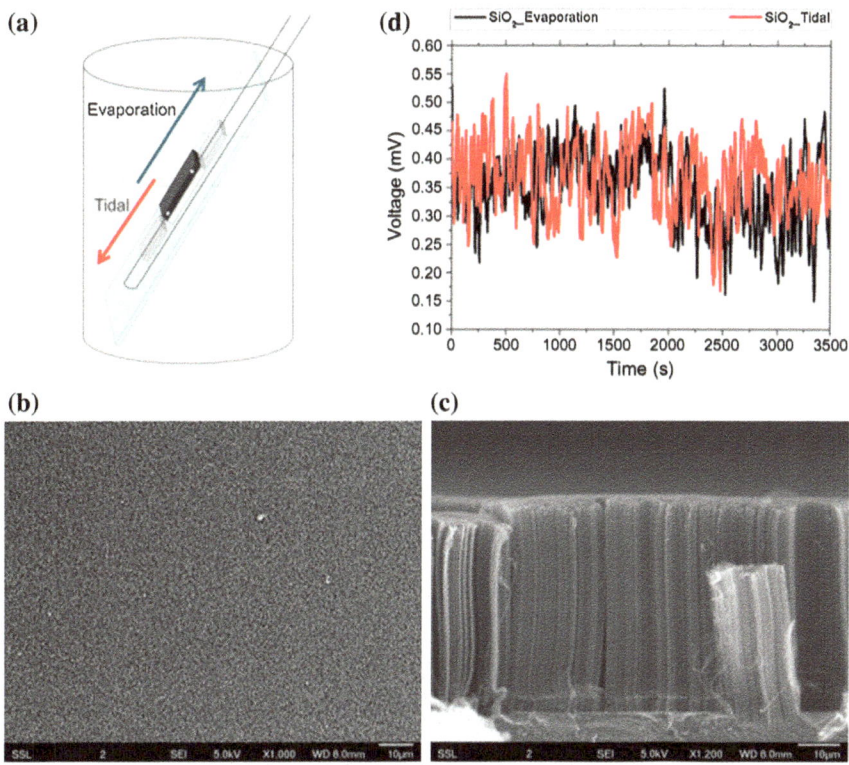

Fig. 3 **a** Schematic of direction of water flow for different methods used. **b–c** SEM of **b** top and **c** side view of as-synthesized CNT array. **d** Potential generated from evaporation and tidal method

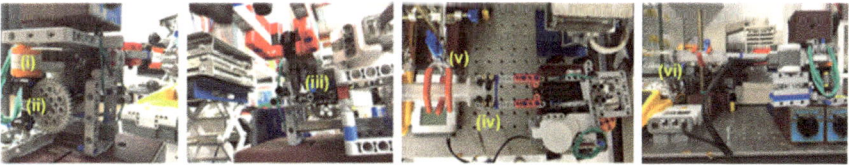

Fig. 4 Self-designed, self-constructed water retracting system (i) Motor (ii) Worm to 40-tooth (iii) Rack and pinion (iv) Claw (v) Retort clamp (vi) Tubing

ratio repeats twice to increase torque and decrease speed. A 16-tooth gear lies on the same axel as the second 40-tooth to create a rack and pinion arm system. A claw that clamps onto the syringe plunger is pulled backwards while the syringe body is held by a retort clamp. A tube is fed from inside the tank to the mouth of the syringe, enabling water to be drawn.

Structures are created on the synthesized CNTs' array in an attempt to further improve the generated potential. Using a laser patterning system (Fig. 1c), 660 nm laser beam is focused onto the sample (placed on a computer controlled stage) through a 50× optical objective. Microsoft Visual Basic is used to generate code to interface with the controller to shift the sample with respect to the laser spot. 13 centralised channels of 2500×100 µm, with gaps of 100 µm are cut from the CNT, forming the channel-like structure. Figure 5a shows an optical image of the patterned channels oriented vertically. A 40° tilted SEM image of the channels is presented in Fig. 5b. Figure 6c shows the side profile of the channels. To determine the effect of channels, potential generated from two different orientations of the channels are investigated. When the channels are oriented horizontally, it generated an average of 6.75E−04 which was much greater than that of vertically oriented channels (3.22E−04).

From the above results, we proposed the following mechanism. Plain CNT has a hydrophobic surface that repels water. Hence interaction between ions in the water and the negatively charged CNT surface is hence minimized (Fig. 6a). For horizontally oriented channels, more CNTs are exposed, hence allowing more interaction with the solution. Furthermore, the sides of the CNT pillars are slightly more hydrophilic than the top surface [5]. This allows water to penetrate into the porous CNTs forest, hence allowing more interactions between the ions in the water with

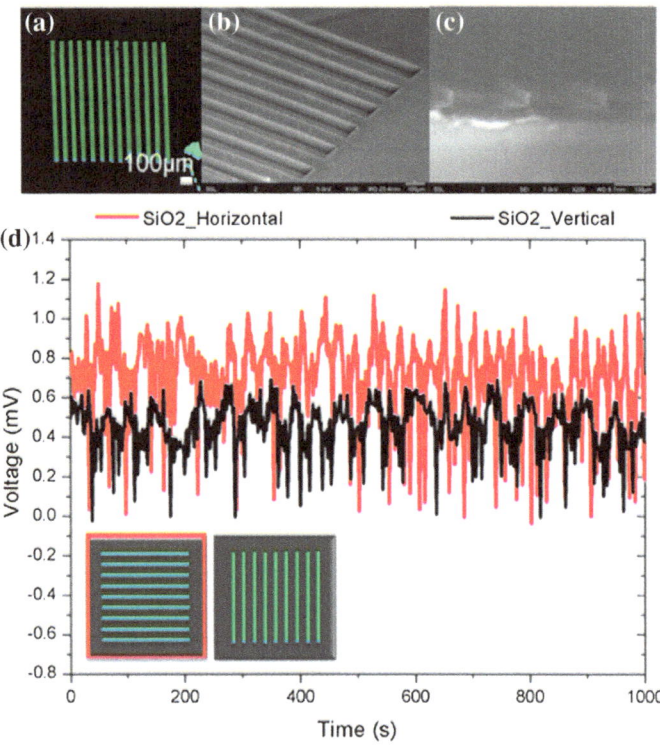

Fig. 5 **a** Optical image of CNT channels, **b** top view of CNT channels, **c** side view of CNT channels, **d** potential generated from horizontal and vertical channels

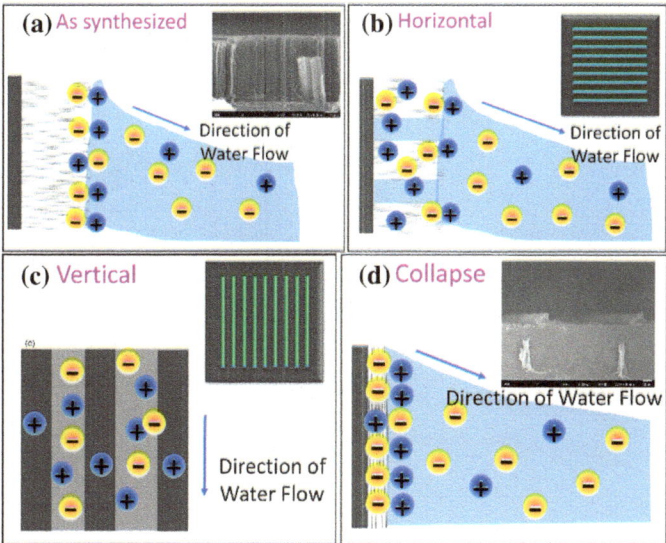

Fig. 6 **a** Schematic of mechanism of as-synthesized, **b** horizontal channels, **c** vertical channels and **d** collapsed sample. Insets are schematic and SEM images of the respective structures

the CNT. As a result, more positively-charged ions are attracted to the negatively-charged CNT. The meniscus of the water in between pillars also aids the separation of the ions by entrapping the ions in the CNT pillars. As a result, higher potential is detected (Fig. 6b). For vertically oriented CNT channels, ions are allowed to pass through the sample without any interaction with the CNT itself. Consequently, there are less ions attracted to the CNT and less separation of ionic charges. Although the meniscus of the water would trap some ions in the CNT, the smaller exposed surface area only allowed a small amount of ions to be trapped. Thus, the sample with vertical channels generates a lower potential compared to its horizontal counterpart (Fig. 6c). As such, if the above proposal is true, then by densifying the CNT channels, reducing porosity of the CNT channels would limit the interaction between the ions and the CNTs by limiting the ability of the water to flow through the CNT forest (Fig. 6d).

Hence, one would expect the compacted sample to generate lower potential. To verify the hypothesis, patterned CNTs sample is adhered to an aluminum stick and submerged into ethanol and withdrawn slowly (Fig. 1c). As the sample is being pulled out, the surface tensions pulls the CNT pillars downwards and causes them to collapse and densify. Expecting lower value from the compacted sample, larger CNTs array synthesis on quartz is used. This will allow the measured potential to fall within the detectable limit of the sourcemeter. Similar measurements are also obtained from horizontally and vertically orientated channels on quartz. The results (Fig. 7a) shows similar trend as that presented in Fig. 4d.

Figure 7b shows generated potential obtained from not collapsed/not compacted channels verse collapsed/compacted channels. Inset are the corresponding side view

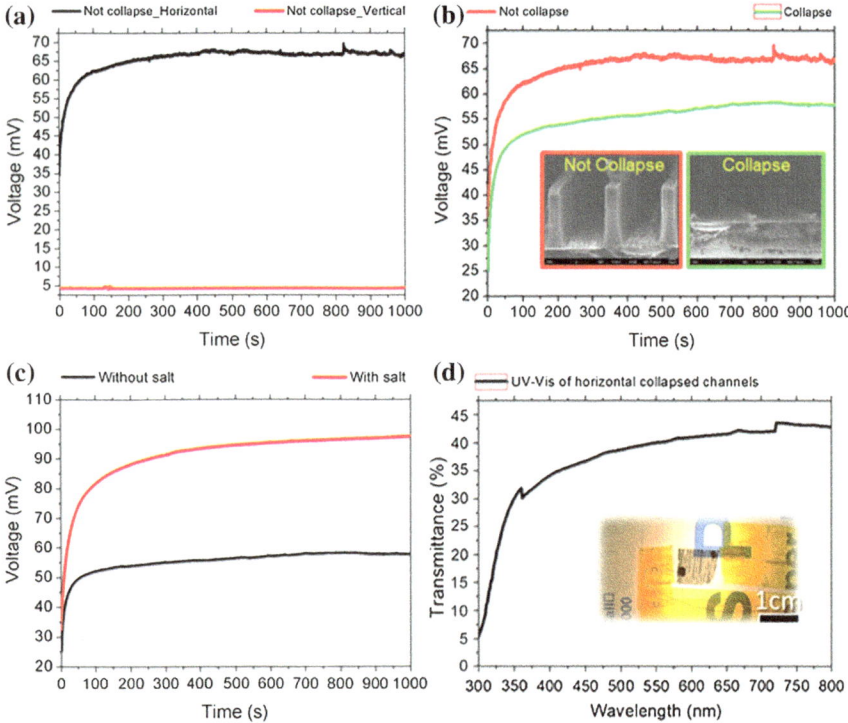

Fig. 7 a Similar trend obtained from horizontal and vertical channels on quartz substrates as compared to those obtained from SiO$_2$ substrates. **b** Graph and SEM images comparing collapsed and not collapsed channels on quartz samples. **c** Potential generated from collapsed quartz samples in salt solution and water. **d** UV-Vis spectrum of the horizontally oriented compacted/collapsed sample on quartz. Inset shows the background words being seen through the sample

SEM images. Furthermore, the role of ions and surface charges on the CNTs can be amplified by replacing water with a more ionic solution. In this case, saturated salt solution is used. Due to the increase of free-flowing ions in the liquid, the saturated salt solution should generate more potential and a higher voltage is recorded. This is supported by the results of the experiment where the potential generated by the sample in saturated salt solution is twice that of the sample in water (Fig. 7b).

An added functionality of the sample arises when the CNT channels on quartz are collapsed/compacted. Given that quartz is a transparent substrate, when collapsed, the compacted CNTs will create a degree of tint. The degree of tint can be tuned by varying the thickness of the CNT channels and gaps between the CNT channels on quartz during the laser patterning process. This would allow the samples to be coated on glass panels to be used in buildings as tinted windows. To determine how translucent the compacted sample is, UV-Vis spectrometer is used to obtain the percentage of transmittance obtained from the sample with respect to different wavelength of incident light (Fig. 7d). Inset shows photo of the background words being partially seen through the sample.

5   Conclusion and Further Advancements

Various CNT microarrays were engineered to determine an optimal design that generates the highest potential. Detailed analysis was conducted to shortlist and identify the optimal microarray and proposed mechanism. The microarrays cut and tested consisted of as synthesized CNT, horizontal and vertical channels, and collapsing of channels. This opens the spectrum of possible applications of the prototype electrodes. Based on prior understanding of zeta potential, the suggested mechanism consists of the separation of ions through the attraction of ions within the liquid medium and the negatively-charged CNT surface, and the interaction between the ionic charges. With its potential as a sustainable source of electricity, the prototype can be commercialised and used in drains or even as tinted windows for buildings to power appliances. Be it rain or shine, our results when coupled with solar panels will allow potential to be generated under any weather conditions, especially in Singapore's context.

Website link to live demonstration of tidal effect: https://www.physics.nus.edu.sg/~physowch/microtide/.

Appendix

See Figs. 8, 9 and 10.

Fig. 8 Raman of CNT sample

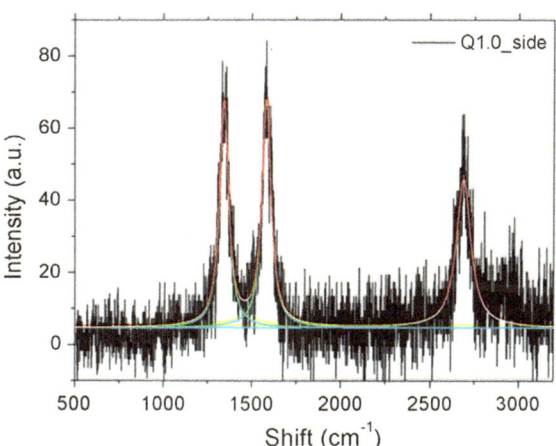

Fig. 9 Schematic of
supports

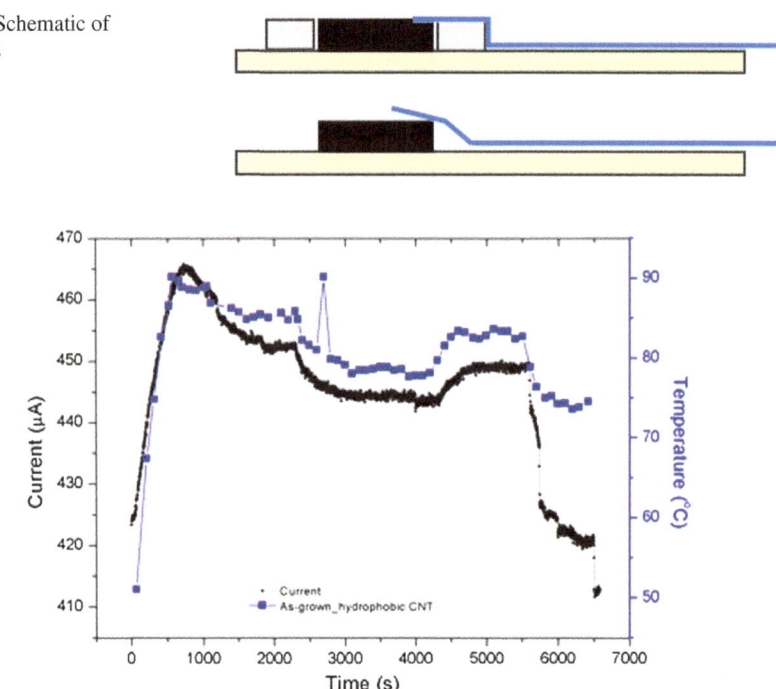

Fig. 10 Current produce by hydrophobic CNT sample at 0.2 V

References

1. Park, S., Vosguerichian, M., & Bao, Z. (2013). A Review of fabrication and applications of carbon nanotube film-based flexible electronics. *Nanoscale, 5,* 1727–1752.
2. Che, Y., Chen, H., Gui, H., Liu, J., Liu, B., & Zhou, C. (2014). Review of carbon nanotube nanoelectronics and macroelectronics. *IOPscience.*
3. Xue, G., Xu, Y., Ding, T., Li, J., Yin, J., Fei, W., et al. (2017). Water-evaporation-induced electricity with nanostructured carbon materials. *Nature Nanotechnology.*
4. Lim, K. Y., et al. (2003). Laser pruning of carbon nanotubes as a route to static and movable structures. *Advanced Materials, 15,* 300–303.
5. Li, P., Lim, X., Zhu, Y., Yu, T., Ong, C.-K., Shen, Z., et al. (2007). *Tailoring wettability change on aligned and patterned carbon nanotube films for selective assembly.* American Chemical Society.
6. Srivastava, A., Srivastava, O. N., Talapatra, S., Vajtai, R., & Ajayan, P. M. (2004). Carbon nanotube filters. *Nature Materials, 3,* 610–614.

The Application of Markov Chains in Multiplayer Online Battle Arena (MOBA) Games

Yongren Xu, Qi Zhang and Zhi Cheng Tan

Abstract The prediction of future states, be it of daily events or complex dynamical systems, has always been a challenging topic in the application of probability theory into real life. The benefit of such predictions based on current conditions, and independent of the past, is far-reaching and manifests itself in multiple areas such as transport, biology and economics-areas closely linked to our everyday life. Yet, the study of this topic has not been easy due to the lack of access to vital data of dynamical systems. In this paper, we propose a revolutionary method of studying such systems, through the application of Markov chains into MOBA online games. Using this complex game as a simulation of dynamical systems and an easy source of data, we aim to predict the movement of a game character based on its present position, using the model constructed with Markov chains theory. Through careful experiments, comparing recorded and expected results, and critical evaluation of the method used, we have proven that Markov chains can be effectively applied to yield an accurate prediction of the future state. And with this, we wish to have shed some light on devising innovative ideas on examining dynamical systems, as well as the potential of improving dynamical systems that will benefit the larger society.

Keywords Markov chains · Dynamical system · Probability

1 Introduction and Purpose

In daily life, people usually need to make the best decisions for themselves, based on the present situation and expected future developments. Scaled up, in some dynamical systems, it would be beneficial to predict future developments as it can help make the

Y. Xu · Q. Zhang (✉) · Z. C. Tan
Anderson Junior College, Singapore, Singapore
e-mail: zhangqi.990705@gmail.com

Y. Xu
e-mail: mxyr007@126.com

Z. C. Tan
e-mail: zhichengmoo@gmail.com

© Springer Nature Singapore Pte Ltd. 2019
H. Guo et al. (eds.), *IRC-SET 2018*,
https://doi.org/10.1007/978-981-32-9828-6_9

Fig. 1 Marked points are
neutral creatures. ▲ This
symbolizes a point of neutral
creatures

system more efficient. For example, in transport systems, it would be advantageous
to be able to predict future traffic conditions, given the current traffic conditions. With
proper predictions, it could potentially reduce or prevent the problem of road congestion damaging many economies and contributing to carbon emissions worldwide.
The system being probabilistic and stochastic, makes it suitable for Markov chains
to model such a system—since one can make predictions of the future based solely
on the system's present state and independent from its history. Unfortunately, data on
real life dynamical systems and those alike are often relatively inaccessible or hard
to collect, significantly hindering the theoretical study and construction of a proper
model on such systems. Recognising the similar randomness underlying dynamical
systems and players' positions in Multiplayer Online Battle Arena (MOBA) games,
we can therefore use Markov chains to model a MOBA game, which serves as a
viable simulation of real-life dynamical systems, and of which we have full access
to essential data.

MOBA games are complex games popular worldwide. In the games, players duel
on a map as heroes. Heroes can raise their prowess by defeating neutral creatures,
located at fixed positions on the map; or gank[1] enemies, impeding their development. However, it can be difficult to locate enemy heroes due to a lack of vision
on the opposing team. Hence we aim to model the movement of heroes on the map
using Markov chains, helping players predict enemy positions and experimenting the
applicability of Markov chains in predicting the future states of dynamical systems.

We decided to use "DOTA 2", a typical MOBA game, as our research case study.
The overwhelming popularity[2] that it enjoys identifies it as a significant game for our
study. A map of "DOTA 2" (version 7.07) is shown below (Fig. 1). For easy reference,
we labeled the positions of neutral creatures with letters A–I, and a simplified map is

---

[1]Gank refers to actively moving around the map in order to kill an enemy hero.

[2]Monetary contributions of its fans worldwide made up the new record of $20,770,460 in compe-
tition prize money for "The International, DOTA 2 Championships".

Fig. 2 Simplified map

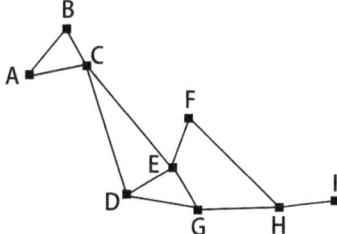

provided (Fig. 2). By killing neutral creatures, the hero earns gold, which can be used to purchase upgrades and strengthen him. It is thus logical for heroes to approach and kill neutral creatures that are highest in value and closest to him.

2 Hypothesis

It is reasonable to assume that the probability of a hero choosing a point of neutral creatures is dependent only on the distance of the point from the hero and the value of the point. From this, various weighting factors can be considered to construct a model, they are

$$V - D, \ V^2 - D, \ V - D^2, \ 2V - D, \ V - 2D, \ \frac{V}{D^3}, \ \frac{V}{D^2}, \ \frac{V}{D}$$

where V represents value of a point and D represents the distance from the point to the hero's current position.

However, we hypothesize that the best fit should be from the choice proportional to value and inversely proportional to distance

$$P_{X \to Y} \propto \frac{V_Y}{D_{XY}}, \tag{1}$$

where X and Y denotes two adjacent points of neutral creatures.

The following assumptions are made to construct a plausible mathematical model:

(1) There is only one hero who kills the neutral creatures at each side of the map, at any given time, and he does this continuously.
(2) The movement time of the hero is negligible vis-a-vis the time used to defeat the neutral creatures.
(3) The refresh period for each neutral creature is 1 min (1 min in actuality).
(4) The refresh period for each neutral creature is unaffected by any factor (actually it may be affected).
(5) It takes 30 s for any hero to defeat one point of neutral creatures.

Due to previous assumptions, the movement of heroes between points of neutral creatures fulfils the Markov property (memoryless/independent of the past) and hence can be modeled by Markov chains.

3 Methodology

3.1 Derivation of Model

A Markov chain is a random process with the property that, conditional on the present, the future is independent of the past [1]. It can be used to model a random walk process through the formula

$$
P_{present} = \begin{pmatrix} P_{11} & P_{12} & \cdots & P_{1i} \\ P_{21} & P_{22} & \cdots & P_{2i} \\ \vdots & \vdots & \vdots & \vdots \\ P_{i1} & P_{i2} & \cdots & P_{ii} \end{pmatrix}^{n} \times \begin{pmatrix} P_1 \\ P_2 \\ \vdots \\ P_i \end{pmatrix}.
\tag{2}
$$

Based on the previous assumptions, the action of a hero only depends on the present position of the hero and is independent of its previous positions. Thus, the action is a Markov process and can be modeled using a Markov chain.

In "DOTA 2", the map (as shown in the Introduction section) is split into two parts—the green side, Radiant, and the red side, Dire. As the map is rotationally symmetrical about the centre, we can use the Radiant side for our study. Suppose that a Radiant hero is hitting neutral creatures and a Dire hero wants to find him.

The possibilities of the Radiant hero's position can be used to form a matrix P, in which $P(X)$ represents the possibility of the hero to be at position X.

$$
P = \begin{pmatrix} P(A) \\ P(B) \\ P(C) \\ P(D) \\ P(E) \\ P(F) \\ P(G) \\ P(H) \\ P(I) \end{pmatrix}
\tag{3}
$$

According to our hypothesis, the probability should be inversely proportional to the distance (D) and directly proportional to the value (V). Therefore, the probability of every position can be represented by

$$P(X \to Y_n) = \frac{\frac{V(Y_n)}{D(X \to Y_n)}}{\frac{V(Y_1)}{D(X \to Y_1)} + \frac{V(Y_2)}{D(X \to Y_2)} + \cdots + \frac{V(Y_n)}{D(X \to Y_n)}} . \qquad (4)$$

All data are organized in Table 1.

Therefore, the position after time t (minutes) can be represented by

$$P = \begin{pmatrix} P'(A) \\ P'(B) \\ P'(C) \\ P'(D) \\ P'(E) \\ P'(F) \\ P'(G) \\ P'(H) \\ P'(I) \end{pmatrix}^T \times \begin{pmatrix} 0 & 0.296 & 0.704 & 0 & 0 & 0 & 0 & 0 & 0 \\ 0.276 & 0 & 0.724 & 0 & 0 & 0 & 0 & 0 & 0 \\ 0.344 & 0.380 & 0 & 0.154 & 0.122 & 0 & 0 & 0 & 0 \\ 0 & 0 & 0.356 & 0 & 0.372 & 0 & 0.272 & 0 & 0 \\ 0 & 0 & 0.166 & 0.219 & 0 & 0.425 & 0.19 & 0 & 0 \\ 0 & 0 & 0 & 0 & 0.747 & 0 & 0 & 0.253 & 0 \\ 0 & 0 & 0 & 0.360 & 0.427 & 0 & 0 & 0.213 & 0 \\ 0 & 0 & 0 & 0 & 0 & 0.351 & 0.231 & 0 & 0.418 \\ 0 & 0 & 0 & 0 & 0 & 0 & 0 & 1.00 & 0 \end{pmatrix}^{\lfloor \frac{t}{0.5} \rfloor} .$$

$$(5)$$

3.2 Experiment

To test the model, we recorded the position of the "Anti-mage" hero at 0.5-min intervals in each game, across 50 games. The "Anti-mage" is chosen as he is typically chosen to hit neutral creatures.

The initial state condition (i.e. position at $t = 0$) from actual experiment is substituted into the model as the initial condition (i.e. into matrix P), in order to calculate the expected probability of the hero being at each of the labeled neutral creature points, at each recorded time. The probability calculated through data is the calculated average value of 5 nearby points, i.e., the moving average (MA):

$$\bar{x}_n = \frac{x_{n-2} + x_{n-1} + x_n + x_{n+1} + x_{n+2}}{5} . \qquad (6)$$

The resulting two probability tables—one containing experimental data; the other containing theoretical values obtained from the model—are then compared, yielding the graphs shown in the appendix for evaluation of the extent of accuracy using the Markov chains model.

Table 1 Probability of movement between points

From	To								
	A	B	C	D	E	F	G	H	I
A	0	0.296	0.704	0	0	0	0	0	0
B	0.276	0	0.724	0	0	0	0	0	0
C	0.344	0.38	0	0.154	0.122	0	0	0	0
D	0	0	0.356	0	0.372	0	0.272	0	0
E	0	0	0.166	0.219	0	0.425	0.19	0	0
F	0	0	0	0	0.747	0	0	0.253	0
G	0	0	0	0.36	0.427	0	0	0.213	0
H	0	0	0	0	0	0.351	0.231	0	0.418
I	0	0	0	0	0	0	0	1	0

4 Results

By summing up the number of instances where "Anti- mage" appears at each neutral creature point, at every point of time (**n**); and the total number of effective records at each point of time (**N**), we obtained the raw data for calculation of probabilities, which is given by **n** over **N**. From these data, we plotted 2 graphs for each of the nine neutral creature points, as shown below (Figs 3–11).

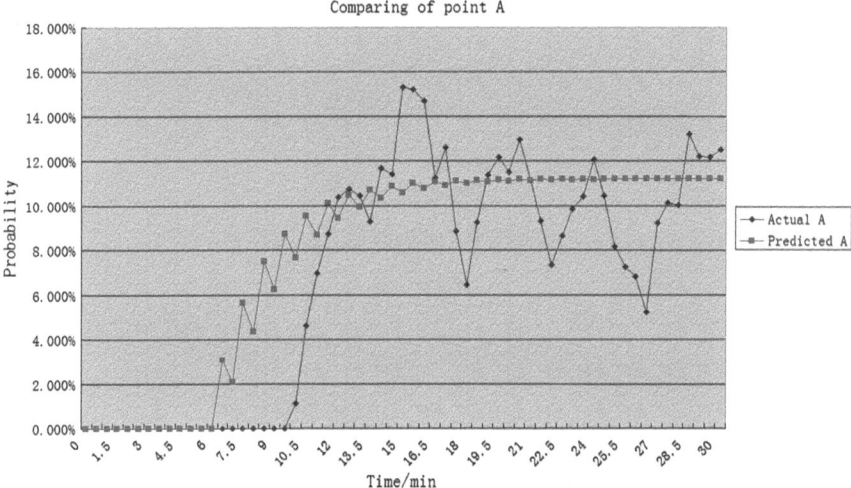

Fig. 3 Comparing the graphs of actual A and predicted A

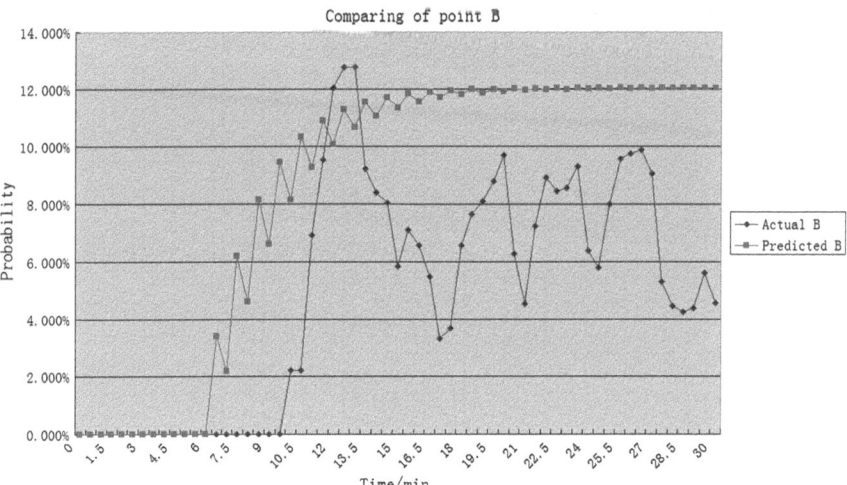

Fig. 4 Comparing the graphs of actual B and predicted B

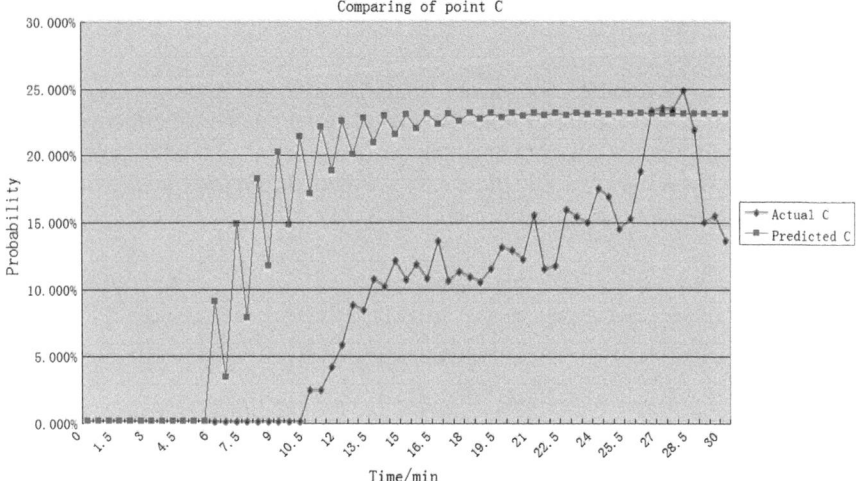

Fig. 5 Comparing the graphs of actual C and predicted C

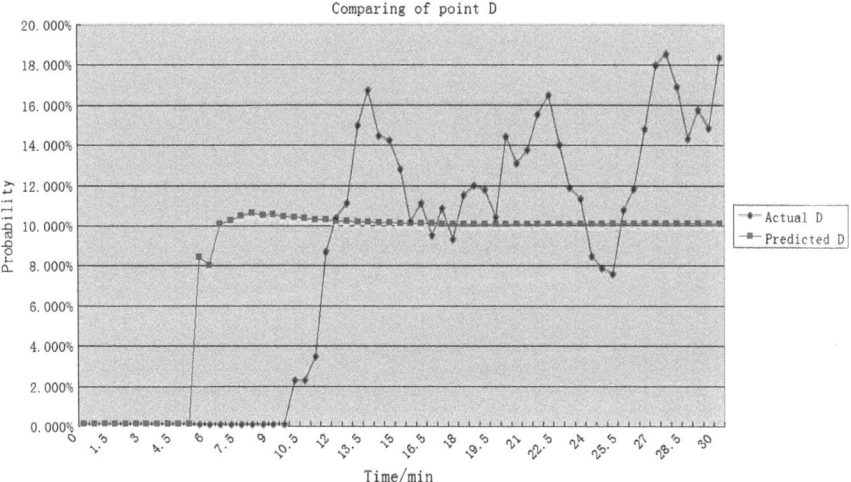

Fig. 6 Comparing the graphs of actual D and predicted D

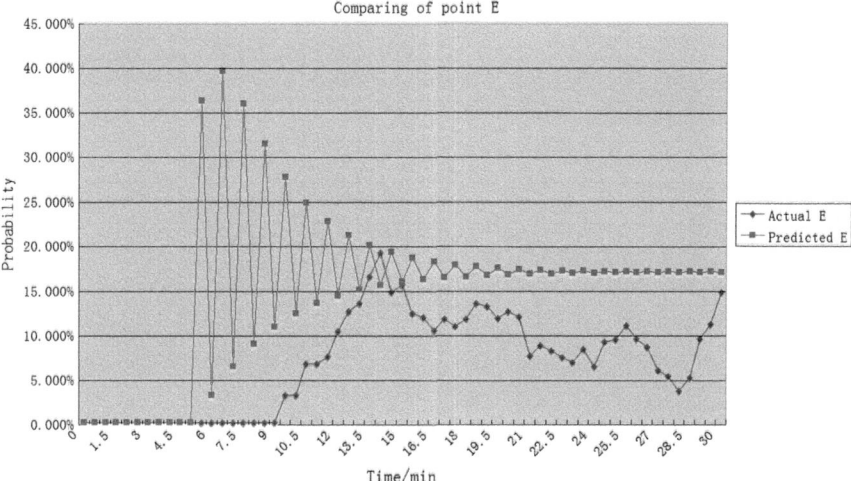

Fig. 7 Comparing the graphs of actual E and predicted E

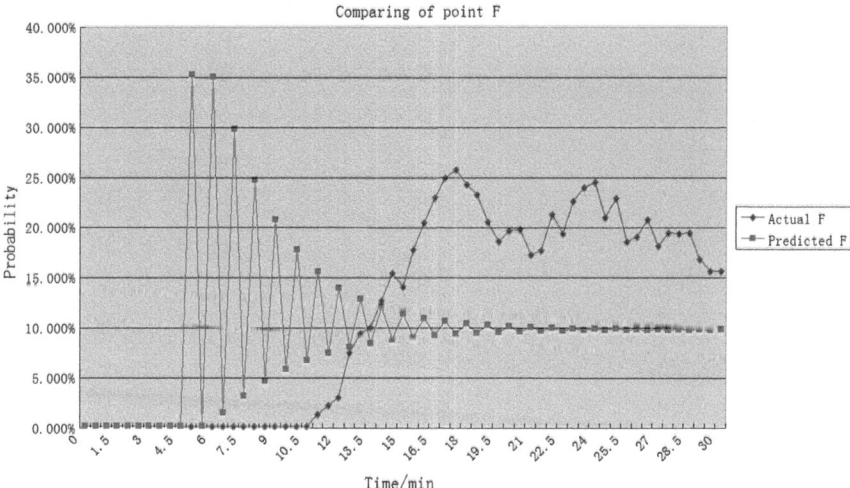

Fig. 8 Comparing the graphs of actual F and predicted F

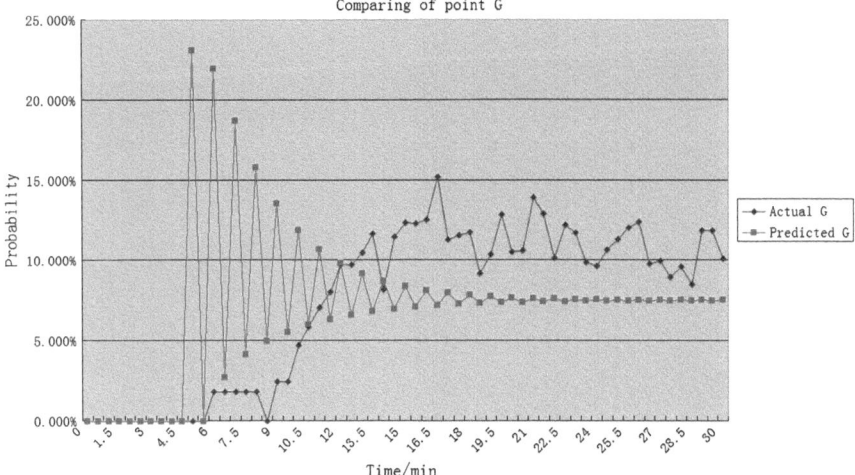

Fig. 9 Comparing the graphs of actual G and predicted G

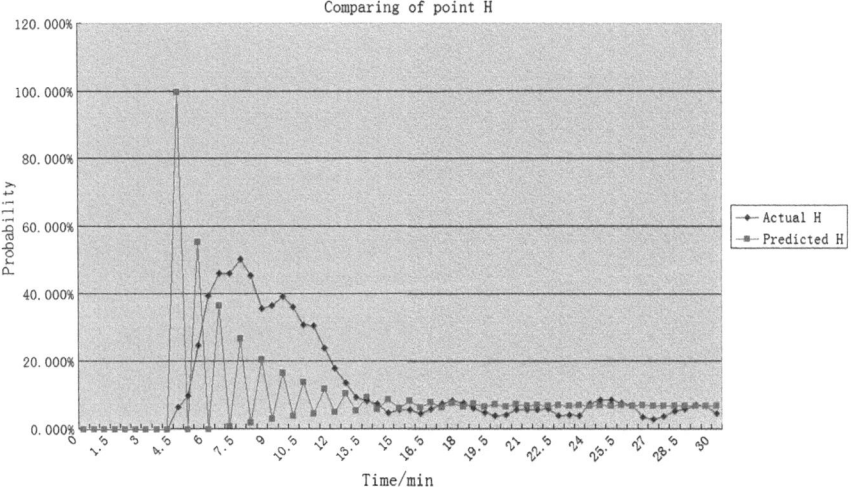

Fig. 10 Comparing the graphs of actual H and predicted H

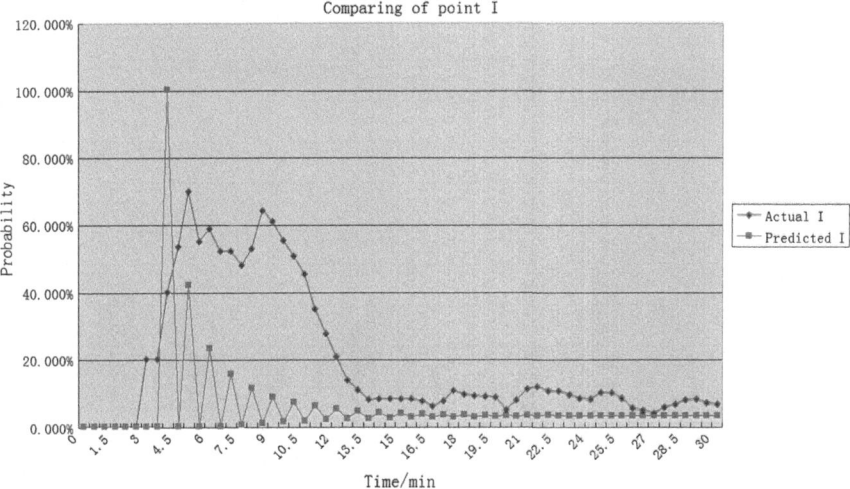

Fig. 11 Comparing the graphs of actual I and predicted I

In the graphs, the horizontal axis represents time in minutes while the vertical axis represents the probability of the hero appearing at the point indicated in each graph. The actual probabilities calculated using the recorded data, as mentioned above, is used to plot the graphs in blue; while the graphs in pink represents the predicted probability from the model.

For all the graphs, during the first 10 min, the differences between the two curves are very large. This is likely due to the hero being too weak at the beginning of the game to combat the neutral creatures. Most players will choose to stay in the lane instead of hitting neutral creatures during the first 10 min. Between 10 and 30 min, it is evident that the current model $\frac{V}{D}$ does not predict the actual situation very well. As such, we repeated the experiment process with the other various weighting factors $V - D$, $V^2 - D$, $V - D^2$, $2V - D$, $V - 2D$, $\frac{V}{D^3}$, $\frac{V}{D^2}$, and taking into consideration only the 10–30 min segments of the graphs when evaluating the different models. Then, graphs of the relative difference between the actual and predicted probabilities for each point, as calculated using each of the formulas stated above are plotted (Fig. 12).

In the graph, the horizontal axis represents the points (A, B, ···, I). The vertical axis represents the relative difference calculated. Different colors of the bars represent the different Markov chains with different weighting factors. The graph is plotted to compare the relative difference for various formulas, across various points.

The graph of the Mean Squared Error of different models is plotted using the actual and expected probabilities (Fig. 13).

In the graph, the horizontal axis represents the formula of various weighting factors. The vertical axis represents the average relative difference calculated across 9 points. Below the graph, data is listed for a better observation. The graph is plotted to compare the average relative difference for the 9 points across the various formulae.

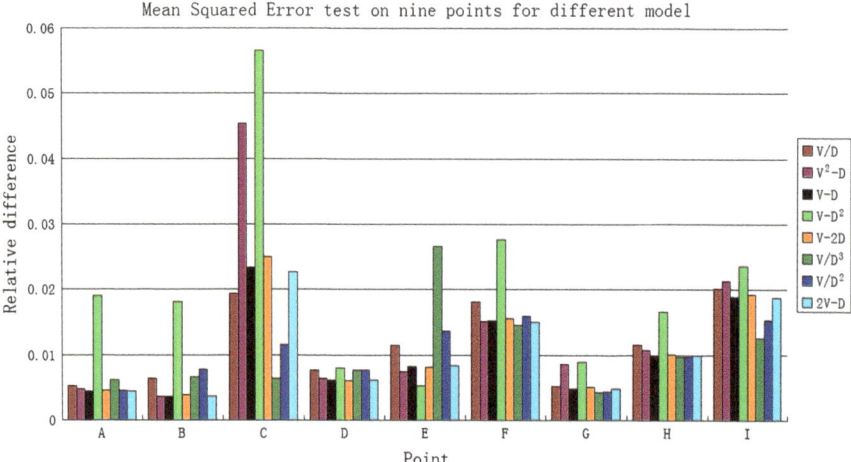

Fig. 12 Bar graphs comparing the prediction accuracy of the different formulas across the 9 different points

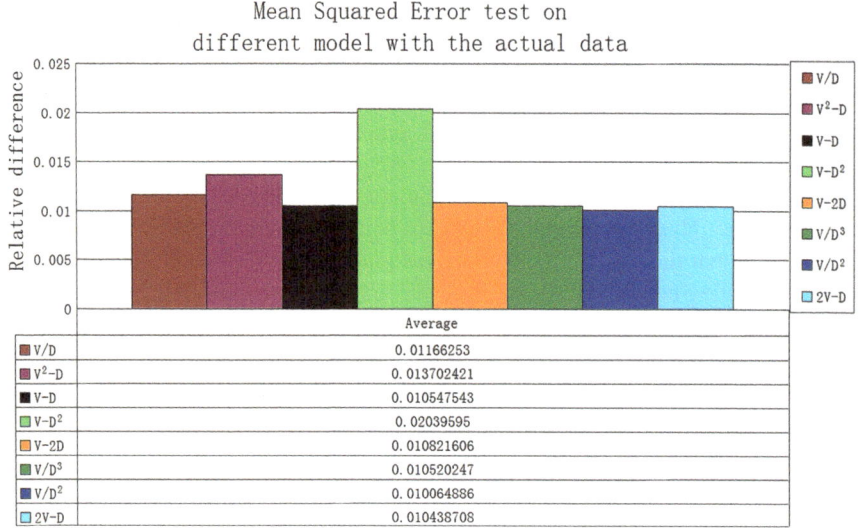

Fig. 13 Bar graphs comparing the average accuracy of prediction of the different formulae across the 9 points

5   Discussion and Conclusion

Interestingly, from our results, it shows that the weighting factor $\frac{V}{D^2}$ gives the most accurate prediction, despite some persistent discrepancies from the actual situation. While this means that our hypothesized weighting factor $\frac{V}{D}$ is not the most accurate prediction, it shows that perhaps in actuality, players regard the distance factor in the formula as proportionally more significant when compared to the value factor of the given point of neutral creatures.

The persistent discrepancies could be attributed to the following:

(A) At point C, the actual curve is mostly under the predicted curve. This shows that players usually undervalue point C. One possible reason is that the hero actually hits the neutral creatures faster than expected. Given the relatively low density of neutral creatures in the upper half of Radiant side (points A, B, C), the supply hardly satisfies the need for refreshed neutral creature points. Thus, players do not tend to go to point C despite its high value.

(B) At point F, the actual curve is mostly above the predicted curve. This shows that players overvalue point F. The same line of reasoning as in A) can be used to explain this phenomenon, as F lies in a region with relatively high density of neutral creature points.

(C) At point I, the actual curve is mostly above the predicted curve. This shows that players overvalue point I. One possible reason is that point I is close to the "safe lane" [2], another major source of gold. This factor, in addition to point I's inherent value, also contributes to the probability of the hero being at I, thus giving point I a higher effective value for the purpose of our model.

In conclusion, it is an effective idea to use Markov chains to model the movement of the hero in MOBA games. This will be useful for a player in planning and executing a successful gank on the enemy. In addition, the model also shows that it is possible to scale it up to real-life dynamical systems, applying Markov chains to real-life problems such as modeling the movement of cars and predicting traffic jams.

Acknowledgements Our deepest appreciation is due to our mentor, Dr. Lee Ching Hua (scientist at IHPC, A*STAR, and Adjunct Assistant Professor at NUS), for inspiring us and guiding us with great patience and enthusiasm.
We are also profoundly indebted to our school teachers: Mr. Poh Wei Leong, Mr. Ho Kian Tong, and Madam Resma Bte Gulzar Mohd, for their enlightening advice; as well as all our friends who have supported us in one way or another.
That said, any errors or inadequacies that remain are entirely our own.

References

1. Grimmett, G. R., & Stirzaker, D. R. (2001). *Probability and random processes* (3rd ed., p. 213). New York: Oxford University Press.
2. Dota2 WIKI. https://dota2.gamepedia.com/Lane. Accessed March 25, 2018.

Investigating the Air Quality in Bus Stops Using IoT-Enabled Devices

Teo Hong Ray, Danish Uzair B. Abdul Hamid Khan Surattee
and A. Muhammed Madhih

Abstract Air pollution is a global problem, with particulate matter being one of the major air pollutants. Small particulate matter smaller than 2.5 μ in diameter (PM$_{2.5}$), are the most harmful. Traffic emissions are a major source of particulate matter in urban cities. This issue is particularly pertinent for bus commuters, who spend extended periods of time exposed to traffic emissions during their daily commute waiting at bus stops. This study employed a mobile Internet of Things (IoT) device to detect PM$_{2.5}$ air quality at bus stops of various conditions. It was of interest in this study to investigate PM$_{2.5}$ air quality with regards to three parameters. Firstly, the position within bus stop. Secondly, the frequency of buses arriving and lastly, the amount of vehicular traffic including cars, motorcycles and lorries. The results of this investigation shows that there are marginal differences in PM$_{2.5}$ air quality between positions in a bus stop. Air quality further away from the road is found to be consistently though marginally better compared to the edge of the road. The variation of frequency of bus arrivals did not provide a strong correlation with the air quality contrary to previous literature. The amount of vehicular traffic also did not have a strong positive linear relationship with air quality contrary to expectations.

Keywords Internet of things · Air quality · Particulate matter · Bus stops · Public transport · Singapore

1 Introduction

Air pollution is a global problem as the world is becoming more modernized, with particulate matter being one of the major air pollutants [1]. Small particulate matter, such as those smaller than 2.5 μ in diameter (PM$_{2.5}$), is the most harmful as they cannot be naturally removed from the body once inhaled [1]. This results in adverse health effects from exposure, ranging from cardiovascular problems such as arteriosclerosis, stroke and myocardial infarction [2], to neurological disorders like white matter disease and exacerbating Alzheimer's and Parkinson's diseases [3], and

T. H. Ray · D. U. B. Abdul Hamid Khan Surattee (✉) · A. Muhammed Madhih
Centre for Smart Systems, Singapore University for Technology and Design, Singapore, Singapore
e-mail: danish.uzair.b.abdul.hamid.k.s.2017@vjc.sg

© Springer Nature Singapore Pte Ltd. 2019
H. Guo et al. (eds.), *IRC-SET 2018*,
https://doi.org/10.1007/978-981-32-9828-6_10

even psychological distress [4]. Although Singapore is known as a 'garden city' and its air quality ($PM_{2.5}$ concentration of 19 $\mu g/m^3$) is significantly better than that of least developed countries (average $PM_{2.5}$ concentration of 49 $\mu g/m^3$) [5], this number is an average across the whole country and certain groups within the population might be exposed to higher $PM_{2.5}$ concentrations.

One such group would be bus commuters, who spend extended periods of time exposed to traffic emissions during their daily commute when waiting at bus stops. These traffic emissions are a major source of particulate matter in urban cities [6]. A study conducted in London found that a bus commuter's total exposure to $PM_{2.5}$ is twice as much as that of a car commuter [7]. Another study in Buffalo, New York suggested that exposure to $PM_{2.5}$ inside a bus stop is 18% higher than exposure outside the bus stop [8]. In Singapore, a country where 46.7% of working people take a bus when commuting to work [9], this problem is one that can have far reaching effects.

In response to the hazards of air pollution, several cities have put in place Internet of Things (IoT) systems to monitor air quality. For example, in Chicago, sensors mounted on lampposts were deployed to measure several air pollutants. Chicago used this data to analyze and forecast air quality incidents [10]. Meanwhile, the city of Dublin uses sensors mounted on bikes in its bike share program to gather citywide air quality data [10]. However, the data gathered from these systems do not focus on specific microenvironments within the city. Thus, it is hard to analyze the causes of air pollution and the effects of potential interventions on air pollution at specific locations.

There have been limited studies conducted with regards to air pollution in bus stops in Singapore. One of these studies analyses the nature of air pollutants and the size of particulate matter in 5 crowded bus stops in Singapore [11]. However, these 5 crowded bus stops are not representative of bus stops in Singapore, and the findings may be unable to account for the $PM_{2.5}$ concentrations of bus stops at other locations.

Hence, this study will use IoT devices to collect and analyze data from many bus stops with varying characteristics in Singapore, in search of factors that would influence the $PM_{2.5}$ concentrations in bus stops. As there are many variables that might affect the $PM_{2.5}$ concentration, 3 parameters were targeted in this study: Position within bus stop, frequency of buses arriving and vehicular traffic including cars, motorcycles and lorries. The $PM_{2.5}$ concentration might vary depending on the position within the bus stop due to the distance away from pollutants, such as the bus exhaust or the road [12]. The frequency of buses arriving is a good gauge of the extent of the stop-start driving behavior of buses, which could heighten $PM_{2.5}$ concentrations [13]. It is also suggested that vehicular traffic emissions from vehicles travelling along the road contribute substantially to $PM_{2.5}$ concentrations in urban areas [14].

This valuable information can be used by the Land Transport Authority (LTA) of Singapore to predict which bus stops would be the most polluted. They would then be able to take on a targeted approach to improve areas such as the ventilation [15] or position [16] of bus stops predicted with high $PM_{2.5}$ concentrations. Ultimately, this research aims to show how IoT devices can be effectively used for environmental monitoring and research.

2 Hypothesis

$PM_{2.5}$ concentrations are higher at the back of the bus stops as it is closer to the bus exhaust, the primary source of pollution from a bus. The edge of the road, which is closer to passing vehicles, will contain higher $PM_{2.5}$ concentrations. The frequency of buses arriving and the volume of vehicular traffic have a positive correlation with the $PM_{2.5}$ concentration at bus stops.

3 Methods and Materials

To test this hypothesis, a field study was conducted using a mobile Air Quality detector (AirQ), which is an Internet of Things (IoT) device, along with the corresponding mobile AirQ smartphone app, jointly developed with Centre for Smart Systems at Singapore University of Technology and Design.

3.1 Mobile AirQ and Smartphone App

The mobile AirQ is a small (77.70 mm × 60.15 mm × 32.10 mm) and lightweight (118 g) device, owing partly to its modern 3d printed casing. It contains a Particulate Matter (PM) sensor (Plantower PMS7003) and a temperature and humidity sensor (Bosch BME280) wired to a development board (Linkit 7697), which is powered by a lithium-ion rechargeable battery (Fig. 1).

3D modelling using Fusion 360 also enables us to create compact structures that help to hold the components in place, preventing them from falling out of position when the mobile AirQ is moved about. This ensures the durability of the mobile AirQ. For example, the PM sensor is firmly held such that the air inlet and outlet is always exposed to surrounding air. Moreover, the design is easy to replicate at a low cost.

The Linkit 7697 board is designed for IoT applications, boasting low cost and low power consumption while providing inbuilt Wi-Fi and Bluetooth connectivity. During the course of this project, the mobile AirQ was found to have a battery life of 3 h. These characteristics make the mobile AirQ easily portable.

During data collection, the mobile AirQ device is first turned on and paired to a smartphone through Bluetooth within the mobile AirQ app. Data streams from the environmental sensor are then processed by the Linkit 7697 and sent via Bluetooth to be displayed on the app, which is updated every 5 s (Fig. 2).

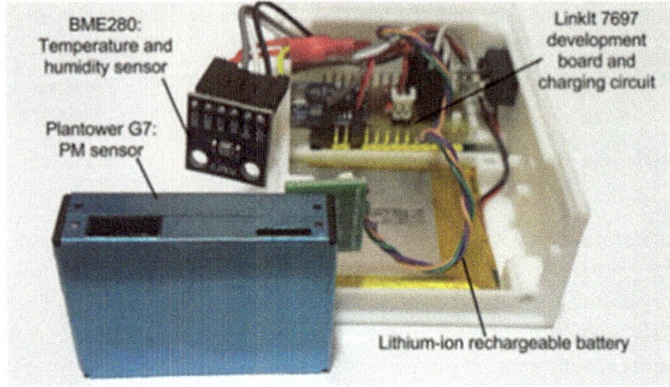

Fig. 1 Diagram showing the different components in the Mobile AirQ

Fig. 2 Diagram shows how data is received and displayed on the Mobile App

The data streams are also able to be exported as a csv file for ease of data analysis (Fig. 3).

The Mobile AirQ and AirQ app is user-friendly and intuitive to use. This enables the device and app to be used by people who wish to conduct environmental studies but lack the technical knowledge of complicated monitoring equipment.

Data collection

Table 1 shows how data were collected for the different variables under investigation.

A good variety of bus stops were chosen, with a good spread of variables such as the busyness of the adjacent road, land use of the surrounding area and size of bus stop (Table 2).

All measurements were taken within 7:30 to 10:00 during the morning peak hour on weekdays, from 15 December 2017 to 31 December 2017. The morning peak hour was chosen as there is a high number of commuters travelling, making it an important time period to study. This would also capture the highest exposure to $PM_{2.5}$ by commuters due to the relatively high traffic density at that time.

	A	B	C	D	E	F	G	H
1	Date	Time	PM 2.5	Temperat	Humidity	PM 10	Latitude	Longitude
2	9/11/2017	3:35:12	20	26.53	45.26	21	1.307104	103.9193
3	9/11/2017	3:35:17	19	26.52	45.27	20	1.307104	103.9193
4	9/11/2017	3:35:22	19	26.54	45.3	21	1.307104	103.9193
5	9/11/2017	3:35:27	20	26.51	45.31	22	1.307104	103.9193
6	9/11/2017	3:35:32	22	26.53	45.32	24	1.307104	103.9193
7	9/11/2017	3:35:37	22	26.52	45.2	23	1.307104	103.9193
8	9/11/2017	3:35:42	21	26.52	45.16	24	1.307104	103.9193
9	9/11/2017	3:35:47	21	26.5	45.23	25	1.307104	103.9193
10	9/11/2017	3:35:52	21	26.53	45.47	26	1.307104	103.9193
11	9/11/2017	3:35:57	20	26.55	45.87	26	1.307104	103.9193
12	9/11/2017	3:36:02	22	26.54	45.61	28	1.307104	103.9193
13	9/11/2017	3:36:08	22	26.53	45.31	27	1.307105	103.9193
14	9/11/2017	3:36:13	24	26.54	45.12	28	1.307105	103.9193
15	9/11/2017	3:36:18	26	26.56	45.16	29	1.307105	103.9193
16	9/11/2017	3:36:23	23	26.54	45.03	25	1.307105	103.9193
17	9/11/2017	3:37:46	25	26.76	48.79	27	1.307533	103.9193
18	9/11/2017	3:37:48	26	26.76	51.47	27	1.307533	103.9193
19	9/11/2017	3:37:51	25	26.78	50.27	25	1.307533	103.9193
20	9/11/2017	3:37:53	26	26.79	48.77	26	1.307533	103.9193
21	9/11/2017	3:37:56	23	26.81	47	25	1.307533	103.9193
22	9/11/2017	3:37:58	23	26.81	46.45	25	1.307533	103.9193
23	9/11/2017	3:38:01	24	26.82	46.12	26	1.307533	103.9193

output (+)

Ready

Fig. 3 CSV file containing output data

Table 1 Methods of data collection for each variable studied

Factor analyzed	How data were collected
Position within bus stop	Two experiments were conducted. The first experiment involved 14 bus stops, where the $PM_{2.5}$ concentrations at the front, middle and back of the bus stop was measured (Fig. 4). The second experiment involved 6 bus stops, where the $PM_{2.5}$ concentrations at the edge of the road and far from the road was measured (Fig. 4)
Frequency of buses arriving	6 bus stops were studied in this experiment. The number of buses stopping at the bus stop was counted over 15 min, and the average rate of buses passing per minute was derived The ambient $PM_{2.5}$ concentration was also measured by standing 10 m away from the bus stop (Fig. 4), where the bus does not come to a stop and starts to move off, thus eliminating the stop-start effect
Vehicular traffic	14 bus stops were studied in this experiment. The number of other vehicles passing by the side of the road nearer to the bus stop was counted over 5 min, and the average rate of vehicles passing per minute was derived

Table 2 Bus stops chosen for investigation

Bus stop name	Bus stop location
Aft Sims Way	Near KPE
Blk 111	Rivervale Plaza
Blk 248A	Compassvale Primary School
Blk 2C	Near Geylang West CC
Chai Chee Ind Pk	Decathlon at Chai Chee
Dhoby Ghaut Stn	Plaza Singapura
Kallang MRT	Kallang MRT
Katong Shop Ctr	Katong Shopping Centre
Maranatha Hall	Before Tanjong Katong Secondary School
Opp Blk 2C	Near Kallang MRT
Opp Playground@ Big Splash	East Coast Park
OUE Bayfront	OUE Bayfront
Paya Lebar Stn	Paya Lebar MRT
Siglap Link	Behind Victoria School

The mobile AirQ was standardized to be placed at 1 m above the ground as that is the average height of one's head when seated within the bus stop. All $PM_{2.5}$ readings were averaged over 5 min to improve the accuracy of the data.

4 Results and Discussion

4.1 Position within Bus Stop

4.1.1 Front, Middle and Back

Comparison between the average $PM_{2.5}$ concentrations at the front (33.24 $\mu g/cm^3$), middle (31.35 $\mu g/cm^3$) and back (32.54 $\mu g/cm^3$) of the bus stops surveyed reveals a marginal variation of $PM_{2.5}$ concentrations between the different positions of the bus stop (Fig. 5) within 1.89 $\mu g/cm^3$. This is mainly due to the diffusion where the $PM_{2.5}$ particles spread across a span of not more than 3 meters in width (Fig. 4).

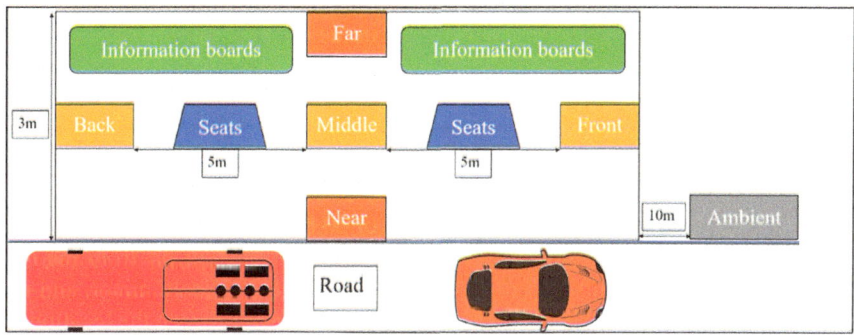

Fig. 4 Locations at the bus stop where data was collected

Fig. 5 The average $PM_{2.5}$ concentrations at the front, middle and back of a bus stop

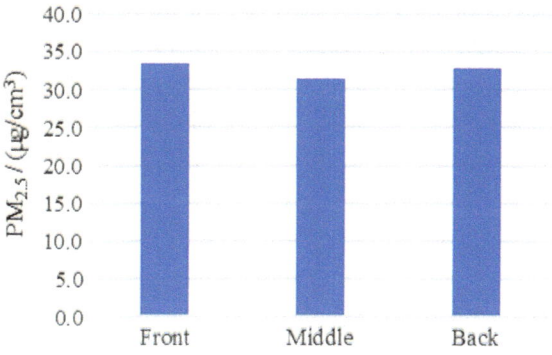

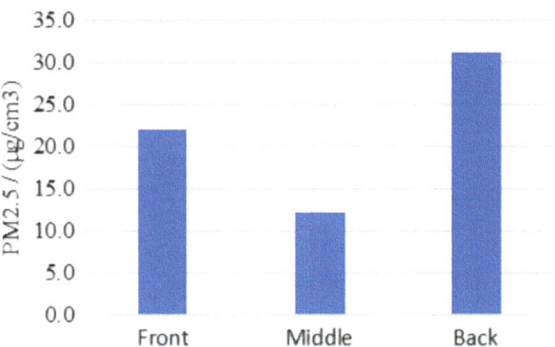

Fig. 6 The PM$_{2.5}$ concentrations at the front, middle and back of Dhoby Ghaut Stn

However, it is interesting to note the huge difference in PM$_{2.5}$ concentrations for Dhoby Ghaut Stn (Fig. 6), with the back being higher than the front by 9.22 µg/cm^3, and the front being higher than the middle by 9.70 µg/cm^3. This causes the standard deviation of the PM$_{2.5}$ concentrations at Dhoby Ghaut Stn to be more than 3 times higher than that at other bus stops. This could be due to the presence of an air-conditioned Mass Rapid Transit (MRT) station exit right behind the bus stop. The presence of air-conditioners could filter the polluted air from the road and release fresh air near the bus stop, causing anomalies in the data collection.

4.1.2   Edge Versus far from the Road

Comparison between the PM$_{2.5}$ concentrations at the edge of the road (average of 32.31 µg/cm^3) and far from the road (average of 30.82 µg/cm^3) reveal that the edge of the road has a consistently higher PM$_{2.5}$ concentration than far from the road (Fig. 7). Nevertheless, this difference of 1.49 µg/cm^3 is relatively minor, and it is suspected that this might be attributed to the short distance between the locations of the two readings of not more than 3 m, and the fact that PM$_{2.5}$ is suspended in the air longer than larger particles [17], allowing it to diffuse more profoundly.

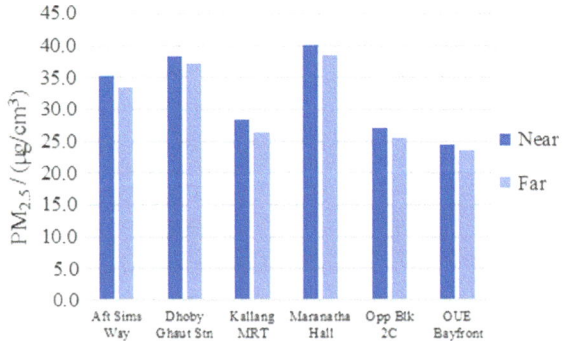

Fig. 7 The average PM$_{2.5}$ concentrations at the edge versus far away from the road

Fig. 8 Difference between the additional PM$_{2.5}$ caused by buses operating at the bus stop against frequency of buses arriving

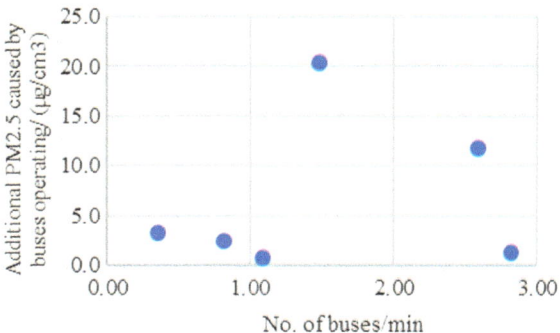

4.2 Frequency of Buses Arriving

To analyze how the frequency of buses arriving affects the PM$_{2.5}$ concentration at bus stops, the average PM$_{2.5}$ concentration at bus stops was calculated by averaging, which is calculated by averaging the PM$_{2.5}$ readings from the front, middle and back of the bus stops. The additional PM$_{2.5}$ concentration was calculated by taking the difference between the average PM$_{2.5}$ concentration in the bus stop and the ambient PM$_{2.5}$ concentration. This is to control all other variables, such as the vehicular traffic or weather condition that could contribute to the PM$_{2.5}$ concentration of bus stops other than the stop-start effect of buses.

Results unexpectedly show that the frequency of buses arriving has an inconsistent effect on the PM$_{2.5}$ at bus stops (Fig. 8). This challenges what was proposed by Buonanno, Fuoco and Stabile in their paper [13], that the stop-start effect of buses would have a large impact on the PM$_{2.5}$ concentrations at bus stops. If found to be true, this is a significant discovery as it would mean that researchers have been wrongly identifying the extent of pollution in bus stops mainly based on the frequency of buses arriving.

However, these results are to be taken with a pinch of salt as the data collection for the ambient PM$_{2.5}$ was rather subjective, due to obstructions that prevent us from standardizing the location for data collection of the ambient PM$_{2.5}$. The additional PM$_{2.5}$ concentrations at OUE Bayfront and Dhoby Ghaut Station are exceptionally higher due to the ambient PM$_{2.5}$ concentrations there being lower than at the other bus stops. Thus, buses are seen to have a significant impact on PM$_{2.5}$ concentrations at such places only.

4.3 Vehicular Traffic

To analyze the effect of vehicular traffic on the PM$_{2.5}$ concentration at bus stops, the ambient PM$_{2.5}$ was used for comparison instead of the average PM$_{2.5}$ as the stop-start driving behavior of buses might unintentionally inflate the PM$_{2.5}$ concentration. Data

Fig. 9 Ambient PM$_{2.5}$ concentration against number of vehicles passing per minute

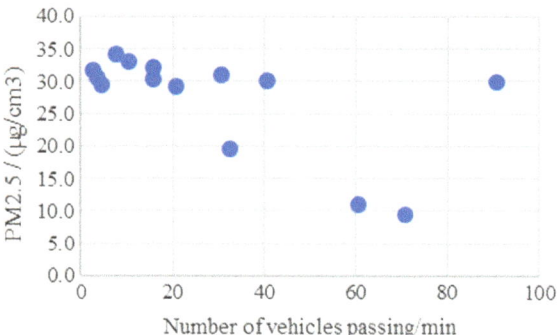

collected shows that the number of vehicles passing per minute is also unrelated to the ambient PM$_{2.5}$ concentration at bus stops (Fig. 9), unlike the predicted graph of a strong linear, positive correlation. This could be due to the strict regulations in Singapore, requiring all light petrol-driven vehicles, including cars, to be fitted with catalytic converters [18] that vastly reduce the amount of particulate matter and other air pollutants released from the exhaust of vehicles.

Something surprising was the two data points collected within the Central Business District (CBD) which had an exceptionally low ambient PM$_{2.5}$ (highlighted in Fig. 9). This could be due to the numerous presence of air-conditioned buildings in the region. As mentioned previously, air-conditioning could possibly filter the indoor air of polluting particles, assuming the units are properly maintained [19]. The clean air within these indoor locations can be released to the outside environment whenever the building doors are open, which is especially so during the morning peak hour, when many office workers are entering their worksites. As OUE Bayfront is located about 50 m from the sea, a morning sea breeze was felt during data collection that further disperses particulate matter, leading to lower PM$_{2.5}$ concentrations. From these two observations, wind might affect the PM$_{2.5}$ concentrations, making it an interesting area for further research.

5   Conclusion

This study is a proof of concept that IoT devices, which are less cumbersome than traditional monitoring equipment, can still be used for environmental monitoring and research. In addition, the results are exciting as some counterintuitive trends have been uncovered, which could change the way we think about urban air pollution and how we tackle it. Moving forward, it is hoped that this study can prompt further research using IoT devices that can validate these findings and explore other factors that might influence commuters' PM$_{2.5}$ exposure at bus stops, such as the presence of greenery or the effect of wind.

References

1. Doreswamy, H. S., & Sudheendra, S. R. (2010). Major air pollutants and their effects. *Mapana Journal of Sciences, 9*(2), 21–27.
2. Bourdrel, T., Bind, M.-A., Béjot, Y., Morel, O., & Argacha, J.-F. (2017). Cardiovascular effects of air pollution. *Archives of Cardiovascular Diseases, 110,* 634–642.
3. Babadjouni, R. M., Hodis, D. M., Radwanski, R., Durazo, R., Patel, A., Liu, Q., et al. (2017). Clinical effects of air pollution on the central nervous system; a review. *Journal of Clinical Neuroscience, 43,* 16–24.
4. Sass, V., Kravitz-Wirtz, N., Karceski, S. M., Hajat, A., Crowder, K., & Takeuchi, D. (2017). The effects of air pollution on individual psychological distress. *Health & Place, 48,* 72–79.
5. Brauer, M. (2016). *PM2.5 air pollution, mean annual exposure (micrograms per cubic meter).* Retrieved from The World Bank: https://data.worldbank.org/indicator/EN.ATM.PM25.MC.M3?end=2015&start=2015&view=map&year_high_desc=false.
6. Yatkin, S., & Bayram, A. (2007). Elemental composition and sources of particulate matter in the ambient air of a Metropolitan City. *Atmospheric Research, 85*(1), 126–139.
7. Rivas, I., Hagen-Zankera, A., & Kumar, P. (2017). Exposure to air pollutants during commuting in London: Are there inequalities among different socio-economic groups? *Environment International, 101,* 143–157.
8. Hess, D., Ray, P., Stinson, A., & Park, J. (2010). Determinants of exposure to fine particulate matter (PM2.5) for waiting passengers at bus stops. *Atmospheric Environment, 44*(39), 5174–5182.
9. Department of Statistics Singapore. (2015). *General household survey 2015.* Singapore: Department of Statistics, Ministry of Trade & Industry, Republic of Singapore.
10. Bousquet, C. (2017, April). *How cities are using the internet of things to map air quality.* Retrieved from DATA-SMART CITY SOLUTIONS: http://datasmart.ash.harvard.edu/news/article/how-cities-are-using-the-internet-of-things-to-map-air-quality-1025.
11. Velasco, E., & Tan, S. H. (2016). Particle exposure while sitting at bus stops of hot and humid Singapore. *Atmospheric Environment, 142,* 251–263.
12. Kaura, S., Clark, R., Walsh, P., Arnold, S., Colvilea, R., & Nieuwenhuijsen, M. (2006, January). Exposure visualisation of ultrafine particle counts in a transport microenvironment. *Atmospheric Environment, 40*(2), 386–398.
13. Buonanno, G., Fuoco, F., & Stabile, L. (2011), Influential parameters on particle exposure of pedestrians in urban microenvironments. *Atmospheric Environment, 45*(7), 1434–1443.
14. Pant, P., & Harrison, R. M. (2013, October). Estimation of the contribution of road traffic emissions to particulate matter concentrations from field measurements: A review. *Atmospheric Environment, 77,* 78–97.
15. Lim, A. (2016, July). *Fans at bus stops to cool you down?* Cool! Retrieved from The Straits Times: http://www.straitstimes.com/singapore/transport/fans-at-bus-stops-to-cool-you-down-cool.
16. University of California—Los Angeles. (2017, November). *Relocating bus stops would cut riders' pollution exposure, study finds: Researchers' healthy solution: Move the sites 120 feet away from intersections.* Retrieved from ScienceDaily: https://www.sciencedaily.com/releases/2017/11/171107113222.htm.
17. Gardiner, L. (2008, June). *Aerosols: Tiny particulates in the air.* Retrieved from Windows to The Universe: https://www.windows2universe.org/earth/Atmosphere/particulates.html.
18. DieselNet. (n.d.). *Emission standards Singapore.* Retrieved from DieselNet: https://www.dieselnet.com/standards/sg/.
19. PURE Living Blog. (2015, June). *Can air conditioners help reduce indoor air pollution?* Retrieved from PURE Solutions: http://www.pureroom.com/Pure_living_blog/can-air-conditioners-help-reduce-indoor-air-pollution/.

Novel Method to Reconstruct a Surface Grid Using Linear Regression Modelling

Si Chenglei, Mao Yu Di, Pang Eng Meng Wyzley and Chen Shunfa

Abstract The main focus of this report is to explore methods to laser scan objects by mapping the locations of the scanned pixels into real-world coordinates through an algorithm with the use of a projected laser plane. This report presents two different approaches: A simplified method to laser-assisted 3D scanning using Geometrical Relations and a proposed Linear Regression Models method that was based on the simplified one and compares the differences between the two. The general form of the regression models was chosen after observation of the main mathematical constructs used in the stack model and plane model of the line-based laser triangulation method. These geometrical concepts, utilizing mainly the equations and intersections of lines and planes were inspiration for the regression model. The process follows that a laser plane is generated using a laser diode and cylindrical lens and cast onto an object. After taking images of the object at different angles, image analysis is carried out using Python 2 and OpenCV2. Then, the Linear Regression Models are trained with pre-existing data before making predictions while the Geometric Relations method uses the measurements of the mechanism in the equipment to map the red pixel location detected in the images to the point cloud of the object.

Keywords 3D scanning · Linear regression models · Generating a point cloud

1 Introduction

Both physical, as well as digital 3D models are useful for a wide variety of applications such as industrial design, prototyping, and even in the production of movies and video games. [1] Our purpose is to develop a precise simple set of procedures with algorithms that when coupled with a cheap set-up can allow us to seamlessly scan objects, find the coordinates of red laser points in an image, and generate a 3D digital model of the object, which is the point cloud. For the sake of comparison,

S. Chenglei (✉) · M. Y. Di · P. E. M. Wyzley
River Valley High School, Mathematics Leaders Academy, Singapore, Singapore
e-mail: sichenglei1125@gmail.com

C. Shunfa
CRADLE, Science Centre Singapore, Singapore, Singapore

© Springer Nature Singapore Pte Ltd. 2019   123
H. Guo et al. (eds.), *IRC-SET 2018*,
https://doi.org/10.1007/978-981-32-9828-6_11

an overall projected cost of a possible setup we can build is about USD$85 (Raspberry Pi 3 model B—USD$45, Picamera—USD$30, stepper motor—USD$5, acrylic cutouts—USD$5) [2] which is less than those found in the market currently at around USD$200–$700 for simple ones and USD$10000–$20000 for much advanced 3D scanners [3]. That being said, our algorithm can be coupled with any image capturing device, a projected laser plane and Python 2 to work. This will open up future avenues for anyone with a smart phone to be able to carry out 3D scanning with a simple algorithm. Currently, some methods that have been adopted include the stack model and plane model of using the line-based laser light triangulation method [4], which gave rise to our method that makes use of the plane equations and intersection of lines and planes. Our report will only focus on different ways of making use of a projected laser plane in image analysis for 3D scanning and how to map the features extracted from the analysis to actual 3D points in space.

2   Engineering Goal

To design a set of algorithms and procedures that will work in tandem with a projected laser plane and a cheap setup that will make 3D scanning less costly. This will make it easier to generate accurate point clouds and 3D models of objects.

3   Methodology and Materials

A brief description of the Geometric Relations (GR) method will be covered first and will be followed by reason for the Linear Regression Models (LRMs) method of choice. The essence of the GR method is to figure out how to map pixel locations in pictures to actual positions of points in 3D space and reconstruct a surface.

3.1   Image Analysis

Let there be a bright red point $P(P_x, P_y)$ on the image. We split the image into channels, use a sliding w by *DimensionX* window (as seen from the dotted line) and run the cv2.minmaxLoc function on the window, which would detect the brightest point $P(P_x, P_y)$ in the window.

Following that, we took 2 features of the image: The distance, X_p of the red pixel from the center of *DimensionX* and Y_p which is that from the center of *DimensionY*.

$$X_p = P_x - \frac{DimensionX}{2} \tag{1.1}$$

$$Y_p = \frac{Dimension Y}{2} - P_y \tag{1.2}$$

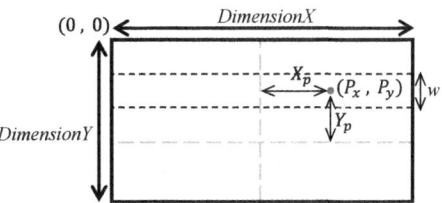

Equation (1.2) is opposite as the origin in an image and the origin that the function will give its output is at the top-left hand corner.

3.2 Geometric Relations (GR) Method

Light reflects from an object and passes through the lens of the camera we use and hits the CCD Array, with length $L = 3.68$ mm. For every unique ray of light that is reflected from the object shines on a unique point on the CCD Array, a discrete pixel is formed on the image. Therefore, it is possible to introduce a dependent variable, l, that is the unknown distance in mm from the center of the CCD Array. We can then say that l is dependent on the distance in pixels of the red pixel from the center of the image, X_p which can be calculated using (1.1) because of the discrete nature of the pixel and light. This relationship is given by the following:

$$\frac{l}{L} = \frac{X_p}{Dimension X} \tag{2}$$

where L is the known length of the CCD Array, *DimensionX* is the known total number of pixels along the x-axis of the image and X_p is the calculated distance in pixels using (1.1) between the middle of the image and the pixel in pixels. By using the concept of similar triangles (Appendix), l, X_p, fL and z can be found to have the following relationship:

$$\frac{z}{fL} = \frac{X_m}{l} \tag{3}$$

where X_m is X_p converted from pixels into mm and is also our final aim. By making l the subject in (3) and substituting into (2), we can write out X_m and Y_{m-} (in a similar fashion) in terms of z:

$$X_m = \frac{L}{Dimension X \cdot fL} \cdot X_p \cdot z \tag{4.1}$$

$$Y_m = \frac{H}{DimensionY \cdot fL} \cdot Y_p \cdot z \qquad (4.2)$$

where H is the known width of the CCD Array, $DimensionY$ is the known total number of pixels along the y-axis of the image and Y_p is the calculated distance in pixels using (1.2) between the middle of the image and the pixel in pixels. The red laser light can be expressed as a plane with the following general equation:

$$ax + by + cz = d \qquad (5)$$

where x, y and z are variable points which lie on the plane and a, b, c and d are constants that have already been determined using a real-life coordinate system. Value of z can be obtained by substituting in (4.1) and (4.2) into (5) to get the following equation:

$$a\left(\frac{L \cdot X_p}{DimensionX \cdot fL} \cdot z\right) + b\left(\frac{H \cdot Y_p}{DimensionY \cdot fL} \cdot z\right) + z = d \qquad (6)$$

Rearranging (6), z can be expressed as the following:

$$z = \frac{d}{\lambda} \qquad (7)$$

where

$$\lambda = a\left(\frac{L \cdot X_p}{DimensionX \cdot fL}\right) + b\left(\frac{H \cdot Y_p}{DimensionY \cdot fL}\right) + 1 \qquad (8)$$

and (8) and (a, b, c, d) are the Plane's constants. Thus, the coordinates of the red pixel can be obtained for plotting inside a 3D coordinate system by using (4.1), (4.2) and (7) after taking the required measurements of our equipment.

Notably, there are 2 things that this original method fails at. Firstly, if the camera does not use a CCD Array, or uses a non-uniform, non-linear type of image sensor [5], the similar triangles property that we use for our calculations falls apart and we are unable to take the sensor length as a measurement. Secondly, the wide variety of measurements we have to take while absorbing the effect of manufacturing defects will result in inaccuracy. By looking at the expression (4) alone, there are a lot of unknowns we have to measure about the setup (a, b, d, L, H, fL), and it may not even be possible for us to manually measure the internal hardware of the camera at times.

3.3 Linear Regression Models (LRMs) Method

In order to combat the limitations of the GR method, we proposed our Linear Regression Models(LRMs) method. Our LRMs method takes advantage of the fact that no matter what type of image sensor is used, the red pixel location within the image is the same due to the mechanics of a camera [6]. Regardless of image sensor type, we can train the parameters in our LRMs to fit the images taken by the camera as long as we have pre-existing data. An intuitive way to think of our LRMs method is that we are finding the specifications of an arbitrary CCD Array in a camera that could have took the exact same images in some arbitrary position. Rather than taking the manual hardware measurements, we train these measurements by telling our models what to output given a certain input.

3.4 Initialising Linear Regression Models (LRMs)

Our LRMs method consists of the input matrix, the input that will be gathered using our code from a single image, in the following form:

$$I = \begin{pmatrix} x_1^1 & x_2^1 & 1 \\ \vdots & \vdots & \vdots \\ x_1^n & x_2^n & 1 \end{pmatrix} \tag{9}$$

where x_1^i denotes the X_p (see (1.1)) of the i-th red pixel detected and x_2^i denotes the Y_p of the i-th red pixel detected. Upon observation of the type of expressions that the x, y and z coordinates hold, namely (4.1), (4.2) and (7), we see that the expression for z, (7) can be manipulated to give the following form:

$$\frac{1}{z} = \frac{a}{d}\left(\frac{L}{DimensionX \cdot fL}\right) \cdot X_p + \frac{b}{d}\left(\frac{H}{DimensionY \cdot fL}\right) \cdot Y_p + \frac{1}{d} \tag{10}$$

where (a, b, c, d) are the Plane's constants, L and H are the length and the height of the CCD Array, fL is the focal length of the camera and X_p and Y_p are shown in (1.1) and (1.2). Observation of (10) can tell us that $\frac{1}{z}$ can be modelled as an output of the following linear regression model:

$$Z = \frac{1}{h_\alpha(I)} \tag{11.1}$$

$$h_\alpha(I) = I\alpha^T \tag{11.2}$$

where $\alpha = \begin{pmatrix} \alpha_1 & \alpha_2 & \alpha_0 \end{pmatrix}$ is the parameters vector to be trained and $Z \in \mathfrak{R}^n$ is the output vector of dimensions, $(z_1 \ldots z_n)$ and all z coordinates that correspond to the red pixels in Image with input I in (9). We can note that (11.1) looks very similar to (7) and (11.2) very similar to (10) with a certain group of constants, $\frac{a}{d}\left(\frac{L}{DimensionX \cdot fL}\right)$ and $\frac{b}{d}\left(\frac{H}{DimensionY \cdot fL}\right)$ are weights to X_p and Y_p and $\frac{1}{d}$ as an intercept term. Two

Fig. 1 Example of an image with a red pixel

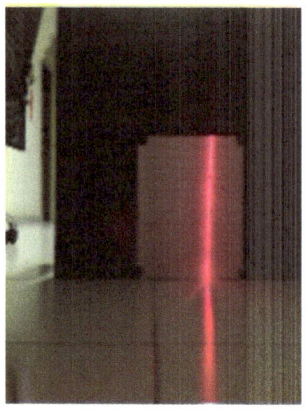

other similar models are constructed based on observations of (4.1) and (4.2) and are modelled based on the following (Fig. 1):

$$X = h_\beta(x_1, Z) = (Z_{x_1})\beta^T \qquad (12.1)$$

$$Y = h_\gamma(x_2, Z) = (Z_{x_2})\gamma^T \qquad (12.2)$$

where $\beta = (\beta_1, \beta_0)$ and $\gamma = (\gamma_1, \gamma_0)$ are parameters to be trained and $Z_{x_k} = \begin{pmatrix} x_k^1 z_1 & 1 \\ \vdots & \vdots \\ x_k^n z_n & 1 \end{pmatrix}$ for $k = 1$ and 2. For both cases however, an intercept term is added.

Figure 2. is a simple flowchart showing our algorithm once we have trained our parameters.

Fig. 2 Flowchart of algorithm

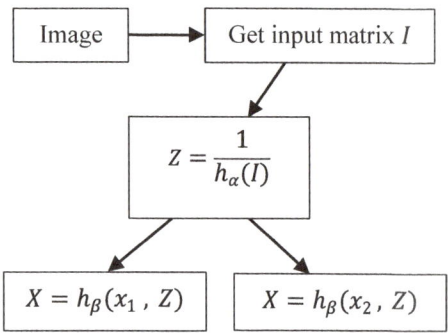

3.5   Training Linear Regression Models (LRMs)

In prediction using regression, it is important that our training data and input follow the same distribution [7]. In this case, our images have to follow the same "distribution". We define this "distribution" by the position of our laser plane with respect to our camera and the resolution of the image. As long as the "distribution" of our input and training data are consistent, the model is valid.

Upon setting our regression models, the parameters then have to be trained using real data. Our training data, $D = \{I, \tau = (\tau_x, \tau_y, \tau_z)\}$ where the input matrix I for the images on the left and is the collection of (x, y, z) ground-truths. For training of parameter α in (11.2), two images (Fig. 3) were taken where a piece of graph paper was fixed a distance away, $z = 60$ and $z = 30$ on a flat surface and stood perpendicular to the ground with origin on the graph paper labelled at the same (x, y) position as the camera. Next, we design a cost function to determine how well fits the truth value of 30 and 60. For this, we adopt the mean-square error (MSE) [8] as our cost function:

$$J_z(\alpha) = \left(I\alpha^T - \tau_z\right)\left(I\alpha^T - \tau_z\right)^T \tag{13.1}$$

where $\tau_z = \left(\frac{1}{30}, \ldots, \frac{1}{30}, \frac{1}{60}, \ldots, \frac{1}{60}\right)$, $\tau_z \in \Re^{2n}$ is the truth vector, and there are n $\frac{1}{30}$'s and n $\frac{1}{60}$'s. We can use the normal equation method to train the parameters. In our case, it is the following:

$$\alpha = \left(I^T I\right)^{-1} I^T \tau_z \tag{13.2}$$

For training of γ in (12.2), the truth-vector τ_y was gathered by finding the difference y-coordinates of the pixels both at the bottom, B and the top of the graph paper, A (Fig. 4) and divide it by the actual length of the paper. This gives us the change in length for every pixel, $\delta = \frac{L}{L'}$. Next, we identified the of the y-coordinate of A laser at bottom of the graph paper, as well as the real-world y-coordinate of

Fig. 3 At z = 60 (Left); z = 30 (Right)

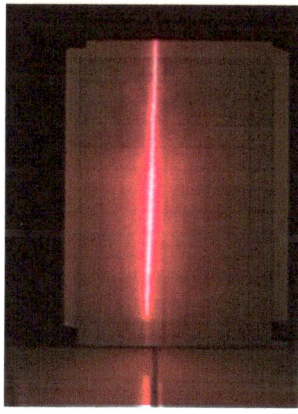

Fig. 4 Example of graph
paper

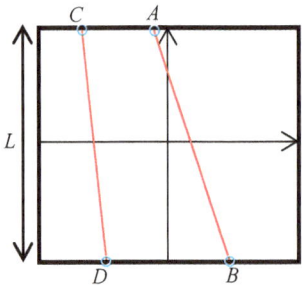

that position to be our starting, s_0. Then, our truth vector, τ_y, was generated in the
following form: $t_i = s_0 + i\delta$ for $i = 0, \ldots, n$. Afterwards, we use the same normal
equation method similar to (12.2) minimise a new cost function:

$$J_y(\gamma) = \left(h_\gamma(x_2, Z) - \tau_z\right)\left(h_\gamma(x_2, Z) - \tau_z\right)^T \tag{14.1}$$

$$\gamma = \left(Z_{x_2}^T Z_{x_2}\right)^{-1} Z_{x_2}^T \tau_y \tag{14.2}$$

where (14.1) is the MSE of our regression model, $h_\gamma(x_2, Z)$ against the truth-vector,
y. For training of β in (12.1), two separate lines, (Fig. 4) AB and CD were taken
at different distances. AB being the line captured at $z = 60$ and CD being the line
captured at $z = 30$. Then, the difference in x-coordinates between point A and B
as well as C and D were computed. Then, we repeat the procedure similar to the
training of where we find the change of length for every pixel, δ_x and generate our
training data in a similar fashion as γ. Then, the same normal equation method is
used to minimise a similar cost function:

$$J_x(\beta) = \left(h_\beta(x_1, Z) - \tau_x\right)\left(h_\beta(x_1, Z) - \tau_x\right)^T \tag{15.1}$$

$$\beta = \left(Z_{x_1}^T Z_{x_1}\right)^{-1} Z_{x_1}^T \tau_x \tag{15.2}$$

3.6 Translation and Rotation of Points

After the eventual cloud of (x, y, z) coordinates are generated for a particular image,
the object was rotated on a makeshift turntable by degree (radians). A total of $\frac{2\pi}{\theta}$
images are taken this way. The points obtained will also have to be translated and
rotated to fit the rotation of the turntable. This procedure consists of the following:

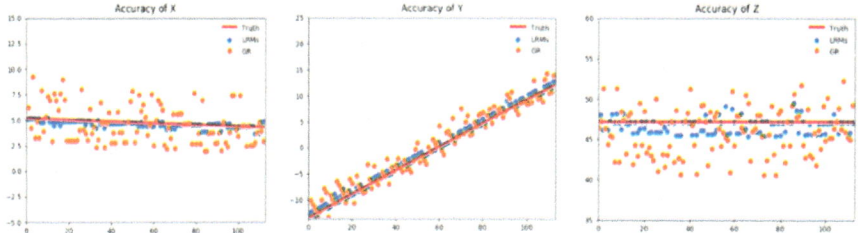

Fig. 5 Test result of X, Y and Z

$$p_{finalr} = \begin{pmatrix} \cos(\varphi) & 0 & \sin(\varphi) & dx \\ 0 & 1 & 0 & dy \\ -\sin(\varphi) & 0 & \cos(\varphi) & dz \\ 0 & 0 & 0 & 1 \end{pmatrix} p_r^T, \varphi = \theta r \qquad (16)$$

where $p_r = \begin{pmatrix} x & y & z & 1 \end{pmatrix}$ and (x, y, z) is a coordinate output from the rth Image taken.

4 Results and Discussion

We tested the accuracy of our LRMs against the GR method by taking a test input at a distance of $z = 47$. The results (Fig. 5) show the accuracy of the LRMs method against the GR method. It can be seen visually that the LRMs method follows more closely to the Truth than the GR method. We then applied hypothesis testing and used a z-test to compare between the mean differences of the two methods with the Truth. The method that has a smaller difference is considered to be more accurate than the other as its output coordinates are closer to the ground truths. Since p-value is smaller than 0.05, we conclude that at 5% significance level, there is sufficient evidence that $\bar{E} < \bar{D}$. Therefore, the LRMs were more accurate in reproducing the point cloud of the object. We then tried scanning two different 3D objects and took 16 images of both input objects, the cardboard pyramid and the cube with $\theta = \frac{\pi}{8}$ in (16). Figure 6 shows the visual difference in results for the original method against the alternative method. Visually, the point cloud produced by the LRMs method follows the original structure better.

5 Conclusions and Recommendations for Future Work

In this report, we have presented 3D scanning with LRMs method along with the geometrical relations method. There is a significant improvement made from the original method as shown in the results obtained as well as in terms of usability,

Fig. 6 Table of comparison

Actual Object	Point Cloud under Geometric Relations Method	Point Cloud under Linear Regression Model Method
Pyramid 		
Cube 	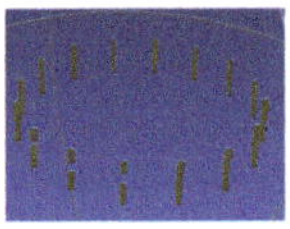	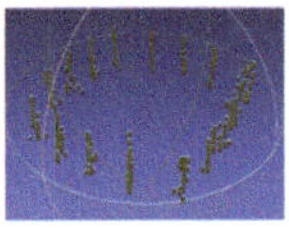

where there is no need for the user to know the specifics of the camera-capturing device, which may be hard to find, unlike for the geometrical relations method where variables must be known. We have constructed an integrated setup coupled with procedures and algorithms which involves rotating the object on a platform linked to a stepper motor controlled via a Raspberry Pi 3 module. It also takes images using the Picamera controlled through a pattern of outputs to the GPIO (General Purpose Input/Output) pins.

While our results are still a far-cry from the state-of-the-art and nowhere near suitable for industrial use, the LRMs approach shows great promise given that we were able to replicate the shape of objects with just 2 images used as training data and a simple linear model.

We suggest training the LRMs with data from slopes instead of a blank graph paper so as to provide more variance of 'z's for the LRMs to be trained. Future works can also include using Anomaly detection methods first to rid the input of noise as well as varying the resolution of the image. More sophisticated models (Neural Networks, Support Vector Machines) can also be used, however, the original geometric basis will then be lost. Other methods of optimizing the cost (e.g. Gradient descent with regularization) can also be used.

Appendix: Similar triangle

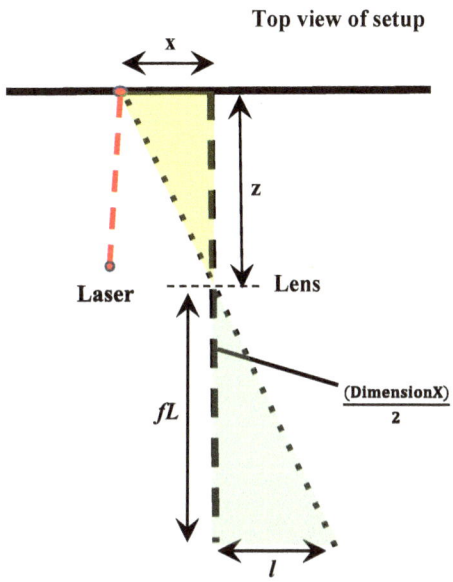

References

1. Modeling, D. (2017, 23 December). Retrieved from Wikipedia The Free Encyclopedia. https://en.wikipedia.org/wiki/3D_modeling.
2. Raspberry Pi 3 Model B SBC. (2012). Retrieved from RS Components Ltd. https://uk.rs-online.com/web/p/processor-microcontroller-development-kits/8968660/.
3. 3D Scanners Comparison. (2018). Retrieved from Aniwaa Pte. Ltd. https://www.aniwaa.com/comparison/3d-scanners/.
4. Bradshaw, G. (1998/1999). *Non-contact surface geometry measurement techniques*. Ireland.
5. Golowczynski, M. (2016, 23 June). *Digital camera sensors explained*. Retrieved from what digital camera. http://www.whatdigitalcamera.com/technical-guides/technology-guides/sensors-explained-11457.
6. Camera. (2017). Retrieved from Wikipedia the Free Encyclopedia. https://en.wikipedia.org/wiki/Camera#Lens.
7. Training, Test and Validation Sets. (2017, 29 December). Retrieved from wikipedia the free encyclopedia. https://en.wikipedia.org/wiki/Training,_test,_and_validation_sets.
8. McCormick, C. (2014, 04 March). *Gradient descent derivation*. Retrieved from Chris McCormick. http://mccormickml.com/2014/03/04/gradient-descent-derivation/.

Demonstration of Switching Plasmonic Chirality via Geometric Transformations for Biosensing Applications

Xiuhua Zhu and Eng Huat Khoo

Abstract The ability to detect and differentiate between left and right-handed biopolymers has important biosensing applications. Although this can be done using circular dichroism (CD), the CD signal is usually weak. To amplify it, plasmonic nanostructures can be used. This paper presents a simulation study of the use of a planar array of cubic nanostructures, demonstrating the switching of chirality through performing geometric transformations. It has been found that when the cubes are rotated about their vertical axis and when the angles φ and θ of incident light are varied, CD can be observed, and performing the opposite action results in the switching of CD. As the use of cubes as compared to other structures, such as the chiral gammadion structure in reported works, has the merit of being easier to fabricate, the methods described in this paper have much potential in being used to detect chiral biopolymers.

Keywords Circular dichroism · Plasmonic nanostructures · Switching of chirality · Simulation study

1 Introduction

The building blocks of life comprise amino acids, nucleic acids and sugars [1], all of which are chiral molecular units [2, 3], where the mirror image of the molecular unit does not superimpose with its original unit [3]. Biopolymers, which are macromolecules formed from these units, also exhibit chirality on molecular and supramolecular scales [4]. The ability to detect chiral biopolymers and distinguish between the left- and right-handed biopolymers is important as it has extensive biosensing applications in biomedical and pharmacological fields [5], such as the

X. Zhu
Raffles Institution, Singapore, Singapore

E. H. Khoo (✉)
Photonics and Electronics, IHPC, A*STAR, Singapore, Singapore
e-mail: khooeh@ihpc.a-star.edu.sg

© Springer Nature Singapore Pte Ltd. 2019
H. Guo et al. (eds.), *IRC-SET 2018*,
https://doi.org/10.1007/978-981-32-9828-6_12

detection of amyloid diseases [4]. However, there is substantial difficulty in differentiating left- and right-handed biopolymers due to their similar physical and chemical properties [6].

One way to differentiate left- and right-handed biopolymers is by making use of their interaction with left- (LCP) and right- (RCP) handed circular polarized light [7]. Biopolymers with the same handedness as the incident circular light waves interact more strongly with the light, thus absorbing a greater fraction of the light, as opposed to biopolymers with opposite handedness as the incident circular light waves [8]. The differential absorption of LCP and RCP light is known as circular dichroism (CD) [7, 9]. Unfortunately, the CD signal is weak [5, 10], and therefore not very sensitive and does not provide a robust method for judging handedness in dilute solutions [5].

To amplify the CD signal, plasmonic nanostructures can be used as a light field enhancement method [11], where localized surface plasmon polaritons interact with the chiral molecules. The localized field allows for a longer time of interaction with the chiral molecules, thereby producing a stronger CD signal to be detected. The most common nanostructure used to amplify the CD signal is the gammadion cross, which is chiral [12]. Reported works have demonstrated experimentally that the sensitivity detection of chiral molecules improved by a factor of 10^6 using gammadion plasmonic structures [4]. In addition, numerous three-dimensional (3D) chiral nanostructures have also been proposed, such as metal helix structures [13–15] and layer-by-layer chiral plasmonic nanostructures [16]. Furthermore, it has been discovered that CD can be observed in not only intrinsic chiral structures, but also in achiral structures with an oblique incident light beam [17]. The use of planar arrays of achiral nanostructures, such as L-shaped nanostructures, has been explored as such nanostructures present an advantage in being easier to fabricate than the aforementioned 3D or planar chiral nanostructures [17].

This paper presents a simulation study of a planar array of achiral cubic plasmonic nanostructures for the detection of chiral molecules. Although naturally achiral by themselves, it is expected that cubes can exhibit chiral properties when geometric transformations, namely rotations, are applied to them or when light is incident on them at oblique angles. In addition, performing the opposite action should result in the switching of the CD spectrum. In this study, the cubes were rotated in the xy-plane and the resultant CD spectra are presented. In addition, the effect of oblique incidence on the nanostructures was also studied, by varying the angle(s) of light incident on the cubic nanostructure array. Compared to gammadion nanostructures and other chiral structures, the benefit of using cubic nanostructures lies in the comparative ease in fabrication, similar to the achiral L-shaped nanostructures.

2 METHODS

The simulations were designed and run using Lumerical's Nanophotonic finite-domain time-difference (FDTD) Simulation Software. The FDTD method is a grid-based numerical modeling technique that approximates the derivatives in the

Maxwell curl equations by finite differences. Material properties are defined at various spatial coordinates and fields are then calculated for small successive time steps where the step size is small compared to the inverse of the frequency of the source [18].

The design of the simulation is as shown in Fig. 1. A gold cubic nanostructure of sides 200 nm was positioned atop a larger borosilicate glass substrate of reflective index 1.517. Water is chosen to be the background medium and the background index is set to 1.333. A fine mesh with cell division of 20 nm in the x-, y- and z-directions was placed around the cube. The simulation region was set to 800 nm in the x- and y-directions to create an array in the xy-plane with a periodicity of 800 nm. A perfectly matched layer (PML) was set in the z-direction to impose absorbing boundary conditions to prevent reflections of energy back into the structure that was being simulated. A monitor was placed 100 nm beneath the cubic nanostructure to measure the intensity of the electric field of transmitted light. Two light sources of plane waves were used to create circularly polarized light, where one of the light source has its polarization angle set to $0°$ and another to $90°$. The phase angles of light sources also differ by $90°$.

To obtain the CD spectrum, two simulations were run, one with LCP light and one with RCP light, where the phase angle of the second light source was $-90°$ and $+90°$ respectively. It should be noted that although CD by conventional definition is the difference in the absorption of LCP and RCP light, Lumerical's FDTD solutions provide only the data on transmittance and not absorbance.

Fig. 1 Design of simulation of cubic nanostructures (not drawn to scale)

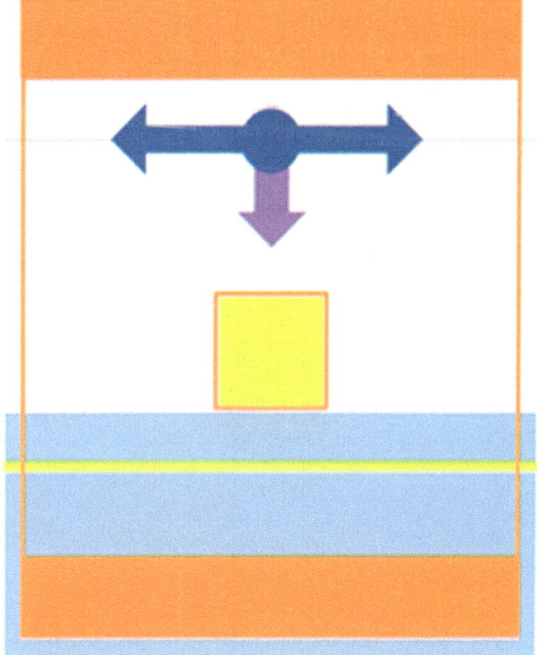

However, given that $I = T + R + A + S$, where I is the total intensity of incident light and T, R, A and S are the transmitted, reflected, absorbed and scattered light respectively, and since reflected LCP and RCP light are approximately equal and scattered light is negligible, CD can be approximated to be the difference in transmitted RCP and LCP light when the incident light is of equal intensity. Therefore, in this paper, CD is defined by

$$CD = T_R - T_L. \tag{1}$$

The array of cubic nanostructures to be simulated as illustrated in Fig. 1 would not produce CD signals as it is achiral. To produce CD signals, geometric transformations were performed to break the symmetry. This was attempted in two ways:

First, the cube was rotated clockwise about its vertical axis in the xy-plane from $0°$ to $75°$ at regular intervals as shown in Fig. 2. Incident light was normal to the top surface of the cube. It was expected that deviation from the $45°$ mark by the same amount in opposite directions, i.e. the cube being rotated $15°$ and the cube being rotated $75°$, as well as $30°$ and $60°$, would produce opposite CD spectra due to their being mirror images. No CD signal was expected to be observed when the cube is rotated $45°$ because, similar to that for $0°$, the cube would be symmetrical and therefore achiral.

Second, the angles at which light is incident on the cubic nanostructure array were varied while keeping the angle of rotation of cube to $0°$. The angles φ and θ (as illustrated in Fig. 3) were varied first separately and then together. In the first trial, angle φ was varied from $-75°$ to $75°$ while angle θ was kept to $0°$. In the second trial, angle θ was varied from $-75°$ to $75°$ while angle φ was kept to $0°$. In the third trial, both angles φ and θ were varied from $-75°$ to $75°$ where $\varphi = \theta$. In all three cases, it was predicted that deviation from the $0°$ mark by the same amount in opposite directions, e.g. angle(s) φ and/or θ of incident light being $15°$ and that being $-15°$, would produce opposite CD spectra.

Fig. 2 Top view of the cube being rotated $0°$, $15°$, $30°$, $45°$, $60°$ and $75°$

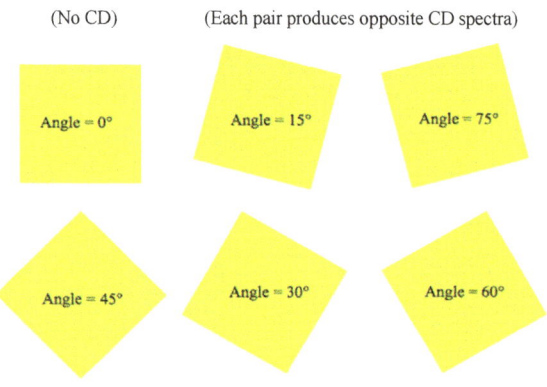

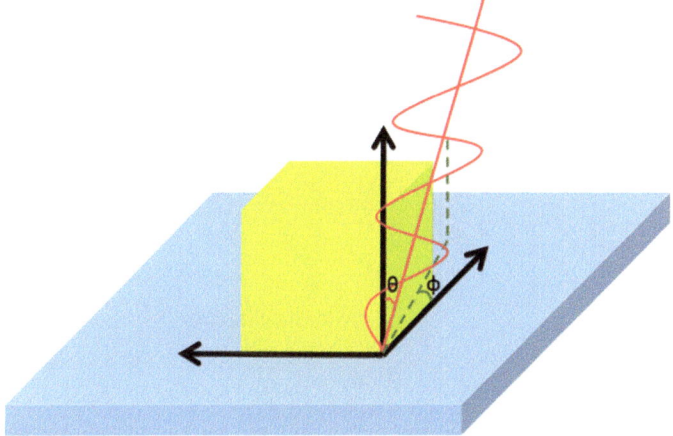

Fig. 3 Schematic diagram showing oblique incidence of light on the cube

3 Results and Discussion

The attempt to produce CD by rotation of the cubic nanostructures was successful and the CD spectra obtained is as shown in Fig. 3. As predicted, the CD spectra of the cube rotated 75° and 60° are the negative of the CD spectra of the cube rotated 15° and 30° respectively, where the CD at every wavelength for 15° and 75°, as well as 30° and 60°, has equal magnitude but opposite signs. The flipping of the sign of CD indicates a switching behaviour resulting from a change in the handedness of the nanostructure. Also, no CD signal is produced at 0° and 45°, which confirms the hypothesis. It is noted that of the three modes, the first provides the most significant switching effect (Fig. 4).

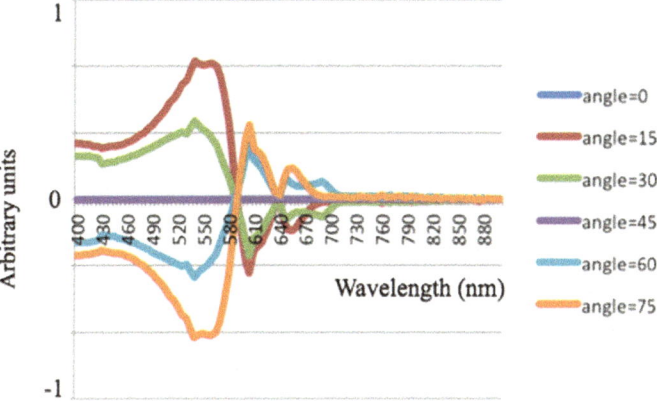

Fig. 4 CD spectra of cubic nanostructure array when cubes were rotated from 0° to 75°

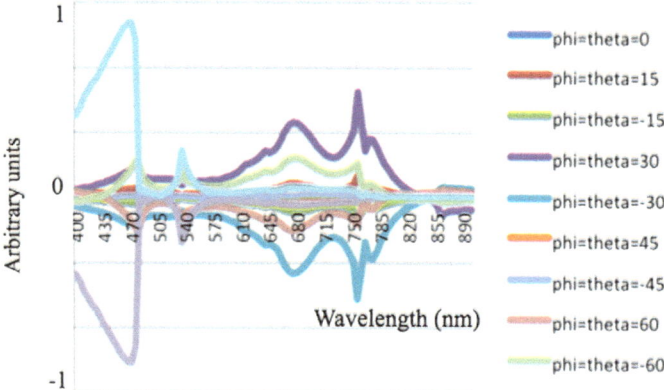

Fig. 5 CD spectra of cubic nanostructure array when both angles φ and θ were varied from $-75°$ to $75°$

On the other hand, varying angles φ and θ of incident light separately was unsuccessful at producing any discernible CD spectra, as the transmittance of RCP light was equal to that of LCP light for every setup. As existing literature indicates that CD could be observed when angles φ and θ are varied separately for other structures such as the L-shaped nanostructures [17], it is likely that the lack of CD here can be attributed to the nature of the cube and more research can be done in this area.

However, CD was once again observed when both angles φ and θ of incident light were varied at the same time. The CD spectra obtained is as shown in Fig. 5. As can be seen from the superimposed graphs, apart from $45°$ and $-45°$ where no CD was observed, the CD spectra of light incident at negative angles ($\varphi = \theta = -15°, -30°, -60°, -75°$) are a vertical reflection of those at positive angles ($\varphi = \theta = 15°, 30°, 60°, 75°$), thus successfully demonstrating the switching of chirality. Among the resonant modes observed, it is noted that the first mode provides the most significant switching effect for $\varphi = \theta = 75°$ and $\varphi = \theta = -75°$.

4 Conclusion

It has been demonstrated that an array of achiral cubic plasmonic nanostructure is capable of producing the effect of switching of CD when geometric transformations namely rotation of the cube and variation of angles φ and θ of incident light are performed. There is therefore much potential in using these methods to distinguish between left- and right-handed biopolymers, especially due to their relative ease in fabrication as compared to gammadion structures and other chiral nanostructures in reported works. For future research, the fabrication of planar cubic nanostructure array with the aforementioned methods to produce CD can be carried out to experimentally verify the findings presented in this paper. In addition, the design of other simple nanostructures to amplify CD signals can be studied as well.

Acknowledgements Zhu Xiuhua would like to extend her sincerest gratitude to her research mentor, Dr. Khoo, for his guidance and support.

References

1. The George Washington University. (n.d.). *Life's Origins*. Retrieved December 26, 2017 from https://www2.gwu.edu/~darwin/BiSc151/Origin/origin.html.
2. Patrick, G. (2015). *Introduction to Drug Synthesis*. S.l.: Oxford Univ Press.
3. Bahar, I., Jernigan, R. L., & Dill, K. A. (2017). *Protein actions: Principles and modeling*. New York: Garland Science.
4. Hendry, E., Carpy, T., Johnston, J., Popland, M., Mikhaylovskiy, R. V., Lapthorn, A. J., et al. (2010). Ultrasensitive detection and characterization of biomolecules using superchiral fields. *Nature Nanotechnology, 5*(11), 783–787. https://doi.org/10.1038/nnano.2010.209.
5. Lu, F., Tian, Y., Liu, M., Su, D., Zhang, H., Govorov, A. O., et al. (2013). Discrete nanocubes as plasmonic reporters of molecular chirality. *Nano Letters, 13*(7), 3145–3151. https://doi.org/10.1021/nl401107g.
6. Pahari, A., & Chauhan, B. (2007). *Engineering chemistry*. Hingham, MA: Infinity Science Press.
7. Ross-Murphy, S. B. (2013). *Physical techniques for the study of food biopolymers*. S.I.: Springer.
8. Newman, J. (2016). *Physics of the life sciences*. New York, S.I.: Springer.
9. Rodger, A., Marrington, R., Roper, D., & Windsor, S. (2005). Circular dichroism spectroscopy for the study of protein–ligand interactions. *Protein-ligand interactions* (pp. 343–364). https://doi.org/10.1385/1-59259-912-5:343.
10. University of Nottingham. (2015, 24 June). *New technique to accurately detect the 'handedness' of molecules in a mixture*. Retrieved December 26, 2017 from https://phys.org/news/2015-06-technique-accurately-handedness-molecules-mixture.html.
11. Maier, S. A., & Atwater, H. A. (2005). Plasmonics: Localization and guiding of electromagnetic energy in metal/dielectric structures. *Journal of Applied Physics, 98*(1), 011101. https://doi.org/10.1063/1.1951057.
12. Png, C. E., & Akimov, Y. (2017). *Nanophotonics and plasmonics: An integrated view*. Boca Raton, FL: CRC Press, Taylor & Francis Group.
13. Gansel, J. K., Latzel, M., Frölich, A., Kaschke, J., Thiel, M., & Wegener, M. (2012). Tapered gold-helix metamaterials as improved circular polarizers. *Applied Physics Letters, 100*(10), 101109. https://doi.org/10.1063/1.3693181.
14. Behera, S., & Joseph, J. (2014). *Tapered-double-helix photonic metamaterial based broadband circular polarizer for optical wavelength range*. In *12th International Conference on Fiber Optics and Photonics*. https://doi.org/10.1364/photonics.2014.m4a.54.
15. Kaschke, J., Gansel, J. K., Fischer, J., & Wegener, M. (2013). Metamaterial circular polarizers based on metal N-helices. *CLEO 2013*. https://doi.org/10.1364/cleo_qels.2013.qtu1a.4

16. Wu, L., Yang, Z., Cheng, Y., Lu, Z., Zhang, P., Zhao, M., et al. (2013). Electromagnetic manifestation of chirality in layer-by-layer chiral metamaterials. *Optics Express, 21*(5), 5239. https://doi.org/10.1364/oe.21.005239.
17. Zhang, Y., Wang, L., & Zhang, Z. (2017). Circular dichroism in planar achiral plasmonic L-shaped nanostructure arrays. *IEEE Photonics Journal, 9*(2), 1–7. https://doi.org/10.1109/jphot. 2017.2670783.
18. Yu, W. (2009). *Electromagnetic simulation techniques based on the FDTD method.* Oxford: Wiley-Blackwell.

Microbial Fuel Cells: Food Waste as a Sugar Source

Phua William, Emma Tan Xiu Wen and Ho Jia Yi Jenevieve

Abstract The recent threat of climate change and rising demands for electricity globally has prompted research into Microbial Fuel Cells (MFCs), a potential source of renewable energy. Another troubling global issue is large amounts of food waste. Use of food waste as substrates in MFCs could convert the food waste into clean energy, tackling both global problems. Hence, this research aims to determine if food waste is a viable substitute for currently used substrates (glucose) in MFCs by replacing glucose (substrate) with banana peels and sugarcane bagasse (inner and outer layers). For each substrate, MFC set-ups were run and the Benedict's Test (quantitative) and Iodine Test were performed. Banana peel was found to be the best substrate among the three food wastes, with the stablest power performance, highest peak and highest average power performance. This was followed by the outer layer of sugarcane and lastly, the inner layer of sugarcane which also had the lowest concentration of reducing sugar. Thus, it was concluded that the substrate's concentration of reducing sugar may affect power performance of the MFC. Furthermore, banana peels and the outer layer of sugarcane were found to be viable alternatives for glucose due to a higher power performance.

Keywords Microbial fuel cell · Food waste · Substrate

1 Introduction

1.1 Background of Research

Currently, climate change poses a great threat to the society we know today. Coupled with rising demands for energy globally in recent years, the issue of climate change has prompted further research into harnessing renewable sources of energy to slow down this phenomenon. Microbial Fuel Cells (MFCs) are one potential source of renewable energy [7, 8].

P. William (✉) · E. T. X. Wen · H. J. Y. Jenevieve
National Junior College, Singapore, Singapore
e-mail: williamphua3@gmail.com

© Springer Nature Singapore Pte Ltd. 2019
H. Guo et al. (eds.), *IRC-SET 2018*,
https://doi.org/10.1007/978-981-32-9828-6_13

Fig. 1 Diagram of the MFC
system [8]

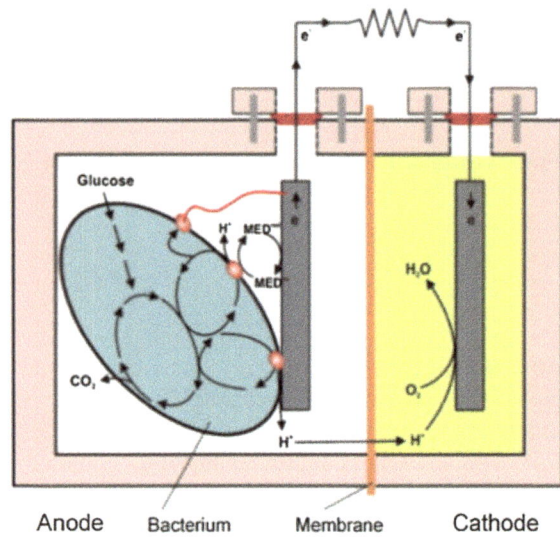

MFCs use microorganisms such as bacteria and yeast as biocatalysts to generate electricity from a variety of substrates including organic waste materials. In the typical mediated dual-chambered MFCs, there is an anaerobic anode chamber, where the biocatalysts oxidise substrates to produce electrons and protons. The electrons are transported to the electrode by the mediator and passed through an external circuit to the aerobic cathode chamber. This current produces electricity. Meanwhile, protons pass through the Proton Exchange Membrane (PEM) to the cathode chamber before joining with the electrons to form water, in the presence of oxygen [8] (Fig. 1).

This research will also explore the use of food waste as substrates in MFCs. Food waste is a global issue that needs to be tackled as well. About 25% of the produced supply of food is wasted in the food supply chain (FSC) [4]. In Singapore alone, 791,000 tonnes of food waste was generated in 2016, of which only 14% is recycled. If food waste can be used as substrates successfully in MFCs, clean energy can be generated while reducing the amount of food waste to be disposed, helping to tackle both global problems. In addition, not much research has been done to investigate the potential use of food waste in MFCs currently.

1.2 Objective

This research aims to find out if food waste is a viable substitute for currently used substrates in MFCs by experimenting using powdered banana peels and the inner and outer layers of powdered juiced sugarcane pulp as the substrate for MFCs.

1.3 Hypothesis

The hypothesis is that food waste is a viable alternative for currently used substrates. All substrates used (inner layer of sugarcane, outer layer of sugarcane and banana peel) will have a comparable performance (equal or better) to that of currently used substrates (glucose), with MFCs using the inner layer of the sugarcane (pith) as substrate having the highest power performance. This is because the pith of the sugarcane is known to have high contents of sucrose which the yeast could feed on for the MFC to generate electricity.

2 Methodology

2.1 Preparation of Food Waste Substrates

Juiced sugarcane pulp was dried in the oven for around 3 h at 100 °C. The inner layer and outer layer of the sugarcane were then separated. Each layer was cut into smaller pieces and blended into fine powder.

Banana peel was dried in the oven for 2 h at 100 °C, before being separated and cut into smaller pieces. The pieces were then blended into fine powder.

2.2 Assembling of Microbial Fuel Cells

Dual-chambered yeast MFC setups were assembled using a Nafion Proton Exchange Membrane (PEM) between the anode and cathode chamber, with each chamber containing an electrode cut from carbon fibre electrode tissue. All MFCs used selected food waste (0.2 g of food waste added to 4.25 ml of buffer) or 0.5% (D+) Glucose solution as the substrate, Methylene Blue (added to buffer) as the mediator and Potassium Permanganate (added to buffer) as the catholyte in a 17:10:32 ratio by volume. Buffer used was 0.1 M phosphate buffer (pH 7.2).

2.3 Testing of Microbial Fuel Cells

For each of the food waste substrates (inner layer of sugarcane, outer layer of sugarcane and banana peel), three MFCs were constructed; two setups with a voltmeter and without a resistor, and one setup with a resistor of 100 Ω and a voltmeter connected in parallel. The setup with the resistor allows for calculation of the amount of current generated by the MFC, while the setups without showed the voltage generated (recorded by the voltmeter) during experimentation. The MFC setup using (D+)

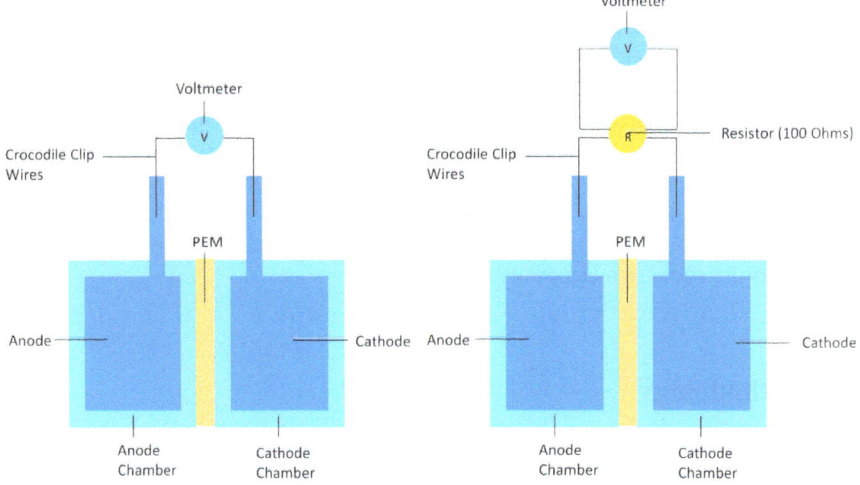

Fig. 2 Diagrams of the MFC setups (with voltmeter and with both resistor and voltmeter)

Glucose solution was with a voltmeter and without a resistor (Fig. 2). Each MFC was run for a duration of 48 h during which it was left undisturbed.

2.4   Food Tests

The Benedict's Test was performed on all 3 substrates. Each substrate was added to buffer in the amount to be used during MFC experimentation. The Benedict's Test was also performed as a quantitative test by testing glucose solutions of 1, 0.5, 0.1, 0.05, 0.01 and 0.005% concentrations and comparing the colour change with that of the substrates researched on.

The Iodine Test was also performed on all 3 substrates.

3   Results and Discussion

3.1   Microbial Fuel Cell Setups Results

Figure 3 shows that all substrates used had an increasing power output in the first 100 min before reaching a power peak. After a period of time, which was varied for each substrate, the power output started to decrease from the power peak until the experiment ended.

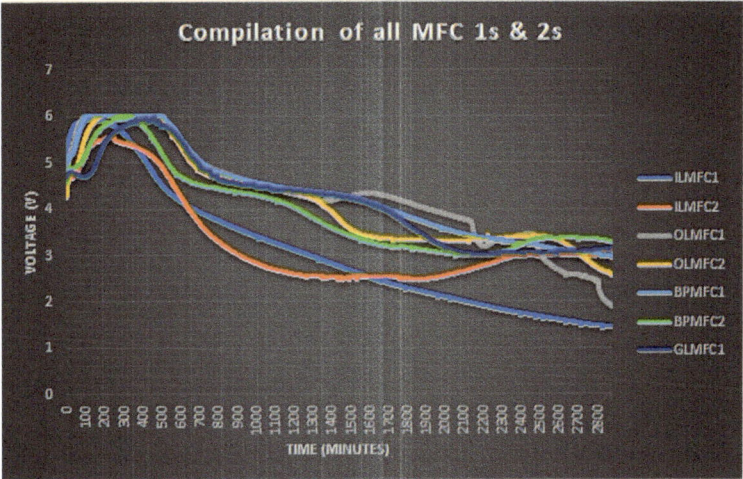

Fig. 3 Graph of Voltage, V against time, minutes MFC was run for MFC setups (without resistors) with inner layer of sugarcane (IL), outer layer of sugarcane (OL), banana peels (BP) and D+ Glucose (GL) as substrates

The highest point of performance of all the MFC setups was 6 V, the highest voltage that the instrument of measurement used could measure. This power peak was observed in the first and second setups of the MFCs using the outer layer sugarcane (OLMFC1 and OLMFC2). For the first setup, this was observed for 507 min and in the second setup, this was observed for 136 min. This power peak was also observed in the first setup of the MFCs using the banana peels during which this was observed for 400 min (BPMFC1). However, during the second setup of the MFCs using the banana peels (BPMFC2), the highest peak was 5.96166 V, at the 315th minute. The peak for the MFC using D+ glucose (GLMFC1) was 5.94888 V at the 466th minute. The highest point of the performance for the MFCs using inner layer sugarcane as a substrate was 5.92971 V, the peak of the first setup (ILMFC1) which occurred erratically from 120 to 142 min. The peak for the second setup (ILMFC2) was 5.53993 V at the 250th minute.

The MFCs using banana peels also attained the highest overall average (across both setups) of 4.82508 V. This was followed by MFCs using the outer layer sugarcane with an overall average of 4.22395 V, then the MFC using (D+) Glucose with an average power output of 4.125874 V and lastly MFCs using the inner layer sugarcane with an overall average of 3.20083 V.

Therefore, since banana peels and the outer layer of the sugarcane have shown comparable power performance, they are viable alternatives for glucose out of the three substrates.

Figure 4 shows that with the added resistor, the MFCs using banana peel still had the highest peak of 0.32907 V (0 min) and average power output at 0.24601 V. This was followed by the MFCs using the outer layer of the sugarcane, with a peak of 0.18211 V (435th minute) and an average of 0.01264 V. Lastly, the MFCs using the

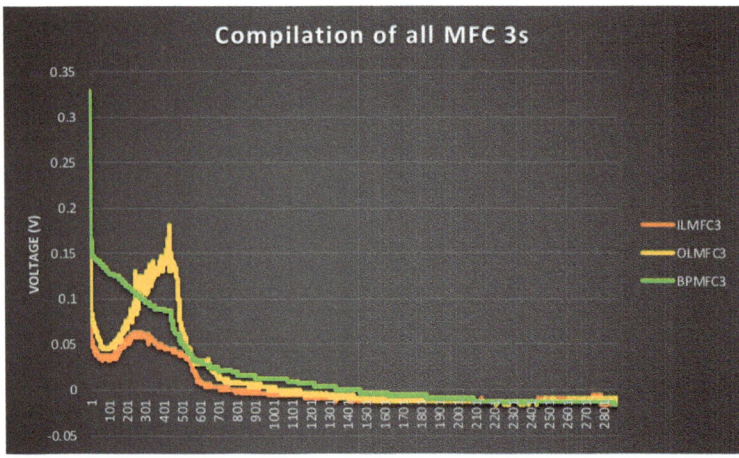

Fig. 4 Graph of Voltage, V against Time, minutes MFC was run for MFC setups (with resistors) with inner layer of sugarcane (IL), outer layer of sugarcane (OL) and banana peels (BP) as substrates

inner layer of the sugarcane had a peak of 0.00799 V (0 min) with an average of 0.00243 V.

The MFC using the outer layer of sugarcane was also shown to have an unnatural power peak before the voltage generated decreased significantly. This could have been due to the drastic changes in the yeast population. The yeast population affects the amount of voltage generated, as a higher yeast population would lead to an increase in the oxidation of the substance in the anode chamber, hence producing more electrons and generating more voltage. In the beginning, there was an increase in the yeast population as the yeast acclimatised to the MFC anode chamber conditions and started multiplying. This will lead to a decrease in the amount of substrate for the yeast to feed on, resulting in insufficient food source to accommodate the growing yeast population. The death rate of the yeast will then catch up, thus causing the yeast population to shrink and lowering the amount of voltage generated.

However, all MFCs with the different substrates were shown to be unstable, and the voltage of all 3 MFCs dropped drastically and went to negative in the later parts of experimentation.

The overall data suggested that banana peel is the best substrate to be used out of the three researched as it had the most stable power performance, with the highest average power performance (across all setups). The outer layer of the sugarcane had the second-best power performance. The inner layer of the sugarcane had the worst power performance with the lowest peak (for all setups) and an unstable performance. In addition, since the power performance of MFCs using banana peels and the outer layer of sugarcane are comparable to that of an MFC using glucose, they have also shown to be viable alternatives for the commonly used substrate glucose.

Table 1 Table with results of the benedict's test and iodine test for the three substrates

Type of substrate	Benedict's test			Iodine test	
	Observation	Presence of reducing sugars	Percentage of reducing sugars present (%)	Observation	Presence of starch
Inner layer of sugarcane	Orange-red precipitate	Yes	0.1–0.5	Solution turned brown	No
Outer layer of sugarcane	Orange-red precipitate	Yes	0.1–0.5	Solution turned brown	No
Banana peels	Orange-red precipitate	Yes	0.1–0.5	Solution turned brown	No

3.2 Food Test Results

All three substrates were shown to contain no starch. All three substrates were also shown to have concentrations of reducing sugar between 0.1 and 0.5%. The inner layer of the sugarcane was observed to have the lowest concentration of reducing sugar amongst the 3 food substrates (Table 1).

3.3 Discussion

The MFC setups using the outer layer and banana waste had better power performances than that of the inner layer, which contradicts the hypothesis. After referring to the food test results, it could be concluded that the differing performances may be due to the composition of the food waste. The outer layer of the sugarcane (rind) consists of small quantities of parenchyma cells and majority of the vascular bundles. The inner layer of the sugarcane (pith) has an abundance of parenchyma cells that store sucrose, and also contains little amounts of vascular bundles [10]. The inner layer (pith) of sugarcane contains most of the sucrose in the sugarcane. However, since sucrose is not a reducing sugar, it did not cause colour-change during the Benedict's Test. In contrast, the outer layer (rind) of the sugarcane contains a lot of fibre which consists of non-starch polysaccharides, such as lignocellulosic fibres [2]. Ethylene, a natural plant hormone that is released when the banana ripens, breaks down complex sugars in the banana peel into simple sugars and also breaks down pectin (a type of fibre) in the banana peel. [5]. This caused the higher concentration of reducing sugars in banana peels.

We can infer that the amount of reducing sugar found in the substrates influenced the power performance of the MFCs with a higher amount of reducing sugar leading to a better power performance. Banana peels, which performed better in the MFCs, were shown to have a higher concentration of reducing sugar than the inner layer of the sugarcane during the Benedict's Test. While the outer layer of the sugarcane appeared to have a higher concentration of sugar during the Benedict's Test, this might not be the case since the banana peel powder was originally brown and the precipitate for it could not be seen as clearly. This is supported by existing research papers stating that banana peels have been shown to have a high percentage of total sugar as invert sugar (22.0%) [1]. Thus, banana peels have a high content of reducing sugar, whereas sugarcane bagasse (combination of the inner and outer layer sugarcane) mainly consists of complex sugars with a percentage of 50% cellulose, 25% hemicellulose and 25% lignin [6].

4   Conclusion

MFC setups were run using three different types of food waste, banana peels and the inner layer (pith) and outer layer (rind) of sugarcane as substrates. Out of the three food types used in experimentation, banana peel was shown to be the best substrate for MFCs as it has the highest peak, highest average power output in power performance (with and without the added resistor) and the most stable performance. The concentration of reducing sugar in the substrate used may have affected the power performance. In addition, banana peel and outer layer of sugarcane were shown to be viable alternatives to currently used substrates in MFCs (glucose) due to comparable power performances in comparison to the results of the MFC using glucose as a substrate.

5   Future Work

Future work regarding the research into the use of other types of food waste as substrate in MFCs, as well as to investigate how the percentage of fibre and complex sugars in the food waste affected the power performance of the MFC can be done. Future work concerning the processing of food waste to facilitate the breakdown of complex sugars such as cellulose in fibres to simple sugars (for example, through the addition of enzymes) before using as a substrate for the yeast can be carried out. Further investigation into why the outer layer (rind) of the sugarcane had such a high concentration of reducing sugar can also be done.

Acknowledgements We would like to thank our school National Junior College, Singapore for the opportunity to conduct this research project under the Science Research and Training Programme (Junior).

References

1. Archibald, J. (1949). Nutrient composition of banana skins. *Journal of Dairy* Science, *32*(11), 969–971. ScienceDirect. http://dx.doi.org/10.3168/jds.s0022-0302(49)92146-3.
2. Han, G., & Wu, Q. (2004). Comparative properties of sugarcane rind and wood strands for structural composite manufacturing. *Forest Products Journal, 54*(12), 283–288.
3. Kummu, M., de Moel, H., Porkka, M., Sierbert, S., Varis, O., & Ward, P. J. (2012). Lost food, wasted resources: Global food supply chain losses and their impacts on freshwater, cropland, and fertiliser use. *Science of The Total Environment, 438*, 477–489. ScienceDirect. doi:https://doi.org/10.1016/j.scitotenv.2012.08.092.
4. Lohar, A. S., Patil, V., Patil, D., & Deelip, B. (2015). Role of mediators in microbial fuel cell for generation of electricity and waste water treatment [Online]. *International Journal of Chemical Sciences and Applications, 6*(1), 6–11.
5. Orwig, J. (2015). People around the world are eating banana peels because they know something that westerners do not. *Business Insider Singapore*, 26 Sept 2015. www.businessinsider.sg/benefits-of-eating-banana-peels-2015-9/?r=US&IR=T.
6. Pandey, A., Soccol, C. R., Nigam, P., Soccol, V. T. (2000). Biotechnological potential of agro-industrial residues. I: Sugarcane bagasse. *Bioresource Technology, 74*(1), 69–80. https://doi.org/10.1016/s0960-8524(99)00142-x.
7. Parkash, A. (2016). Microbial fuel cells: A source of bioenergy. *Journal of Microbial and Biochemical Technology, 8*(3), 4. https://doi.org/10.4172/1948-5948.1000293.
8. Rahimnejad, M., Adhami, A., Darvari, S., Zirepour, A., & Oh, S.-E. (2015). Microbial fuel cell as new technology for bioelectricity generation: A review. *Alexandria Engineering Journal, 54*(3), 745–756. https://doi.org/10.1016/j.aej.2015.03.031.
9. Serrano-Ruiz, J. C. (2016). *New microbial technologies for advanced biofuels: Toward more sustainable production methods*. Apple Academic Press.
10. Siqueira, G., Milagres, A. M. F., Carvalho, W., Koch, G., Ferraz, A. (2011, March 16). Topochemical distribution of lignin and hydroxycinnamic acids in sugar-cane cell walls and its correlation with the enzymatic hydrolysis of polysaccharides. *Biotechnology for Biofuels, Vol. 4*. https://doi.org/10.1186/1754-6834-4-7.

Cross-Talk Between Cell Mechanics and Biochemical Signalling Pathways via Modulation of Nucleocytoplasmic Shuttling of Transcription Factors: A Novel Approach to Study Fibroblasts in Breast Cancer Environments

Kia Hui Lim

Abstract Tumor progression in breast cancer has formerly only been linked to increased integrin signaling as a result of increased extracellular matrix rigidity. In this study, we further examined the role of cell mechanics, namely cell geometry (2D) and matrix rigidity (3D) in modulating the cell signaling response to the inflammatory cytokine TNF-α secreted by cancer cells and the conditioned media from metastatic breast cancer cell line MCF-7. Signal activation was measured by quantifying the nucleocytoplasmic shuttling of transcription factors (p-65, MKL1, SMAD3) in NIH3T3 mouse embryonic fibroblasts via a custom code in software ImageJ and novel methods—Micropatterning and Engineering cells in collagen matrix were employed to achieve the varied 2D and 3D environments of cells. Cell mechanics was shown to impinge on the nuclear morphology of cells—Circular and unpatterned cells tend to have circular nuclei, while rectangular cells and cells grown on rigid matrixes tend to have elongated nuclei. Upon exposure to MCF-7 conditioned media, TNF-α—p-65 signaling pathways were further amplified in cells with circular cells and cells grown on soft matrix; TGF-β—MKL1 pathways in cells grown on soft matrix were particularly highly activated; TGF-β—SMAD3 signaling pathways in all cells were activated, though the pathway was more activated in environments with decreased matrix rigidity. Increase in SMAD3 nuclear levels also led to a proportionate increase in Vimentin levels in cells grown in 3D matrix. Our novel understanding of the cross-talk between cell mechanics and biochemical signaling pathways is important for understanding the behavior of cancer-associated-fibroblasts (CAFs) in breast cancer and the discovery of the highly activated signaling pathways induced by CAFs giving rise to tumorigenesis under various cell environments is vital for the identification and engineering of new therapeutic targets.

Keywords Cellular mechanics · Extracellular matrix rigidity · Cell geometry · Nuclear morphology · TNF-α treatment · MCF-7 conditioned media ·

K. H. Lim (✉)
NUS Yong Loo Lin School of Medicine, Singapore, Singapore
e-mail: limkiahui@gmail.com

© Springer Nature Singapore Pte Ltd. 2019
H. Guo et al. (eds.), *IRC-SET 2018*,
https://doi.org/10.1007/978-981-32-9828-6_14

Transcription factors · Signaling pathways · TNF-α—p-65 · TGF-β—MKL1 ·
TGF-β—SMAD3 · Cancer-associated-fibroblasts · Tumorigenesis

1 Introduction

Cells embedded within tissues regularly sense and respond to external stimuli from
their microenvironment. These include physical forces such as tension forces, shear
stress and compression, in addition to biochemical signaling molecules such as hormone insulin and cytokines Tumor Necrosis Factor (TNF)-alpha and Transforming
Growth Factor (TGF)-beta. While the mechanisms by which these classical biochemical molecules interact with cells have been well established, the mechanical
regulation of cells has only been well characterized recently. The forces from the
ECM acting on cells embedded within are reciprocated by the cellular forces generated by actomyosin contractility and microtubular forces. This balance of forces,
which result in cells obtaining a particular geometry, also helps in the maintenance
of tissue homeostasis. The loss of tissue homeostasis could lead to the development
of cancer.

The role of cell-ECM interactions and cell morphology in promoting cancer progression through the regulation of cell survival, differentiation and proliferation has
become apparent in the recent years. Breast cancer remains as the most prevalent
cancer amongst women, impacting over 1.5 million women each year. Malignant
transformation in breast cancer has been characterized by the progressive stiffening
of the stromal tissue—an increased matrix rigidity in breast tissue had been found
to stimulate proliferation of the mammary cells and promote tumor-like behaviors
[1] via the elevation of integrin-mediated focal adhesions and increased integrin
signaling [2].

Both physical forces and biochemical signaling molecules have shown to have
had an effect on transcriptional levels of genes that code for proteins which play a role
in the cellular response to stimuli [3–5]. TNF-α and TGF-β secreted by surrounding
tumor cells have shown to induce the transformation of a subpopulation of normal
stromal fibroblasts to cancer-associated-fibroblasts (CAFs), which aid the metastasis and angiogenesis of the tumor cells via secretion of matrix metalloproteinase
(MMP)9 and (MMP)2 [6, 7].

In this study, we examined the role of cell mechanics, namely cell geometry
(2D) and matrix rigidity (3D), on the nucleo-cytoplasmic shuttling of transcription
factors (p-65, MKL1, SMAD3) and protein Vimentin in NIH3T3 mouse embryonic
fibroblasts, in the presence of TNF-α and conditioned media from metastatic breast
cancer cell line MCF-7. Our results have shown that the heterogeneous mechanical
microenvironments which fibroblasts exist in give rise to their different responses
to the chemical signals from tumor cells. The novel understanding of the nucleo-
cytoplasmic localization of these several transcription factors in fibroblasts under
various mechanical microenvironments heightens the potential for understanding the
behaviors of cancer-associated-fibroblasts (CAFs). We aim to identify the activated

signaling pathways in CAFs under different mechanical environments likened to that of the different stages of breast cancer to aid in future engineering of more effective pharmaceuticals in breast cancer treatment through targeting the highly activated signaling pathways within the fibroblasts. Current target therapeutics for breast cancer targets namely the growth-promoting protein HER2, which is found in negligible overexpression levels in early-stage and triple negative breast cancer.

2 Methods and Materials

2.1 Cell Culture and Treatment

NIH3T3 fibroblast cells and MCF-7 cells were cultured in Low Glucose DMEM (Life Technologies/Thermo Fisher Scientific, Carlsbad, CA) supplemented with 10% (vol/vol) FBS (GIBCO/Life Technologies/Thermo Fisher Scientific) and 1% (vol/vol) Antibiotic-Antimycotic (Life Technologies/Thermo Fisher Scientific), at 37 °C in 5% CO_2. Cells were split via the aspiration of media, a rinse with 1 × PBS and the use of trypsin for 5 min at 37 °C. Cells were finally partitioned by the addition of cell culture media described above. Conditioned media from MCF-7 cells was obtained from the supernatant from the tissue culture flasks after cells were cultured for 4 days. 500 μL of conditioned media was added to each dish. Cells were stimulated with 20 ng/mL of TNF-alpha (Dissolved in cell culture media) (Sigma) for 30 min.

2.2 Engineering Cells in 3D Collagen Matrix

Collagen I (Rat Tail) was equilibrated with 10 × PBS and 1N NaOH and mixed with 30,000 fibroblast cells in prepared cell culture media (refer to Sect. 3.1) to obtain collagen concentrations 1 and 0.5 mg/mL. Neubauer Chamber was used to determine the average number of cells per mL. 300 μL of this mix was then deposited onto the surface of hydrophobic dishes (ibidi) and allowed to solidify for 3 h after which 0.5 mL of cell culture media/ conditioned media was added and cells were cultured overnight. TNF-alpha treatment was carried out for 30 min after 24 h.

2.3 Micropatterning

Polydimethylsiloxane (PDMS) elastomer (SYL-GARD 184; Dow Corning, Midland, MI) was prepared at a 1:10 ratio of curative to precursor. The PDMS was then poured onto microfabricated silicon wafers containing an array of micro-wells of the desired geometry and cured at 80 °C for 2 h. The solidified PDMS was subsequently peeled off the silicon and overturned. The surface of the PDMS stamps (containing protrusions with the desired geometry) was treated with high power oxygen plasma (Plasma Cleaner Cat. No. PDC-002; Harrick Scientific Products, Pleasantville, NY) for 10 min and 50 mg/mL of fibronectin solution was allowed to adsorb onto the surface. The fibronectin-containing surface was brought into contact with the surface of hydrophobic dishes (ibid), whereby islands of fibronectin with the desired geometry were deposited. PDMS stamps were carefully removed and discarded. This resulted in an array of fibronectin islands of a defined geometry on the culture dish. The remaining area of the dish was passivated with 2 mg/mL Pluronic F-127 (Sigma-Aldrich, St. Louis, MO) to ensure that the cells do not migrate out of the two-dimensional (2D) fibronectin islands. After 10 min, excess acid was washed away twice with $1 \times$ PBS. 65,000 cells were seeded for 45 min. Unadhered cells were removed with a wash of $1 \times$ PBS. Upon addition of cell culture media (refer to Sect. 3.1), dishes were incubated for 3 h. Un-patterned cells were obtained by culturing NIH3T3 cells on dishes uniformly coated with fibronectin.

2.4 Fluorescent Staining on 3D Matrix

Media was aspirated from plates and 3.7% Formaldehyde solution (diluted from 37% formaldehyde stock solution with $1 \times$ PBS) was added. After 20 min, cells were washed thrice with 1 mL of PBS and glycine solution and subsequently permeabilized with 0.5% Triton-X (Sigma-Aldrich) in $1 \times$ PBS for 20 min. Cells were then washed thrice with 1 mL of PBS and glycine solution. 10% goat serum with $1 \times$ IF wash buffer was added for blocking and left for 3 h. For immunostaining, SMAD3 mouse antibody, MKL1 goat antibody, p-65 rabbit antibody, vimentin antibody (1:200 in 5% blocking buffer; Abcam, Cambridge, UK) were used. The plates were then left overnight at 4 °C. Cells were washed thrice with $1 \times$ IF wash for 15 min each before secondary antibodies (1:500 in 5% goat serum and IF wash) were added. After 3 h, the cells were rinsed once with $1 \times$ IF wash and twice with $1 \times$ PBS, each for 15 min. DNA was then stained using Hoechst-33342 (1 mg/mL) and F-Actin was visualized using Phalloidin for 30 min. IF wash buffer contains 250 nM Tween in $1 \times$ PBS.

2.5 Fluorescent Staining on 2D Micro-Patterned Substrates

Cells were rinsed twice with $1 \times$ PBS, followed by fixation with 3.7% paraformaldehyde (Sigma-Aldrich) in $1 \times$ PBS for 15 min. Cells were washed thrice with $1 \times$ PBS and permeabilized with 0.5% Triton-X in $1 \times$ PBS for 15 min. After which, cells were washed thrice again with $1 \times$ PBS. 1 mL of 5% blocking buffer BSA was then added and left for 3 h. For immunostaining, SMAD3 mouse antibody, MKL1 goat antibody, p-65 rabbit antibody (1:200 in 5% blocking buffer BSA; Abcam, Cambridge, UK) were used. Plates were then left overnight at 4 °C. Cells were washed thrice with 1 × PBS and secondary antibodies (1:500 in blocking buffer) were added. They were then incubated for 1 h. DNA was then stained using Hoechst-33342 (1 mg/mL) and F-Actin was visualized using Phalloidin for 30 min.

2.6 Imaging, Digital Image Analysis and Statistical Analysis

Images of single cells were taken using A1R Laser Scanning confocal microscope (Nikon, Melville, NY). Imaging conditions were kept similar in all of the experiments. Image analysis was carried out using a custom code in the software ImageJ [8] (National Institutes of Health, Bethesda, MD) and the statistical analysis was carried out in R [9]. A code was written using the ImageJ 1 × macro language, where sequential analysis of numerous images was engaged in a for increment loop. Within different channels, images were smoothened and the threshold of images were taken. Values of pixels along the Z axis were also included to enable for calculations of nuclear morphometrics in a 3D space. The results table generated also comprised of the measurements of the nuclear levels of various transcription factors via the selection of the threshold nucleus as the Region of Interest as well as the 3D nuclear morphometrics. Data from 30 to 50 cells were taken for each sample.

3 Results and Discussion

3.1 Cellular Mechanics Impinge on Nuclear Morphology

Cells in the 3D collagen matrix had smaller nuclei (smaller nuclear volumes) as compared to cells grown on micro-patterned substrates. Cells seeded on circular micro-patterns of area 500 μ^2 also had smaller nuclei compared to the cells grown on rectangular micro-patterns of area 1800 μ^2, though circular cells had proportionately larger nuclei. The un-patterned (UP) cells had intermediate volume with a large standard deviation. In the 3D matrix, cells in the more rigid matrix (1 mg/mL collagen) had smaller nuclei (Fig. 2). Amongst samples with cells cultured in a 2D matrix, the un-patterned cells were flattest (lower height), followed by the rectangular cells and circular cells (Fig. 3). Cells in 3D matrix have a wide distribution of heights, with cells in the more rigid matrix having generally flatter nuclei. Additionally, the circular cells have a smaller projected area as compared to the un-patterned and rectangular cells. Cells the 3D collagen matrix have a wide distribution of nuclear area, with cells in the softer matrix (0.5 mg/mL collagen) having a generally larger nuclear area (Fig. 4). The nuclear Aspect Ratio (nuclear height/width ratio), a measure of how elongated the nucleus is was recorded. An Aspect Ratio of 1 indicates that the nucleus is circular. We find that the un-patterned and circular cells to have a relatively circular nucleus whereas the rectangular cells have an elongated nucleus, with the Aspect Ratio deviating greater from 1 (Fig. 5). Cells cultured in 3D collagen matrix reflected a decrease in nuclei circularity, with the cells in a more rigid matrix having slightly more elongated nuclei (Fig. 1).

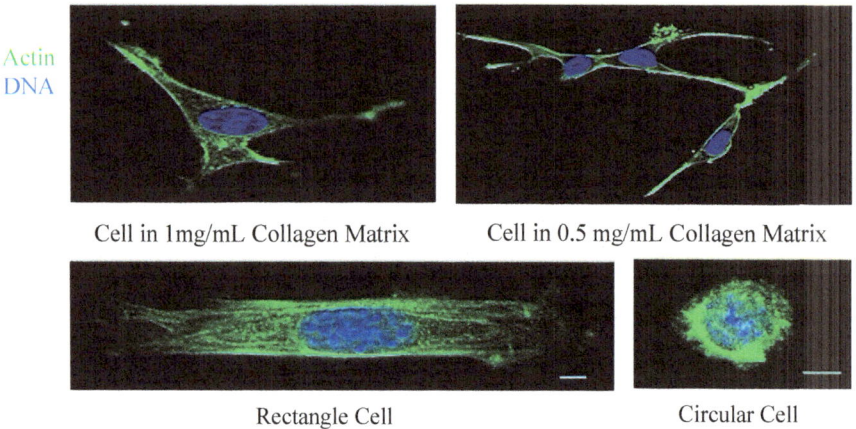

Cell in 1mg/mL Collagen Matrix   Cell in 0.5 mg/mL Collagen Matrix

Rectangle Cell   Circular Cell

Fig. 1 Fluorescence image

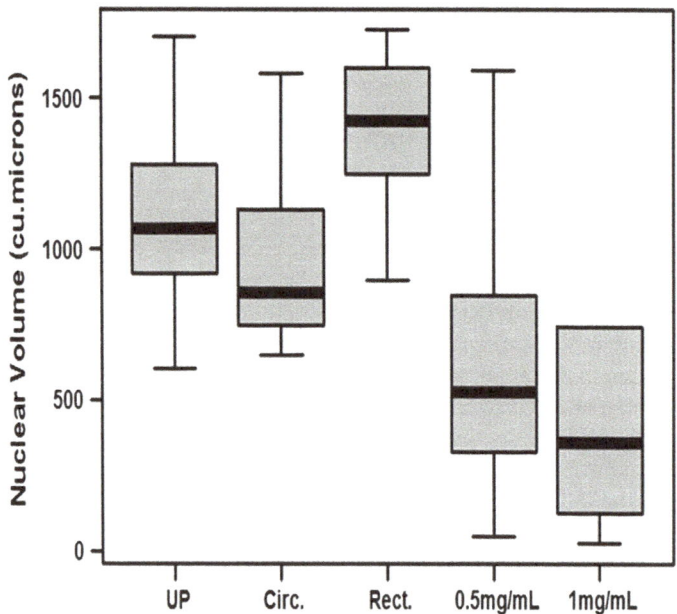

Fig. 2 Nuclear Volume of cells in different mechanical environments

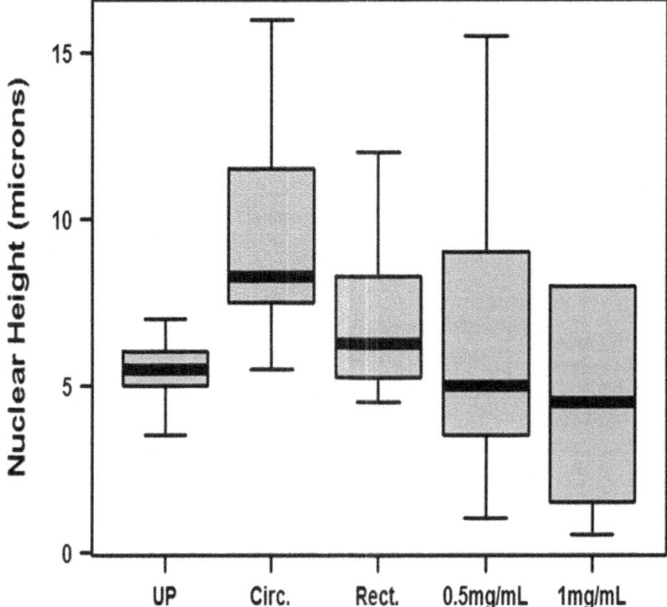

Fig. 3 Nuclear height of cells in different mechanical environments

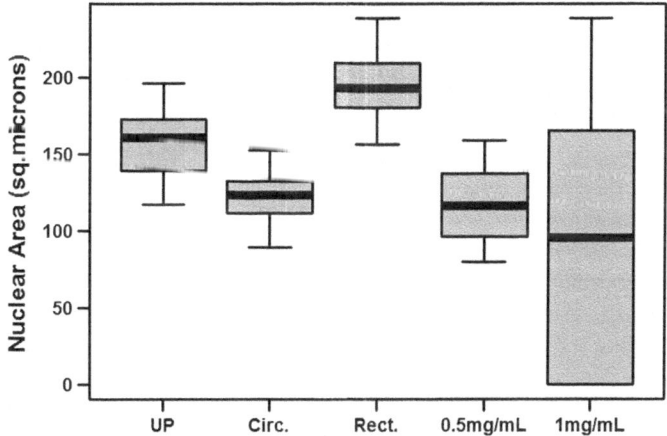

Fig. 4 Nuclear area of cells in different mechanical environments

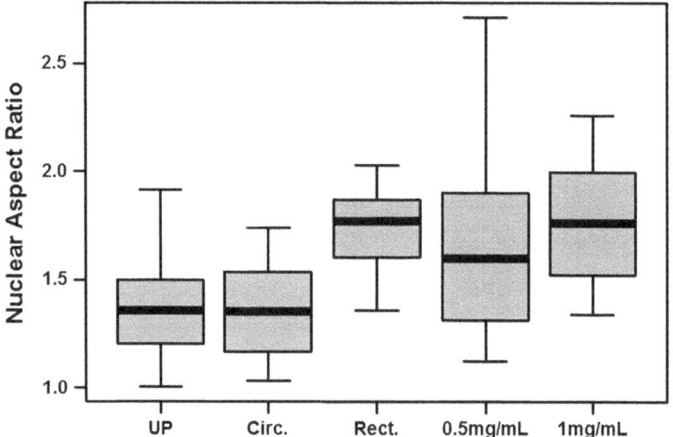

Fig. 5 Nuclear aspect ratio of cells in different mechanical environments

Collectively these results show that the cellular mechanics can modulate the nuclear morphology. These results are in line with previously reported observations [3, 10], which suggests that nuclear shape is modulated by substrate rigidity and cell geometry-induced changes in actomyosin tension.

3.2 Cellular Mechanics Modulate Cellular Response to TNF-Alpha Stimulation

Studies have shown that inflammatory cytokine TNF-alpha which is highly secreted by breast cancer cells induces the p-65 signaling pathway in nearby fibroblasts, whereby p-65/NFkB translocates into the nuclei [11, 12] and transcription of anti-apoptotic and proinflammatory genes occurs [6]. In order to further understand the effects of cells in different mechanical states, namely different geometric shapes and matrix stiffness, on the cellular response to TNF-alpha, we measured the nuclear localization of one of its chief effector molecules p-65.

We find that p-65 translocated from the cytoplasm to the nucleus upon stimulation with TNF-α in cells under all conditions. The changes in cellular mechanics did however play a role in modulating the extent of nuclear abundance of concentration of p-65 in the nucleus observed from the significant increase in nuclear localization of p-65 in rectangular cells as compared to circular cells post TNF-α stimulation. When cells are embedded in the 3D collagen matrix, we find that the cells in softer substrates generally have higher nuclear p-65 levels prior to and post TNF-α stimulation even though a more significant increase in p-65 nuclear levels was observed in cells embedded within a rigid matrix. These results suggest that nuclear levels of p-65 pre-and-post TNF-α stimulation is dependent on cellular mechanics (Fig. 6).

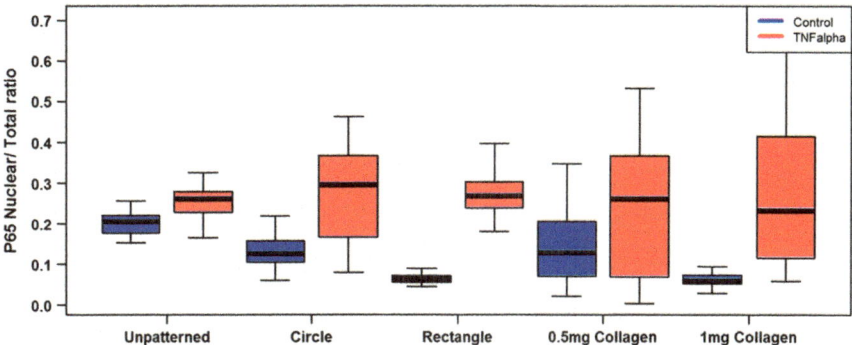

Fig. 6 p-65 nuclear abundance in cells subjected to TNF-α. Overall increase in p-65 nuclear localization across cells in all mechanical environments. Most significant increase in p-65 nuclear levels in rectangular cells and cells in rigid matrix post TNF-α stimulation

In addition to p-65, we also looked at MKL1 which is a cofactor for the ubiquitously expressed transcription factor Serum Response Factor (SRF). MKL1 is a known effector molecule of the cytokine TGF-beta, which is highly secreted by breast cancer cells and known to be antagonistic to TNF-alpha [6]. Translocation of MKL1 into the nucleus has shown to significantly promote the migration of cancer cells [13], as the overexpression of MKL1 potentiated the induction of matrix metalloproteinases, MMP9, transcription by TFG-beta. MMP9 transcription was activated through the recruitment of ASH2, a component of the H3K4 methyltransferase complex by MKL1.

We find that the MKL1 is abundant in the nucleus in both un-patterned and rectangular cells prior to TNF-alpha stimulation, while nuclear levels of MKL1 in circular cells were relatively low. Upon TNF-alpha stimulation, MKL1 exits the nucleus in un-patterned and rectangular cells but minimal decrease in nuclear levels of MKL1 is observed in circular cells. When cells are embedded in the 3D collagen matrix, we find that the cells in more rigid substrates have higher nuclear levels of MKL1 prior to TNF-alpha stimulation and nuclear levels dropped more noticeably in the more rigid matrix after stimulation (Fig. 7).

Collectively, these results have showed how cells modulate the expression genes participating in two antagonistic pathways. Higher cytoplasmic levels of MKL1 were maintained post TNF-alpha simulation, in which higher nuclear levels of p-65 was observed. MKL1 was also observed to have translocated into the nucleus when p-65 translocated from the nucleus into the cytoplasm, thereby maintaining a modular control over the genes that are expressed (Figs. 8 and 9).

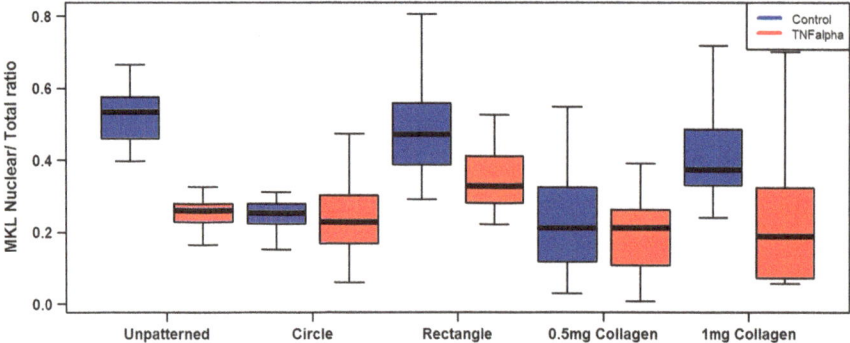

Fig. 7 MKL1 nuclear abundance in cells subjected to TNF-alpha. Overall increase in MKL1 cytoplasmic localization across cells in all mechanical environments. Most significant increase in MKL1 cytoplasmic levels in rectangular and cells in rigid matrix post TNF-α stimulation

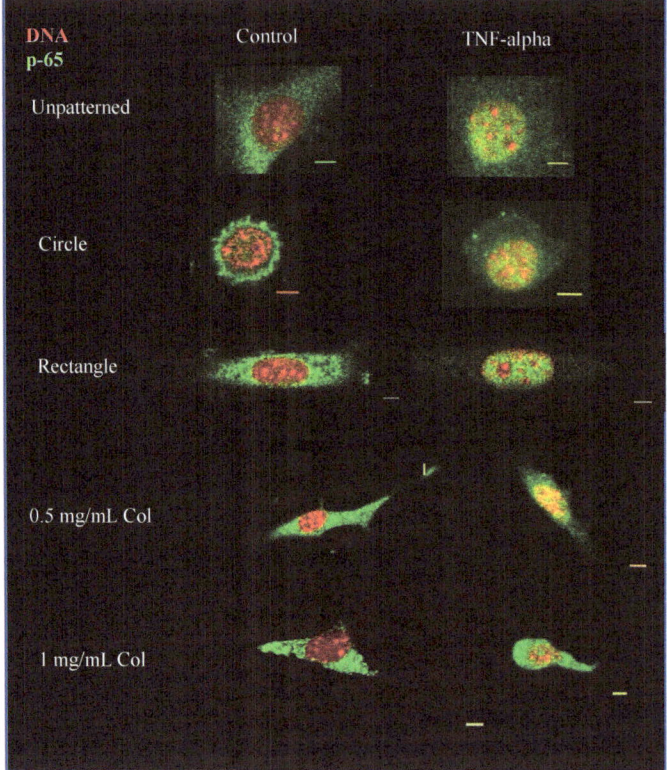

Fig. 8 Fluorescence image of cells subjected to TNF-alpha stained for p-65 and DNA

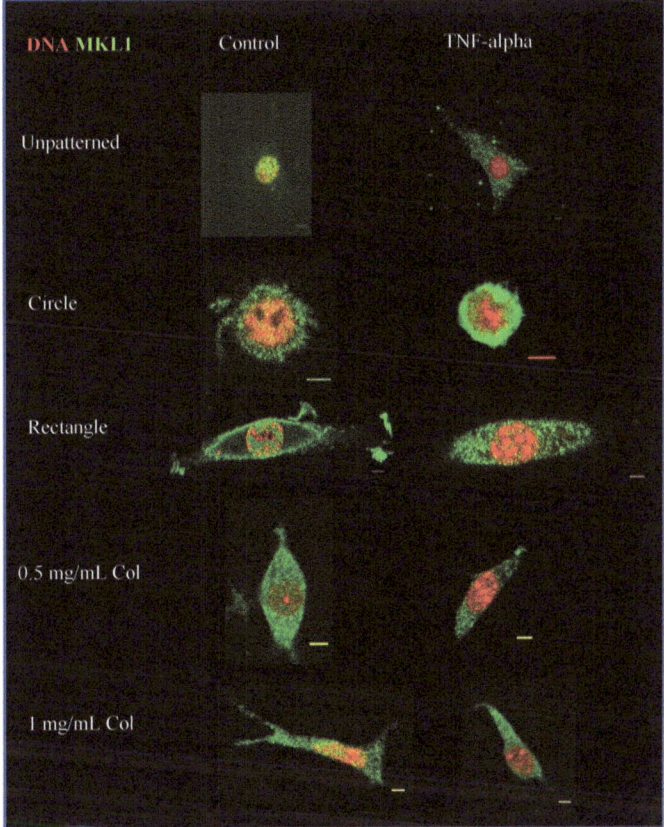

Fig. 9 Fluorescence image of cells subjected to TNF-alpha stained for MKL1 and DNA

3.3 Cellular Mechanics Modulate Cellular Response to Stimulation with MCF-7 Conditioned Media

SMAD3 and MKL1 are transcription factors involved in the TGF-beta pathway and p-65 as mentioned above participates in the TNF-alpha pathway. Cancer cells are known to release both these cytokines [14, 15] and the fibroblasts from the surrounding tissue are susceptible to changing their expression profiles due to the presence of these biochemical signals.

Upon supplementation with conditioned media, we find that the p-65 translocated from the cytoplasm to the nucleus within the un-patterned and circular cells grown in the conditioned media as well as cells grown in the softer matrix. Surprisingly, no significant increase in p-65 nuclear levels was observed in the rectangular cells and cells grown in the more rigid 1 mg/mL collagen matrix (Fig. 10).

SMAD3, another transcription factor in the TFG-beta signaling pathway, translocated from the cytoplasm to the nucleus in cells grown in the conditioned media under

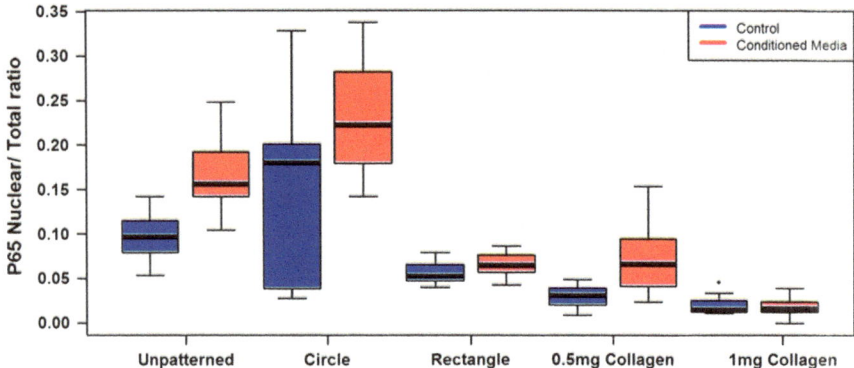

Fig. 10 p-65 nuclear abundance in cells subjected to MCF-7 conditioned media. Overall increase in p-65 nuclear localization across cells in all mechanical environment; remained constant in rectangular cells and cells in rigid matrix Most significant increase in p-65 nuclear levels in circular cells and cells in soft matrix post stimulation with MCF-7 conditioned media

all conditions (Fig. 11). Activation of SMAD3 thereby allows for the regulation of transcriptional responses that are favorable to metastasis via increased production of matrix metalloproteinases, MMP2 [16], which degrade the ECM.

MKL1, the effector molecule of highly expressed TGF-beta from the breast cancer cells, translocated from the cytoplasm to the nucleus in cells grown in the conditioned media under all conditions. This translocation is more prominent in the un-patterned, rectangular cells and cells grown in the softer matrix (Fig. 12).

Collectively, these results reflect the complex nature of the tumor microenvironment and show that the cancer conditioned media contains both TNF-alpha and TGF-beta signaling molecules. More importantly, they show that the mechanical

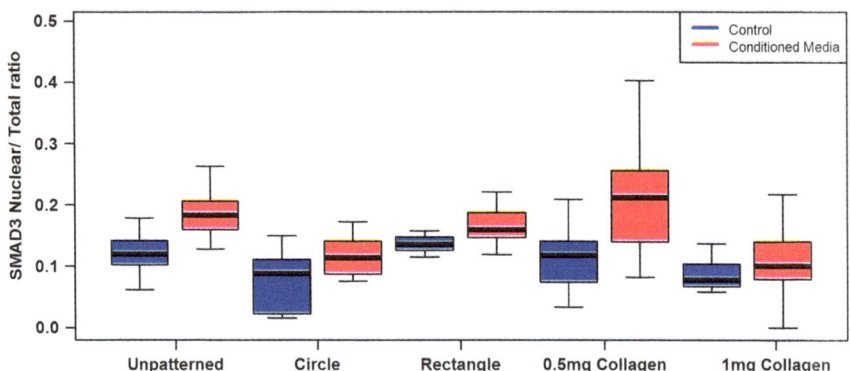

Fig. 11 SMAD3 nuclear abundance in cells subjected to MCF-7 conditioned media. Overall increase in SMAD3 nuclear localization across cells in all mechanical environments. Most significant increase in SMAD3 nuclear levels in cells in soft matrix post stimulation with MCF-7 conditioned media

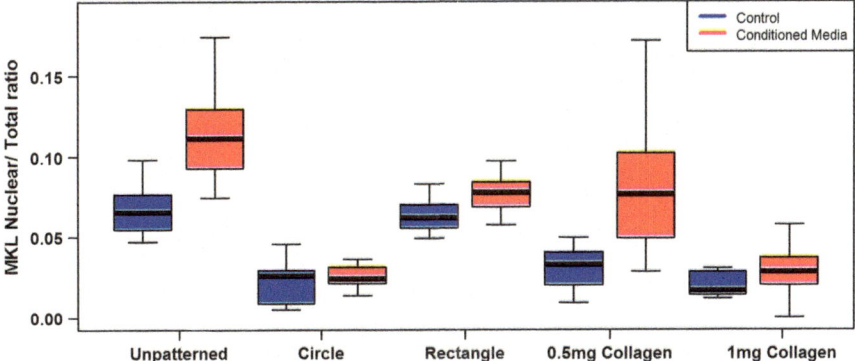

Fig. 12 MKL1 nuclear abundance in cells subjected to MCF-7 conditioned media. Overall increase in MKL1 nuclear localization across cells in all mechanical environments; remained constant in circular cells. Most significant increase in MKL1 nuclear levels in cells in soft matrix post stimulation with MCF-7 conditioned media

state of the cell plays a crucial role in mediating cellar response, specifically that of the nuclear translocation of transcription factors (Figs. 13, 14 and 15).

3.4 Cellular Mechanics Modulate Vimentin Protein Levels

Vimentin, a protein known to maintain cellular integrity, is ubiquitously expressed in mesenchymal cells such as fibroblasts [17]. The binding of TGF-beta secreted from breast cancer cells to serine/threonine kinase receptors in the surrounding fibroblasts, results in the phosphorylation of SMAD2 and SMAD3. This begets the formation of a heteromeric complex with SMAD4 which translocates into the nucleus and modulates the transcription of selected genes such as VIM, thereby contributing to the up-regulation of vimentin expression in the fibroblasts. VIM is a target gene for SMAD3 [7] that codes for the protein Vimentin. In order to observe if the changes recorded in SMAD3 nuclear abundances when exposed to the cancerous conditioned media were also reflected in the protein levels, the Vimentin levels in cells grown in varied matrix rigidities with exposure to conditioned media from breast cancer cells were measured. We find that the Vimentin protein levels do indeed reflect the nuclear abundance of SMAD3—Vimentin levels in cells grown in 3D matrix have increased (Fig. 16) proportionately with their SMAD3 nuclear levels (Fig. 11), in response to increased concentrations of TGF-beta molecules present in the media.

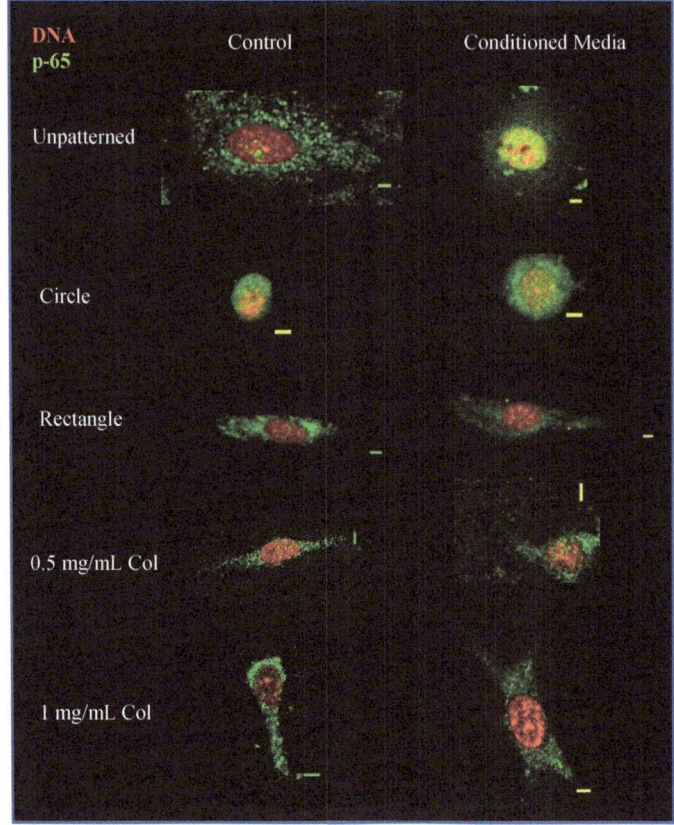

Fig. 13 Fluorescence image of cells subjected to MCF-7 conditioned media stained for p-65 and DNA

4 Conclusion and Future Work

Cells experience a complex combination of external stimuli that they respond to in order to maintain homeostasis within their tissue microenvironments. This study has shown that the cells in different mechanical states (varied cell geometry and matrix rigidity) can differentially process the same signal by differentially modulating the nucleocytoplasmic shuttling of transcription factors, evident for biochemical signaling molecule TNF-α and the complex mix of biochemicals and cytokines released by the metastatic breast cancer cell line MCF-7. Upon exposure to cancer conditioned media, TNF-α—p-65 signaling pathways were further amplified in cells with circular cells and cells grown on soft matrix. TGF-β—MKL1 pathways in cells grown on soft matrix were particularly highly activated. TGF-β—SMAD3 signaling pathways in all cells were activated, though the pathway was more activated in environments with decreased matrix rigidity. Increase in SMAD3 nuclear levels also

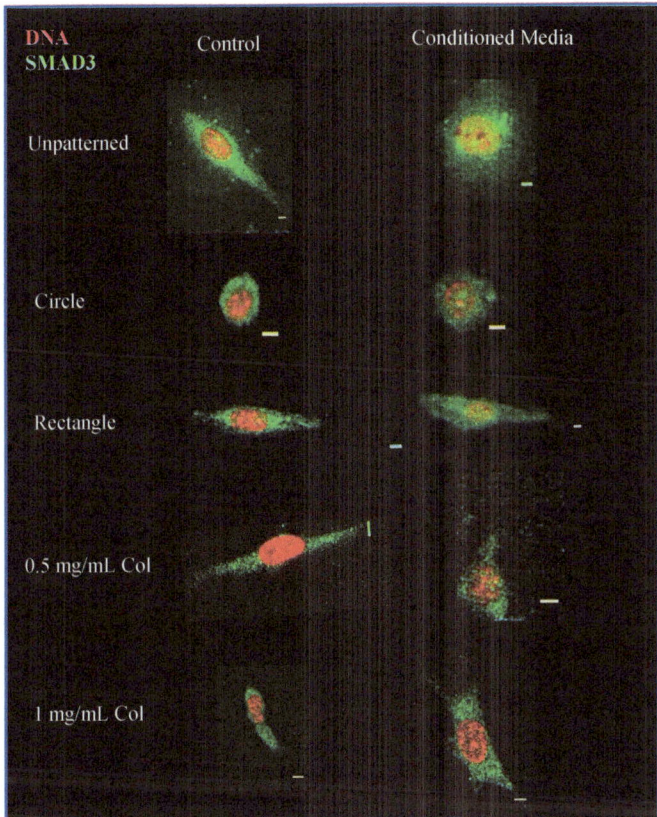

Fig. 14 Fluorescence image of cells subjected to MCF-7 conditioned media stained for SMAD3 and DNA

led to a proportionate increase in Vimentin levels in cells grown in 3D matrix. Our crucial findings indicate that transformations of normal dermal fibroblasts to CAFs during tumorigenesis could depend on the mechanical states of the cell. Our results show that advanced tumorigenesis in breast cancer can also occur in cells within soft ECM, likened to the mechanical environment of the early-stage breast cancer through these different activated biochemical signaling pathways of transcription factors, an important target for new targeted therapeutics, as current target therapeutics for breast cancer namely targets the growth-promoting protein HER2, found in negligible overexpression levels in early-stage and triple negative breast cancer. We postulate that the enhanced nuclear localization of these metastasis-contributing transcription factors in soft matrixes may be due to the increased presence of unstable integrin-mediated focal adhesions resulting in the probable increase in internalization of the soluble cytokine receptors TNFRSF and TGFBR1 via endocytosis [18].

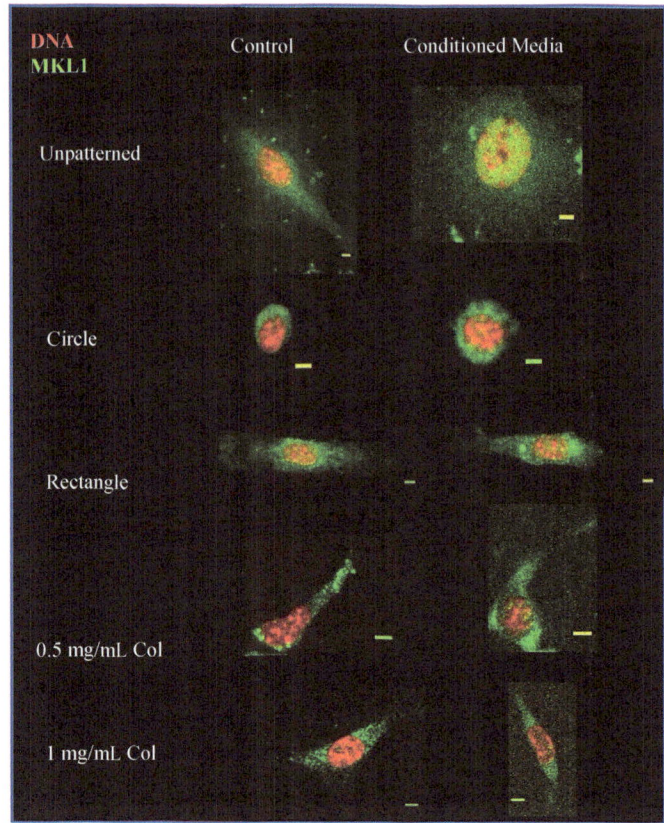

Fig. 15 Fluorescence image of cells subjected to MCF-7 conditioned media stained for MKL1 and DNA

This, along with our findings that cells in softer matrixes generally have proportionately larger nuclei, leads to the probable reduction in diffusion distance between the effector molecules and nucleus within cells, thereby facilitating their nuclear localization in soft matrixes. Further studies involving 3D co-culture models [19] of tumor-fibroblast interactions can help further understand this transformation of normal stromal fibroblasts to CAFs.

Acknowledgements I would like to specially thank Ms. Saradha Venkathachalapathy along with other members of G. V. S. Lab of Mechanobiology Institute, NUS, as well as NUS High School of Math and Science for their unwavering support and guidance throughout this project.

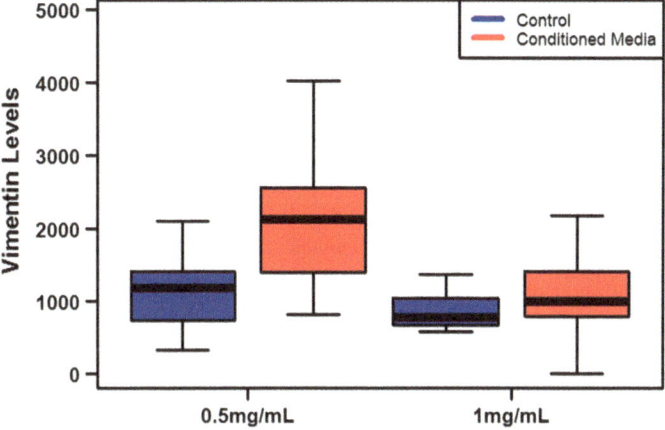

Fig. 16 Vimentin levels in cells subjected to MCF-7 conditioned media in different mechanical environments. Proportionate increase in Vimentin levels with SMAD3 nuclear localization

Appendices

Micro-patterned Fibronectin

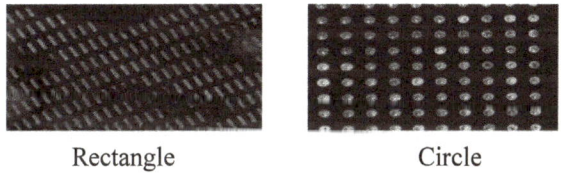

Rectangle   Circle

NIH3T3 Embryonic Mouse Fibroblasts on Patterns

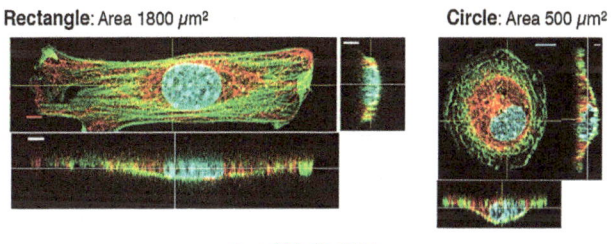

Actin Tubulin DNA

References

1. Kostic, A., Lynch, C. D., & Sheetz, M. P. (2009). Differential matrix rigidity response in breast cancer cell lines correlates with the tissue tropism. *PLoS ONE, 4*(7). https://doi.org/10.1371/journal.pone.0006361.
2. Levental, K. R., Yu, H., Kass, L., Lakins, J. N., Egeblad, M., Erler, J. T., et al. (2009). Matrix crosslinking forces tumor progression by enhancing integrin signaling. *Cell, 139*(5), 891–906. https://doi.org/10.1016/j.cell.2009.10.027.
3. Jain, N., Iyer, K. V., Kumar, A., & Shivashankar, G. V. (2013). Cell geometric constraints induce modular gene-expression patterns via redistribution of HDAC3 regulated by actomyosin contractility. *Proceedings of the National Academy of Sciences, 110*(28), 11349–11354. https://doi.org/10.1073/pnas.1300801110.
4. Pahl, H. L. (1999). Activators and target genes of Rel/NF-κB transcription factors. *Oncogene, 18*(49), 6853–6866. https://doi.org/10.1038/sj.onc.1203239.
5. Meyer, C. J., Alenghat, F. J., Rim, P., Fong, J. H., Fabry, B., & Ingber, D. E. (2000). Mechanical control of cyclic AMP signalling and gene transcription through integrins. *Nature Cell Biology, 2*(9), 666–668. https://doi.org/10.1038/35023621.
6. Stuelten, C. H. (2005). Breast cancer cells induce stromal fibroblasts to express MMP-9 via secretion of TNF-α and TGF-β. *Journal of Cell Science, 118*(10), 2143–2153. https://doi.org/10.1242/jcs.02334.
7. Wu, Y., Zhang, X., Salmon, M., Lin, X., & Zehner, Z. E. (2007). TGFβ1 regulation of vimentin gene expression during differentiation of the C2C12 skeletal myogenic cell line requires Smads, AP-1 and Sp1 family members. *Biochimica Et Biophysica Acta (BBA)—Molecular Cell Research, 1773*(3), 427–439. https://doi.org/10.1016/j.bbamcr.2006.11.017.
8. Rueden, C. T., Schindelin, J., Hiner, M. C., Dezonia, B. E., Walter, A. E., Arena, E. T., et al. (2017). ImageJ2: ImageJ for the next generation of scientific image data. *BMC Bioinformatics, 18*(1). https://doi.org/10.1186/s12859-017-1934-z.
9. The R Project for Statistical Computing. (n.d.). Retrieved from http://www.R-project.org/.
10. Lovett, D. B., Shekhar, N., Nickerson, J. A., Roux, K. J., & Lele, T. P. (2013). Modulation of nuclear shape by substrate rigidity. *Cellular and Molecular Bioengineering, 6*(2), 230–238. https://doi.org/10.1007/s12195-013-0270-2.
11. Erez, N., Truitt, M., Olson, P., & Hanahan, D. (2010). Cancer-associated fibroblasts are activated in incipient neoplasia to orchestrate tumor-promoting inflammation in an NF-κB-dependent manner. *Cancer Cell, 17*(2), 135–147. https://doi.org/10.1016/j.ccr.2009.12.041.
12. Sero, J. E., Sailem, H. Z., Ardy, R. C., Almuttaqi, H., Zhang, T., & Bakal, C. (2015). Cell shape and the microenvironment regulate nuclear translocation of NF-κB in breast epithelial and tumor cells. *Molecular Systems Biology, 11*(3), 790. https://doi.org/10.15252/msb.20145644.
13. Leask, A., & Abraham, D. J. (2004). TGF-β signaling and the fibrotic response. *The FASEB Journal, 18*(7), 816–827. https://doi.org/10.1096/fj.03-1273rev.
14. Coussens, L. M., & Werb, Z. (2002). Inflammation and cancer. *Nature, 420*(6917), 860–867. https://doi.org/10.1038/nature01322.
15. Massagué, J. (2008). TGFβ in Cancer. *Cell, 134*(2), 215–230. https://doi.org/10.1016/j.cell.2008.07.001.
16. Millet, C., & Zhang, Y. E. (2007). Roles of Smad3 in TGF-β signaling during carcinogenesis. *Critical Reviews™ in Eukaryotic Gene Expression, 17*(4), 281–293. https://doi.org/10.1615/critreveukargeneexpr.v17.i4.30.
17. Satelli, A., & Li, S. (2011). Vimentin in cancer and its potential as a molecular target for cancer therapy. *Cellular and Molecular Life Sciences, 68*(18), 3033–3046. https://doi.org/10.1007/s00018-011-0735-1.
18. Cendrowski, J., Mamińska, A., & Miaczynska, M. (2016). Endocytic regulation of cytokine receptor signaling. *Cytokine & Growth Factor Reviews, 32*, 63–73. https://doi.org/10.1016/j.cytogfr.2016.07.002.

19. Jaganathan, H., Gage, J., Leonard, F., Srinivasan, S., Souza, G. R., Dave, B., et al. (2014). Three-dimensional in vitro co-culture model of breast tumor using magnetic levitation. *Scientific Reports, 4*(1). https://doi.org/10.1038/srep06468.

A Study of Chitosan Films as a Feasible Material for Wound Dressing

Geraldine Foo Yawen

Abstract Chitosan was used in this investigation to develop a film suitable for bandaging (wrapping around infected area) wounds to prevent infections. Chitosan is known for its good biodegradability which proves the point of making an environmentally-friendly film that is usable. Previously found by researchers to be useful as a food packaging material [1], it was hypothesised that it would make a good antibacterial bandage film. In this research, various properties like the strength, elasticity, antibacterial properties, air permeability and the effect of water on the film was studied to determine the reliability of the film. Thus, the uses of chitosan as a feasible bandage material was studied.

1 Introduction

In recent years, researchers have found various substitutes for non-biodegradable polymers such as plastics. One of which, is the use of Chitosan film to replace food packaging material like cling wrap. Chitosan, a protein derived from chitin found in exoskeletons of insects, shells of crustaceans and various fungi, has been studied by many researchers for its tremendous medicinal properties. Chitosan is known for its good biodegradability and capacity to form membranes, including its excellent film forming capacity [1]. Therefore, this research aims to find out the other uses of chitosan film.

Plastics are used in several products due to its long-lasting durability. Plastics are recyclable but not all produced end up being recycled. Due to its long biodegradation process, those of plastics that are disposed of have to be burnt in order to get rid of. This eventually leads to global warming. With global warming developing into a much larger problem faced today, a biodegradable film will help to reduce the amount of non-biodegradable materials that have to be burnt, aiding in the lessening of air pollution. As researchers have found the chitosan film to conceive antibacterial properties [1], it is proposed that it would make a suitable material for a bandage.

G. Foo Yawen (✉)
Fairfield Methodist School (Secondary), Singapore, Singapore
e-mail: geraldinefyw@gmail.com

© Springer Nature Singapore Pte Ltd. 2019
H. Guo et al. (eds.), *IRC-SET 2018*,
https://doi.org/10.1007/978-981-32-9828-6_15

Various properties like the usability and reliability of bandages were tested and examined. The strength, elasticity, antibacterial properties, air permeability and the effect of water on the film was studied in this research. In this research, chitosan was used to make a biodegradable film capable of other uses such as the bandaging of wounds to prevent infections.

2 Materials and Methods

2.1 Synthesis of Chitosan Film

1 g of chitosan powder was first dissolved in 2 wt% ethanoic acid (HOAc) to prepare the solution. The mixture was then stirred continuously at room temperature for 24 hours until a homogeneous solution was obtained. The solution was filtered through a synthetic cheesecloth to remove undissolved material. The resulting solution was then casted on a flat acrylic sheet and left to dry in the open at room temperature. After taking the film off the acrylic sheet, the resulting film was washed with 2 wt% sodium hydroxide (NaOH) to neutralise HOAc.

2.2 Modifications Made to Standard Film

Cross-linking: To test for the possible changes in tensile strength of the film, the film was soaked in a 2 wt% sulfuric acid (H_2SO_4) cross-linking solution for 1 h to test for the possible changes in tensile strength of the film. The film was named C—1 g—CL. Amount of powder added: Amount added was interchanged using 1 and 2 g to test for possible changes in various areas tested. The films were named C—1 g, C—2 g.

2.3 Incorporation of Other Chemicals

Glycerol which is known as a good plasticizer for chitosan films [2], was incorporated into some films as a softener to test for possible changes in the elasticity of the film. The films were named C—1 g—Gly and C—2 g—Gly.

Plant essential oils exhibit antibacterial properties (In this case, Lavender, Peppermint and Tea tree essential oil were used as they are commonly used to treat the skin.) was added to enhance the antibacterial strength even further. The films were named C—LV, C—PM, C—TT respectively.

3 Testing of Synthesised Material

3.1 Tensile Strength Test

A film was cut out to a dimension of 4 cm by 3 cm for testing. The film was secured on both ends of the longer side with bull clips. As weights are added, some of the films slipped out of the clip before the breaking point due to the lack of friction. Duct tape was then used to secure the film to the clip. It can be taped on the contact area of the clip and the film. The clip above was then hung onto the retort stand clamp. Until the film tears, additional 50-gram weights were added to the hook that was used to hook onto the bottom bull clip. After each weight was hung, the new length of the film was measured. Cling wrap and food plastic bag which are commonly found in households for protecting wounds from water while showering was also used as the control set-up.

Stress was calculated using the formula

$$Stress = Force/Cross\text{-}sectional\ Area$$

The thickness of each film was measured using the micrometre screw gauge. The mean of the readings at the four corners of the film was taken.

Strain was calculated using the formula

$$Strain = Change\ in\ Length/Original\ Length$$

3.2 Antibacterial Property Test

The antibacterial properties of the films were tested by disc diffusion method. Gram-negative *Escherichia coli (E. Coli)* was used in this experiment. Using aseptic techniques, colonies of *E. Coli* were inoculated into the broth and mixed well. Using the micropipette, 10 μL of bacteria broth was pipetted onto a new agar plate and spread equally using a L-shaped glass rod so as to ensure consistency of bacteria concentration on each plate. The plates with bacteria were then left to dry for 20 min before the films and filter discs were placed onto it.

The made film and homogenous solution of the film was used for the test. The essential oil films were suspected to have the best antibacterial strength considering the fact that many researchers have concluded that essential oils have antibacterial properties. The film was first hole punched to obtain a circle of diameter 0.6 cm. 10 μL of the homogenous solution was micropipetted onto the hole punched filter paper as well. This is to compare the difference in the film's antibacterial properties before (homogenous solution) and after (film) casting. The film and solution-soaked filter discs were placed onto the agar plates and placed inside the incubator machine at 37 °C to allow for bacteria growth. The bacteria was allowed to grow for 3–4 days before observations were made. The resulting diameter of the zone of inhibition was measured using a pair of Vernier Calipers. Each measurement was taken three times and the mean of the three readings was taken. Cling wrap was also used as the control set-up. The wider the inhibition zone, the more antibacterial it is against *E. Coli.*

3.3 Air Permeability Test

The air permeability of the films were tested by using them as a substitute for cling wrap and food plastic bag. A small piece of apple was cut out and placed into a small plastic cup. A film big enough to cover the opening of a plastic cup was then "wrapped" over the opening. To ensure an airtight space, silicon sealant was used to secure the film onto the opening of the cup. Cling wrap and the food plastic bag was made the control set-up. The set-up was left overnight in room temperature before the film was taken off. The physical appearance of the apple was used to determine which film allowed more air to enter than another. The extent of oxidation was deduced by the browning of the apple.

3.4 Effects of Water on Film

A 3 cm by 3 cm dimension film was tested for the effects of water on the film. Effects include the changes in size and solubility of the film. The film was submerged in tap water for 8 hours. Tap water instead of distilled water was used so as to stimulate showering conditions. After 8 hours, the film was taken out from the water and patted dry with filter paper before measurements were taken. The area of the film was then measured.

4 Experimental Data and Analysis

4.1 Tensile Strength Test

The tensile strength of all the films made including the control food plastic bag and cling wrap was compared as shown in Fig. 1. Controls are only for reference to determine reliability between made films and plastics that were already made available. Results were analysed and compared between the films made.

Comparing only the films, the strongest film would be the one which has the greatest stress. As seen in Fig. 1, the C—PM film is the strongest film.

The most flexible film would be the one which has the greatest strain. As seen in Fig. 1, the C—2 g—Gly film is the most stretchable film even though it breaks

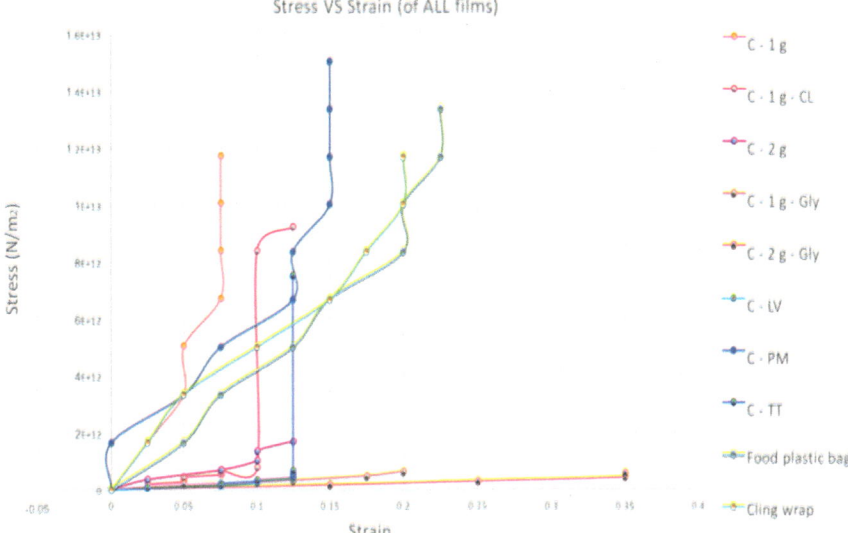

Fig. 1 Stress and Strain (of ALL films)

easily. Glycerol is a softener added to allow the resulting film to be stretched to a wider range. It is hence concluded that adding glycerol during the synthesis of the film enhances the film's ability to stretch.

According to Fig. 1, the film(s) which conceives both good strength and elasticity would be C—1 g and C—1 g—CL as the stress and strain of both films are comparable.

The difference between using 1 g of chitosan powder and 2 g of that was compared as shown in Fig. 2. All films tested did not go through the process of cross-linking so as to ensure accuracy of the results. According to Fig. 2, both types, with and without glycerol, 1 g of chitosan powder dissolved in 2 wt% HOAc is stronger than 2 g of that. At the same time, it was observed from the graph also that both types, with and without glycerol, 2 g of powder made the film more stretchable even though the film was not as strong as when 1 g was used. Hence, it was concluded that 1 g of powder used produced a stronger film while 2 g of powder used produced a more elastic film.

4.2 Antibacterial Property Test

Table 1 shows the result of the zone of inhibition of the filter paper with solution and the film after *E. Coli* was allowed to grow.

The diameter of the zone of inhibition of each film tested is shown in Fig. 3. The bacteria strength was deduced by the diameter of the zone of inhibition. The larger

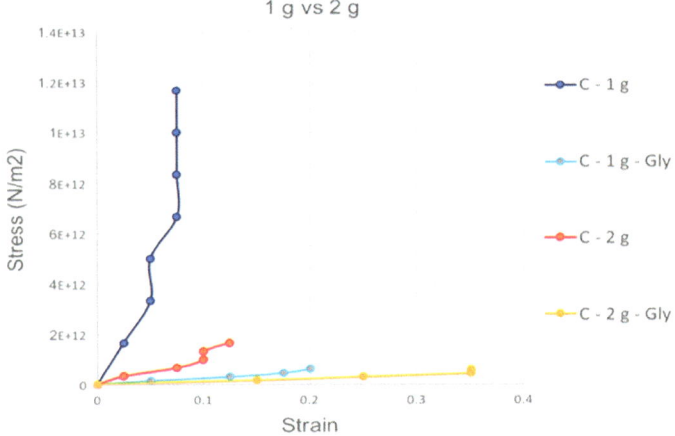

Fig. 2 1 g (of powder) versus 2 g (of powder)

Table 1 Solution versus Film

Type of film	Solution on filter paper	Film
C—LV		
C—PM		
C—TT		
C—1 g		–
Cling wrap (control set-up)	–	

'–' denotes as experiment not carried out

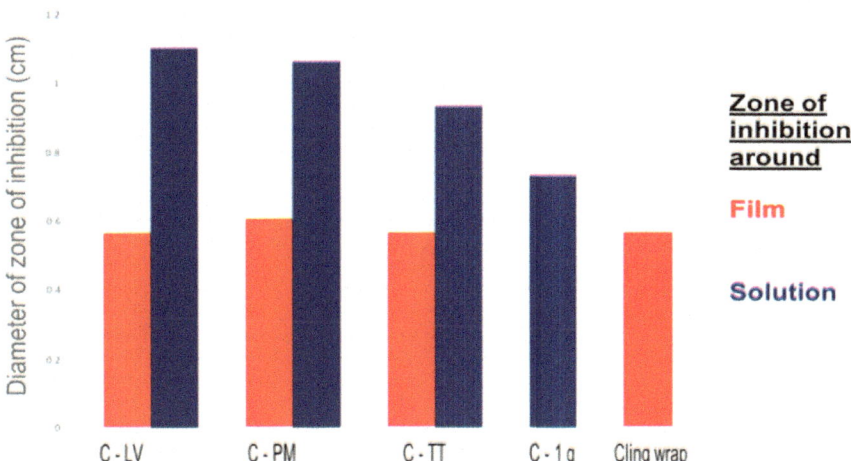

Fig. 3 Bacteria growth deduced by diameter of zone of inhibition

the diameter, the stronger the film is in the protection against bacteria (*E. Coli* in this experiment). The amount of chitosan powder used was standardised to 1 g.

4.2.1 Solution on Filter Paper Versus Film

According to Fig. 3, the zone of inhibition around the solution is clearly visible whereas that of the film is not visible. Hence, it was concluded that only the solution conceived antibacterial properties.

4.2.2 Effects of Essential Oil

According to Fig. 3, comparing the solutions, it can be observed that the chitosan solutions where essential oils were added had better antibacterial strength than the normal chitosan solution. Thus, it was concluded that adding essential oil to the solution enhances the antibacterial strength.

4.3 Air Permeability Test

The amount of chitosan powder used for the films were kept consistent to 1 g to ensure accuracy in the results. Based on the results, the food plastic bag and cling wrap had the allowed the largest amount of oxygen to enter as the apples were the brownest

Table 2 Physical appearance of apples left overnight

C—LV	C—PM	C—TT
C—1 g	Cling wrap	Food plastic bag

of the batch. The films however had the apples which had little change in colour. Therefore, it was concluded that the films may not be the best for the bandaging of wounds due to its inability to allow oxygen flow for the wound to respire. Instead, it is more suitable for the storage of raw food (Table 2).

4.4 Effects of Water on Film

Films were tested to see the changes water made on them. Changes include the expansion or contraction and solubility of the film. The area before the test is 9 cm^2. Based on the results, all the films tested above except the C—1 g film are suitable to be used a bandaging material as they are all able to expand when they are in contact with water. This will prevent exposure of the wound to water when the film expands. It was also predicted that the films were able to absorb water which is useful in terms of the absorption of moisture from the skin. The C—PM film expanded the most of all and would thus make a favourable material (Fig. 4).

5   Conclusion

For a suitable wound dressing material, the film has to be strong, antibacterial, allow oxygen to pass and it has to be insoluble in water. As seen in the results, although chitosan powder itself may not a suitable material, but with the various modifications made to the standard film, it has proven itself to be a suitable material. With the various modifications carried out for more in-depth investigation, the Peppermint film would make a good bandaging material as it is strong and stretchable, and has the second widest zone of inhibition in the antibacterial test as well. Lastly, it allows oxygen to pass adequately and expands the most in water of the tested films.

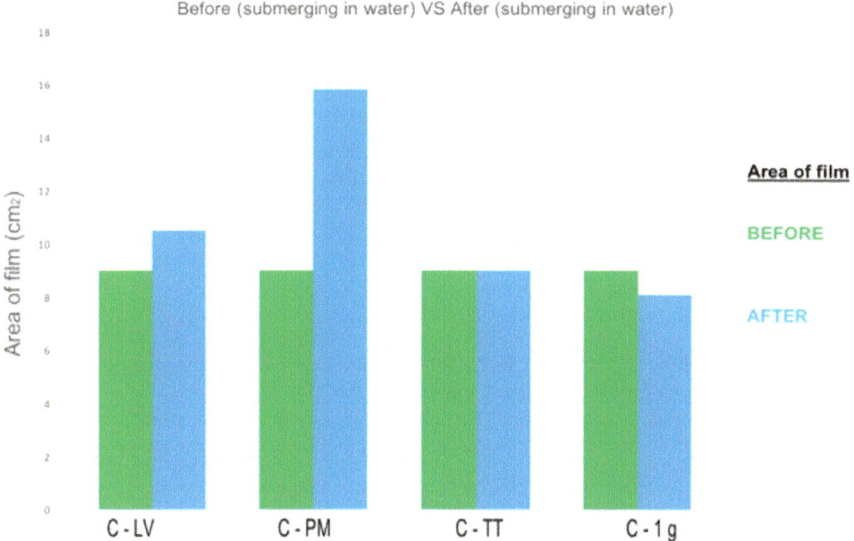

Fig. 4 Effects of water on film (before versus after submerging in water)

Henceforth, general chitosan Peppermint film is an appropriate material for use as a wound dressing material.

6 Limitations

One limitation is that the chitosan powder comes with a high price. The cost before the film is even made proves to be a challenge to introduce to the market. The solution could be to carry out self-extraction of chitin from crustaceans like crabs and shrimps through chemical extraction.

The other limitation of the experiment is that the films are inconsistent in their thickness. This could have affected the results. The solution could be to compress the resulting solution that is already casted on the acrylic sheet with another acrylic sheet to get a flat surface throughout.

7 Future Research

Exploring different concentrations of cross-linking reagents to strengthen the film and develop an even more reliable and durable film in hopes that it would eventually sell in the market. In terms of the synthesis process, cross-linking the essential oil films to strengthen them, exploring a range of concentrations for the amount of chitosan powder added to test for the differences in the various areas tested, testing for the antibacterial strength of the films using gram-positive bacteria to confirm the antibacterial properties, as well as coating the physical film with antibacterial agents to ensure that the final product of film itself is antibacterial. Lastly, replacing essential oils with only their active ingredients to eliminate other ingredients in the essential oils that may potentially affect the results.

References

1. Tripathi, S., Mehrotra, G. K., Dutta, P. K. (2009). Physicochemical and bioactivity of cross-linked chitosan–PVA film for food packaging applications. *Elsevier B.V.*, 372–376.
2. Domjan, A., Bajdik, J., Pintye-Hodi, K. (2009). Understanding of the plasticizing effects of glycerol and PEG 400 on chitosan films using solid-state NMR spectroscopy, *American Chemical Society*, 4667–4673.

Demonstration Kit for Battery-Less RF Energy Harvesting Device

Wu Yongxin, Zhang Junwu and See Kye Yak

Abstract Radio Frequency (RF) Wireless Power Transfer (WPT) is a technology that allows devices to be powered wirelessly. This project aims to develop a functional Radio Frequency (RF) energy harvesting demonstration kit, and to assess the feasibility of building a RF energy harvesting kit. The objective of the kit is to harvest and convert RF energy from a transmitter into DC energy to power up a sensor device, or end device. The main focus is on configuring and evaluating the power management circuit and the end device in the kit. The performance of the kit was evaluated. Testing results showed that the demonstration kit was able to power up the end device when RF energy is transmitted to the receiving antenna in the demonstration kit. When the transmitter is placed 94.0 cm away from the receiving antenna, the demonstration kit can be powered up with a minimum power of 21.90 dBm delivered to the transmitter, and the end device can be powered on continuously with a minimum power of 23.30 dBm delivered to the transmitter.

Keywords Radio frequency · Wireless power transfer · Demonstration kit

1 Introduction

Radio Frequency (RF) Wireless Power Transfer (WPT) is a promising technology to prolong the battery life of low-power sensors which is in huge demand in growing sectors such as Internet of Things (IoT). WPT allows devices to be powered wirelessly, and can either be achieved through having a dedicated antenna to transmit

This project was sponsored by Nanyang Technological University (NTU) as part of the Nanyang Research Programme.

W. Yongxin (✉)
Raffles Institution (Junior College), Singapore, Singapore
e-mail: yonx30@gmail.com

Z. Junwu · S. K. Yak
EMERL, School of Electrical and Electronic Engineering, Nanyang Technological University, Singapore, Singapore

© Springer Nature Singapore Pte Ltd. 2019
H. Guo et al. (eds.), *IRC-SET 2018*,
https://doi.org/10.1007/978-981-32-9828-6_16

183

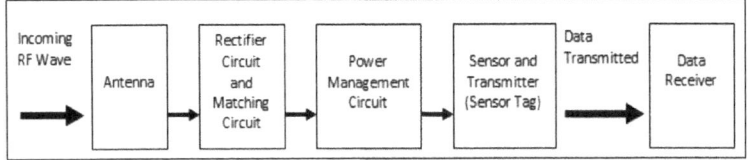

Fig. 1 Overview of demonstration kit components

power, or it can be achieved through harvesting ambient RF energy from existing broadcasting antennas.

The aim of this project is to develop and evaluate a functional RF energy harvesting kit. The kit consists of a rectenna, which comprises a receiving antenna to harvest RF energy and rectifying circuit to rectify and boost the received power to DC, a power management circuit, and the sensor tag which measures the ambient temperature and then sends it wirelessly to a receiver. Figure 1 shows a simple diagram of the demonstration kit.

This project is divided into two parts: The development of the sensor tag and power management circuit.

2 System Setup

As mentioned in the introduction, the demonstration kit comprises of several components shown in Fig. 1. The functions of each component are as follows.

2.1 Receiving Antenna

Firstly, an antenna harvests incoming RF energy. The antenna is comprised of 4 individual patch antennas in an array of 2 by 2. Two antennas are wired in series to form a column, which is then wired in parallel to another column of two antennas in series. This configuration is shown in Fig. 2.

2.2 Rectifier Circuit

The rectifier circuit converts the energy from the antenna into DC current, and boosts the input voltage from the antenna to a level usable by the power management circuit. The rectifiers used are single stage differential Dickson Charge Pumps.

Each antenna is connected to one rectifier, which is then wired in the configuration explained in the antenna section.

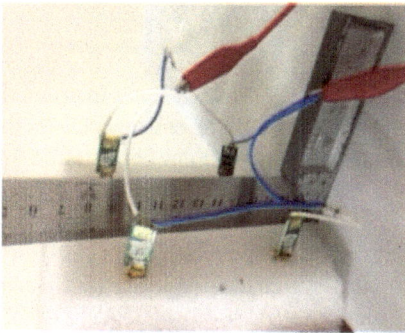

Fig. 2 Receiving antenna array and attached rectifier circuit configuration

A matching circuit for the antennas should be used if necessary. The matching circuit should have impedance that is the complex conjugate of the impedance of the antenna, to maximise power transfer from the antenna to the rectifier circuit [1]. A matching circuit was not used for this setup as testing of the antenna found that its impedance is sufficiently negligible.

2.3 Power Management Circuit

The power management circuit manages the DC energy from the rectifier circuit, and provides sufficient and usable energy to the sensor tag for it to operate.

A Texas Instruments BQ25570 Power Management Integrated Circuit is used in this project for this purpose. It manages the power through several functions. Firstly, received DC energy from the rectifier circuit has to be above 330 mV to start charging the storage element during cold start, or above 100 mV to start charging a storage element during continuous operation [2]. Once input voltage exceeds the minimum requirements, the BQ25570 then receives and boosts the input energy's voltage via a boost converter, which is stored within the storage element. A capacitor is used, and the type and capacitance of the capacitor is determined through testing. Once the charged capacitor reaches a sufficient voltage, the BQ25570 allows the storage element to start powering up the sensor tag. A buck converter regulates the voltage of energy from the capacitor if necessary, so as to provide energy with an acceptable voltage to power up the sensor tag.

2.4 Sensor Tag

The sensor tag used is a Texas Instruments eZ430-RF2500. It consists of an end device and access point. The end device uses the energy from the power management circuit to measure the ambient temperature and its input voltage, and then sends the data wirelessly to the access point.

A second eZ430-RF2500 serves as an access point. The access point is attached to a computer and receives the data from the end device, where it is then displayed on screen through the eZ430-RF2500 Sensor Monitor application. In the default factory setting, the access point sends its data to the access point every 1 s. The Sensor Monitor application displays the data from both the end device and access point.

3 Development of Rectenna

The rectenna development is split into two parts. Firstly, the rectifier circuit also has to be optimised to minimise power loss while boosting the voltage of the received power. Secondly, the antenna design has to be selected and optimised to maximise power absorption and efficiency.

The rectifier circuit has to convert the incoming RF energy from the antenna into DC energy, and boost its voltage to a level usable by the power management circuit, which is at least 100 mV when the circuit is in continuous operation. A single stage differential Dickson Charge Pump, shown in Fig. 3, is used for each antenna.

The diodes used must have a low turn on voltage, in order to minimise power loss from the diode and allow the rectifier to rectify input energy with smaller voltages, thus maximizing its efficiency and output. Reference [1] states that a SMS7630 diode is ideal for this application, and is hence used in the rectifier circuit. Reference [1] also mentions that capacitors with a capacitance larger than 10 pF are sufficient for low power rectifier purposes.

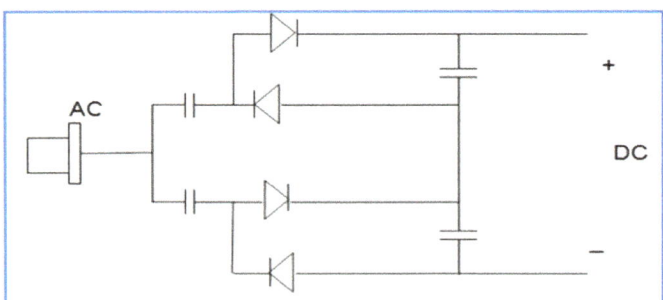

Fig. 3 Single stage differential Dickson charge pump

A square patch antenna is selected for the antenna design due to the ease of fabrication as well as the low profile of the antenna design. These properties allow the antenna to be easily produced, which increases the feasibility of creating the demonstration kit, and also keeps the antenna and subsequently demonstration kit smaller. However, patch antennas have lower efficiency, which necessitates more power transmitted to the antenna to deliver sufficient energy to the power management circuit and power the sensor device [1]. The antenna design was optimised using CST Studio Suite software.

Due to time constraints, optimization of the rectenna was not done. A rectenna provided by the research mentor is used for the final demonstration kit setup.

4   Development of Sensor Tag

The ez430-RF2500 end device needs to operate on energy from the power management circuit and send a signal wirelessly to an access point. To test the functionality of the device, the access point USB is connected to a computer. The end device is then powered up using the provided battery board, and the ez430-RF2500 Sensor Monitor application is run on a computer. Once the end device is connected to the access point, the Sensor Monitor application displays the temperature of both end device and access point, as well as the input voltage of the end device.

The firmware of the end device is edited to reduce its power consumption. Configuration of the end device is done using the source code provided with the ez430-RF2500 Sensor Monitor application, and a compatible code compiler. Code Composer Studio v7 is used for this project and the library rts430.lib is imported from Code Composer Studio v4 for compatibility. The end device code is in the file Main_ED.c. The code is edited to increase the temperature update interval of the end device from 1 to 5 s to reduce power consumption.

This is done by changing the command to measure and send temperature to run only every 5 loops of the end device's inbuilt interrupt function, which is coded to run in 1 s intervals. The following lines of code are edited:

```
static volatile uint8_t sSelfMeasureSem = 0;
→ static volatile uint8_t sSelfMeasureSem = 4;
if(sSelfMeasureSem); → if(sSelfMeasureSem == 5);
sSelfMeasureSem = 1; → sSelfMeasureSem + +;
```

The edited code is then compiled, debugged and uploaded to an end device via the USB FET emulator provided. The end device is powered by shorting the jumpers on the battery board, and access point USB stick is connected to a computer. The configured end device now updates its temperature every 5 s on the Sensor Monitor application, while the access point's update interval remains at 1 s. A successfully

1 Data Update from End
Device ($001)

5 Data Updates from
Access Point ($HUB0)

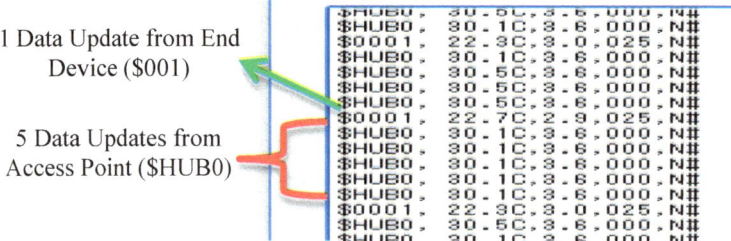

Fig. 4 Configured sensor tag readings on sensor monitor application console

edited end device updates once for every 5 updates from the access point, as shown in Fig. 4.

5 Development of Power Management Circuit

To accomplish the function of the BQ25570 as outlined in Chap. 2, the power management circuit is tweaked in three ways.

5.1 Configuration of Jumpers

The appropriate jumpers on the BQ25570 are shorted to enable the desired functions listed above. As stated in [2], the EN pin is connected to GND to enable the integrated circuit. VOUT_EN is connected to BAT_OK to enable the storage element to start and stop powering the sensor tag at VBAT_OK_HYST and VBAT_OK_PROG values respectively.

5.2 Configuration of User Programmable Values

The user programmable VOUT, VBAT_OK_HYST and VBAT_OK_PROG values are configured to a suitable level by changing the resistors on the BQ25570.

The VOUT value is configured to fit within the acceptable input voltage levels of the sensor tag end device. Reference [3] recommends the supply voltage Vcc to the sensor tag during operation to be as follows:

$$1.8V \leq Vcc \leq 3.6\,V.$$

A VOUT value of 3 V, within the recommended range, is chosen for this project. Reference [2] states the VOUT value follows Eq. (1):

$$VOUT = VBIAS\left(\frac{ROUT2 + ROUT1}{ROUT1}\right). \qquad (1)$$

VBIAS is stated in [2] as 1.21 V, as per the BQ25570 electrical specification table. Cross referencing [2] with [4] shows that ROUT1 = R10 and ROUT2 = R9 on the BQ25570.

Substituting the VOUT = 3 V and VBIAS = 1.21 V into Eq. (1) yields the following:

$$3V = 1.21V\left(\frac{ROUT2 + ROUT1}{ROUT1}\right)$$

$$ROUT2 = \frac{179}{121}ROUT1 \qquad (2)$$

The recommended operating conditions in [2] states that the resistance sum of ROUT1 and ROUT2 should be as follows:

$$11M\Omega \le ROUT1 + ROUT2 \le 15M\Omega. \qquad (3)$$

Combining the Eqs. (2) and (3) gives the following conditions for the resistors:

$$11M\Omega \le \frac{300}{121}ROUT1 \le 15M\Omega. \qquad (4)$$

The following resistors with 1% tolerance were sourced to configure the VOUT value. Accounting for a 1% error in the resistor value gives the following:

$$ROUT1 = (5.23 \pm 0.05)M\Omega.$$
$$ROUT2 = (7.68 \pm 0.08)M\Omega.$$

$$11M\Omega \le ROUT + ROUT2 = (12.91 \pm 0.1)M\Omega \le 15M\Omega.$$

Substituting the above values of ROUT1 and ROUT2 into Eq. (1) yields:

$$VOUT = 1.21\left(\frac{(5.23 \pm 0.0523) + (7.68 \pm 0.0768)}{(5.23 \pm 0.0523)}\right)$$
$$= (2.99 \pm 0.06)V.(2d.p.)$$

The VBAT_OK_HYST and VBAT_OK_PROG values should then be configured to allow for a sufficient charge to be stored in the storage element before it is connected to the sensor tag to power it.

Reference [2] states that the VBAT_OK_HYST and VBAT_OK_PROG values follows the equation below:

$$VBAT\_OK\_PROG = VBIAS\left(\frac{ROK1 + ROK2}{ROK1}\right). \tag{5}$$

$$VBAT\_OK\_HYST = VBIAS\left(\frac{ROK1 + ROK2 + ROK3}{ROK1}\right). \tag{6}$$

As stated in [2], VBIAS is 1.21 V as per the BQ25570 electrical specification table. Cross referencing [2] with [4] shows that ROK1 = R8, ROK2 = R7 and ROK3 = R6 on the BQ25570. Reference [2] also lists the following conditions:

$$VBAT\_OK\_HYST \leq VBAT\_OV = 4.2V.$$

$$VBAT\_OK\_PROG \geq VBAT\_UV = 1.95V.$$

The recommended operating conditions in [2] states that the sum of ROK1, ROK2 and ROK3 should be as follows:

$$11M\Omega \leq ROK1 + ROK2 + ROK3 \leq 15M\Omega. \tag{7}$$

Since VOUT is set at (2.99 ± 0.06)V, VBAT_OK_PROG should be equal to or higher than VOUT. This is to ensure that the voltage output is constant at (2.99 ± 0.06)V, as there is no boost converter between the storage element and output pins to boost the VOUT voltage. A value of VBAT_OK_PROG = 3 V is chosen. The VBAT_OK_HYST value should be high enough to ensure enough energy in the storage element is stored to operate the sensor tag when VBAT discharges from VBAT_OK_HYST to VBAT_OK_PROG. VBAT_OK_HYST is arbitrarily chosen to be 3.6 V. Testing is later done to determine if this value is sufficient and acceptable.

Substituting the VBAT_OK_PROG = 3 V and VBAT_OK_HYST = 3.6 V into Eqs. (5) and (6) respectively yields the following:

$$3V = 1.21V\left(\frac{ROK1 + ROK2}{ROK1}\right). \tag{8}$$

$$3.6V = 1.21V\left(\frac{ROK1 + ROK2 + ROK3}{ROK1}\right). \tag{9}$$

Combining the Eqs. (7), (8) and (9) gives the following conditions for the resistors:

$$ROK1 = \frac{121}{179}ROK2 = \frac{121}{60}ROK3. \tag{10}$$

$$11M\Omega \leq \frac{360}{121}ROK1 \leq 15M\Omega. \tag{11}$$

The following resistors with 1% tolerance were sourced to configure the VOUT value. Accounting for a 1% error in the resistor value gives the following:

$$ROK1 = (4.3 \pm 0.04)M\Omega.$$
$$ROK2 = (6.34 \pm 0.06)M\Omega.$$
$$ROK3 = (2.15 \pm 0.02)M\Omega.$$
$$11M\Omega \leq ROK1 + ROK2 + ROK3$$
$$= (12.79 \pm 0.1)M\Omega \leq 15M\Omega.$$

Substituting the above values of ROK1, ROK2 and ROK3 into Eqs. (5) and (6) yields:

$$VBAT\_OK\_PROG = 1.21\left(\frac{(4.3 \pm 0.04) + (6.34 \pm 0.06)}{(4.3 \pm 0.04)}\right)$$
$$= (2.99 \pm 0.06)V.$$

$$VBAT\_OK\_HYST = 1.21\left(\frac{(4.3 \pm 0.04) + (6.34 \pm 0.06) + (2.15 \pm 0.02)}{(4.3 \pm 0.04)}\right)$$
$$= (3.60 \pm 0.1)V.$$

5.3 Selecting a Storage Element

A storage element has to be selected and connected to the VBAT and GND pins on the BQ25570. A capacitor is chosen for the storage element as the increased energy storage capacity of other storage options are not required for this system as both harvested energy and energy consumption of the sensor tag are relatively low. The capacitor chosen should have a low ESR to reduce power loss when charging and a low leakage to reduce voltage loss while the device is being charged or idle. For ease of testing, the capacitor should be easy to connect and disconnect from the BQ25570 setup. Hence through hole capacitors which can be clipped are preferred over surface mounted capacitors that have to be soldered. Lastly, the capacitor should have a maximum voltage that is higher than the VBAT_OV value of 4.2 V, accounting for tolerance. The first capacitor tested is an electrolytic capacitor of capacitance 2200 μF and voltage rating 50 V.

5.4 Optimising the Power Management Circuit

The configured board is connected to a Rohde & Schwarz SMB100A signal generator to simulate power from the rectenna. VOUT, VBAT_OK_HYST and VOUT_OK_PROG values of the board are tested. The signal generator is set at 2.45 GHz to match the antenna in the complete demonstration kit. The power of the

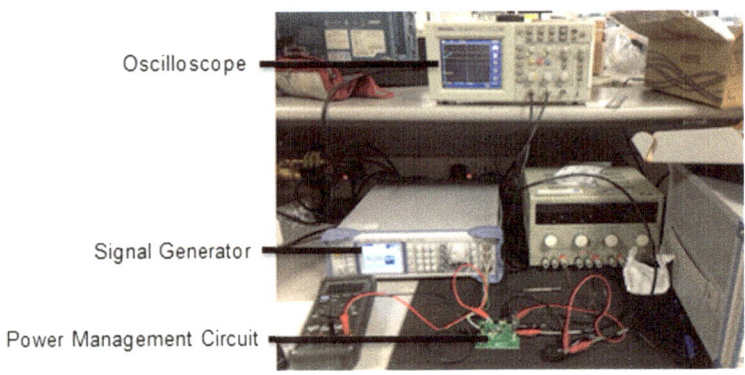

Fig. 5 Setup to evaluate power management circuit

signal generator is set at 6.00 dBm. The rectifier is then connected to VIN and GND of the BQ25570 board to provide power. The capacitor, or storage element, is then connected to VBAT and GND. The voltage levels of VBAT and VOUT are measured using a Tektronix TDS2022 Two-Channel Oscilloscope. The setup is shown in Fig. 5.

When the power source is turned on, VBAT starts rising as the onboard capacitor is charged. Once the voltage of VBAT reaches the programmed VBAT_OK_HYST level of 3.6 V, the output at VOUT is enabled and connected to VBAT. The VOUT voltage should be observed to rise from 0 V to the programmed VOUT level, 2.99 V, once VBAT reaches VBAT_OK_HYST. Further charging of the BQ25570 increases VBAT until it reaches VBAT_OV, after which VBAT is maintained at VBAT_OV.

VOUT rises to 3.04 V when VBAT reaches the VBAT_OK_HYST level of 3.76 V. Further charging increases VBAT until the VBAT_OV value of 4.36 V, after which VBAT is maintained. Figure 6 shows the oscilloscope voltage-time graph of this.

The VOUT value is within the range of the calculated value of (2.99 ± 0.06) V. The recorded VBAT_OK_HYST level is higher than the calculated value of (3.60 ± 0.1) V. This may be due to a faulty resistor or a different VBIAS value from the data

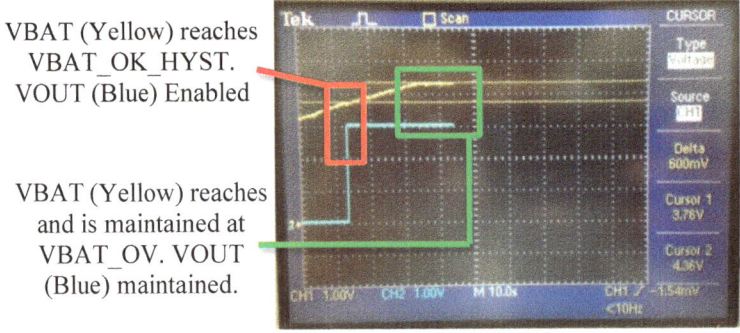

Fig. 6 Oscilloscope voltage-time graph of VOUT and VBAT

VBAT (Yellow) oscillates
between VBAT_OK_HYST
(Upper Cursor) and
VBAT_OK_PROG (Lower
Cursor).
VOUT (Blue) is enabled
briefly when
VBAT_OK_HYST < VBAT
< VBAT_OK_PROG.

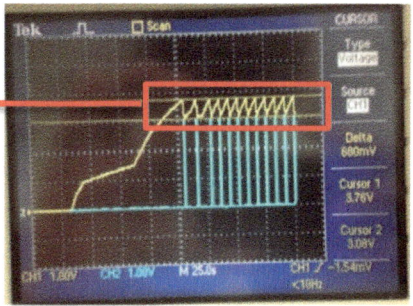

Fig. 7 Oscilloscope voltage-time graph of oscillation of VBAT between VBAT_OK_PROG and VBAT_OK_HYST

sheet. As the values are still within a reasonable range from the calculated value, and are well below VBAT_OV, it was deemed acceptable and was not changed.

To test the VBAT_OK_PROG value, the input power is turned off or disconnected, and loading is applied to VOUT via resistors connected between VOUT and GND. The resistance values of the resistors are unimportant, as long as it provides sufficient loading on VOUT to decrease the voltage of the capacitor fast. VOUT will be disabled and VOUT value decreases to 0 when VBAT < VBAT_OK_PROG. Once VOUT is disabled, VBAT value should then rise again, if input energy to the BQ25570 is available, until it reaches VBAT_OK_HYST again and VOUT is connected to VBAT. The cycle then repeats itself in the pattern shown in Fig. 7.

The sensor tag end device should then be attached to the BQ25570 to check if the configured values are appropriate and sufficient to successfully power the end device and allow it to send at least one measurement to the access point. The end device's battery board is removed and replaced with wires to allow it to be connected to the BQ25570's output pins.

Once connected, the Sensor Monitor application should be started on a nearby computer with an access point attached. The BQ25570 should be set up in a similar fashion as before, with the addition of the sensor tag end device connected to the VOUT and GND pins on the board. The energy input to the BQ25570 is connected to start charging up the BQ25570 and capacitor. Once VBAT reaches VBAT_OK_HYST, the capacitor starts powering the end device.

As seen in Fig. 8, The BQ25570 and capacitor are unable to provide sufficient energy to start up the end device, and no reading is received and displayed by the Sensor Monitor. The VBAT value alternates between VBAT_OK_HYST and VBAT_OK_PROG as the capacitor is charged up, and then discharged by the end device.

To resolve this, VBAT_OK_HYST may be configured to a higher value, so that the voltage in the capacitor is higher when it starts powering the end device. Alternatively, the capacitor's capacitance can be increased by changing it or adding capacitors in parallel, thus increasing the amount of charge available at the same VBAT_OK_HYST voltage.

VBAT (Yellow) oscillates
between VBAT_OK_HYST
(Upper Cursor) and
VBAT_OK_PROG (Lower
Cursor).
VOUT (Blue) enabled briefly
then disabled in a repeating
cycle, but is insufficient to
power up the end device.

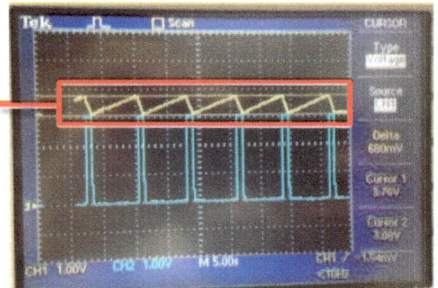

Fig. 8 Oscilloscope voltage-time graph of VBAT and VOUT with end device attached, using 2200 μF capacitor

Configuring the VBAT_OK_HYST value requires new resistors and re-soldering them onto the BQ25570, and is time consuming. Changing the capacitance is hence used.

The capacitance is changed by changing the capacitor to a 4700 μF electrolytic capacitor rated at 16 V. The test is then repeated again to evaluate if the BQ25570 and capacitor is able to power up the sensor tag. Figure 9 shows that the new capacitor contains sufficient charge at VBAT_OK_HYST to power up the end device. There is a short dip of VBAT, indicated by the circle in Fig. 9, due to the end device powering up. Afterwards, VBAT starts rising further as the end device consumes much less energy while running compared to its transient start up power draw. This result indicates that the power management circuit is appropriately configured, and can be assembled with the rest of the demonstration kit.

VBAT (Yellow) reaches
VBAT_OK_HYST, slight
dip in VBAT when
VOUT (Blue) enabled.

VBAT (Yellow) reaches
and is maintained at
VBAT_OV. VOUT (Blue)
Maintained

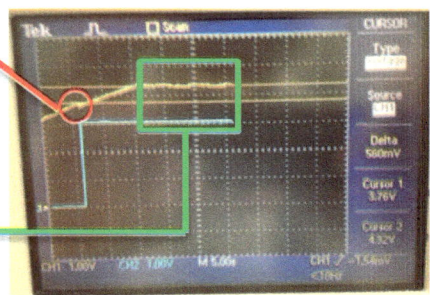

Fig. 9 Oscilloscope voltage-time graph of VOUT and VBAT with end device attached, using 4700 μF capacitor

6 Demonstration Kit Assembly and Testing

To evaluate the performance of the demonstration kit, the following setup is arranged, as shown in the diagram in Fig. 10.

A Rohde & Schwarz SMB100A Signal Generator is connected to an Amplifier Research Model 40S1G4 Signal Amplifier, which amplifies the signal from the Signal Generator to usable levels. The Signal Amplifier is then connected to an Electro-Metrics EM-6961 1—18 GHz Horn Antenna to serve as the transmitting antenna to transmit RF energy. The receiving antenna, and array of 4 2.45 GHz antennas, is placed 94.0 cm away at an angle of 90°, perpendicular to the transmitting antenna. The rectenna is then connected via crocodile clips to the BQ25570 VIN and GND pins. The BQ25570 is configured as stated in Chap. 5.4 with the 4200 μF capacitor attached. An oscilloscope is used to monitor the voltage of VOUT and VBAT. The sensor tag end device is connected to VOUT and GND of the BQ25570. A computer with an access point USB connected is set up, with the access point being 95.0 cm away from the end device.

To evaluate the operation of the demonstration kit, the signal generator is turned on and its frequency and power level are set at 2.45 GHz and −22.00 dBm respectively. The frequency matches that of the receiving antenna. The signal amplifier is turned on and its gain is set at 100%, which gives a minimum gain of 46.0 dB as stated in [5].

The oscilloscope displays the VBAT and VOUT values of the BQ25570 while being charged. The minimum power required to charge VBAT to VBAT_OK_HYST as well as minimum power required to maintain end device operation at VBAT = VBAT_OK_HYST are recorded. Table 1 shows, for these two conditions, the power output of the signal generator, and the transmitter output which is the signal generator's output boosted by the signal amplifier. The potential difference across the power management circuit's input and GND pins is measured using a multimeter to

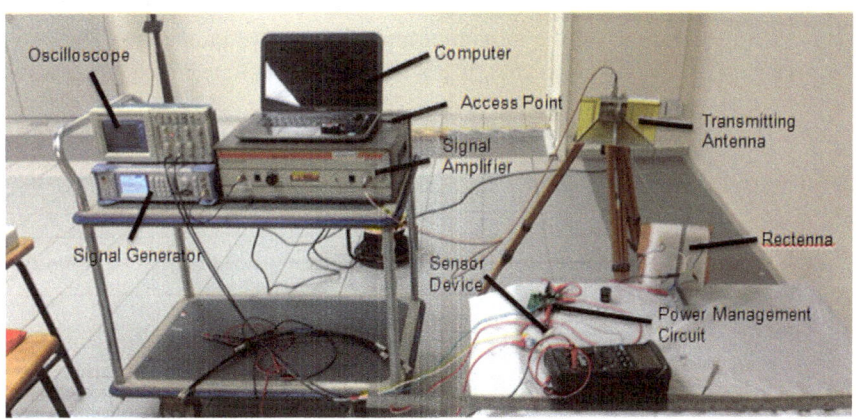

Fig. 10 Picture of actual complete demonstration kit testing setup

Table 1 Results of demonstration kit testing

Conditions	Signal generator power output (dBm)	Transmitter power output (dBm)
Minimum power required to charge VBAT to VBAT_OK_HYST	−24.10	21.90
Minimum power required to maintain end device operation at VBAT = VBAT_OK_HYST	−22.70	23.30

be approximately 0.100 mV at the minimum power required to maintain end device operation.

7 Conclusion

A demonstration kit composed of a receiving antenna, a rectifier circuit, a power management circuit, a sensor tag end device to transmit data, a sensor tag access point to receive data and a sensor monitor application to process and display the data is assembled.

Testing of the setup shows that the kit can operate with a minimum of 21.90 dBm transmitted by a horn antenna from 94.0 cm away, and can power the sensor tag to transmit data. At this level, VBAT levels and continuous sensor tag end device operation can be maintained with 23.30 dBm.

The potential difference of about 100 mV at the minimum transmitted power required to charge kit to VBAT_OK_HYST suggests that the voltage of the energy delivered to the power management circuit input is at the BQ25570's minimum required level. Hence, increasing the input voltage delivered to the power management circuit or decreasing the minimum input voltage requirements of the power management circuit will allow charging of the kit at lower transmitted power levels.

8 Future Work

Based on the performance results of the demonstration kit, the following work is proposed for the demonstration kit:

1. Increase number of stages in the rectifier circuit to increase the voltage delivered to the power management circuit input and allow charging of the circuit at lower transmitted power. Increasing the number of stages will also reduce efficiency of the rectifier circuit due to added components. Hence, testing should be done to

determine the optimum number of stages to decrease the minimum transmitted power required.

2. The power management circuit and storage element may be configured to provide just enough power to power the sensor tag to transmit one measurement before deactivating. The final VBAT_OK_HYST and capacitor capacitance used allows the power management circuit to power the sensor tag for a long time before being discharged. Consequently, however, the storage element takes longer to charge then required if the goal is to enable the sensor tag to transmit once before being disconnected from the storage element.

   The power management circuit itself can be changed from a BQ25570 to a circuit with a lower cold start or continuous operation minimum voltage. This will allow the device to operate off even lower transmitted power levels.

   The storage element can also be changed to reduce ESR and leakage to further reduce power loss and increase efficiency.

3. The Sensor Tag end device firmware can be further edited to reduce the power consumption. The standard end device has a function that waits for an acknowledgement from the access point before sleeping. Disabling this function can reduce power consumption, but will result in the end device not being able to detect if its connection to the access point has been lost, and hence will reduce the robustness of the system. A new sensor tag with even lower power consumption and required input voltage can also be sourced to further increase the efficiency of the system.

Acknowledgements I would like to extend my utmost gratitude to my project mentor, Dr Zhang Jun Wu, for guiding and assisting me throughout every single aspect of the project. I would also like to thank the laboratory technician of the NTU Electronics System Reliability Laboratory, Mr Wee Seng Khoon, for teaching me experimental safety procedures and helping me with various laboratory materials.
I would like to thank my Project Supervisor, Associate Professor See Kye Yak, for accepting me into this NRP Project, and managing the administrative matters for the project. Finally, I would like to thank my NRP Teacher Supervisors Dr Lena Lui and Mr Wong Tze Yang for accepting me into the NRP Programme and handling the administrative matters of the project.

References

1. Wu, Z. J. (2015). *Compact rectenna array on-package for wirelessly powered small device*, PhD thesis, Nanyang Technological University, Singapore.
2. Ultra Low Power Harvester Power Management IC with Boost Charger (Rev. E). (2017). [Online]. Available: http://www.ti.com/lit/ds/symlink/bq25570.pdf.
3. eZ430-RF2500 Development Tool User's Guide, January 2015. [Online]. Available: http://www.ti.com/lit/ug/slau227f/slau227f.pdf. Accessed December 29, 2017.
4. User's Guide for bq25570 Battery Charger Evaluation Module for Energy Harvesting (Rev. A), 2017. [Online]. Available: http://www.ti.com/lit/ug/sluuaa7a/sluuaa7a.pdf.
5. Model 40S1G4 M1 through M4 40 Watts CW 0.7 GHz–4.2 GHz. [Online]. Available: https://www.arworld.us/post/40S1G4.pdf. Accessed January 03, 2018.

A Novel Approach to Desalination Using Modified Bio-wastes as Adsorbents

Ong Jia Xin

Abstract There is a dire need for an economical alternative to convert sea water into fresh water owing to prevalent methods demanding high power and large-scale infrastructures. In this regard, adsorption is considered as an attractive choice due to its renewable and cost-effective nature. In this study, banana stem waste (plantain pith) has been exploited for its capacity to efficiently remove sodium chloride from aqueous solution (sea water). However, removal of Na^+ remains as a challenging task because of its high solubility in water. In the current work, cellulose-based polymer hydrogels are synthesized to increase the adsorption capacity of the extracted cellulose. Results show that cellulose, PVA, PAA and chitosan show similar adsorption capacities which supports our hypothesis. Though the maximum adsorption achieved is not very significant yet, these studies depict that modified bio-waste could serve as viable second, in place of synthetic polymers for desalination process.

Keywords Adsorption · Desalination · Cellulose · Polymer · Hydrogel

1 Introduction

1.1 Background

There is a dire need for an economical alternative to convert sea water into fresh water owing to prevalent methods demanding high power and large-scale infrastructures [1]. In this regard, adsorption desalination is considered as an attractive choice due to its renewable and cost-effective nature [2–4]. Moreover, exploration of industrial wastes as potential adsorbents for purposes of water purification have been of interest in recent times [5, 6]. In this study, banana stem waste (plantain pith) has been exploited for its capacity to efficiently remove sodium chloride from aqueous solution (sea water). Banana is widely grown across the world, with an average production of 120–150 million tons per year. Yet, only 12% of the plant's weight is the fruit,

O. J. Xin (✉)
NUS High School of Math and Science, Singapore, Singapore
e-mail: h1310128@nushigh.edu.sg

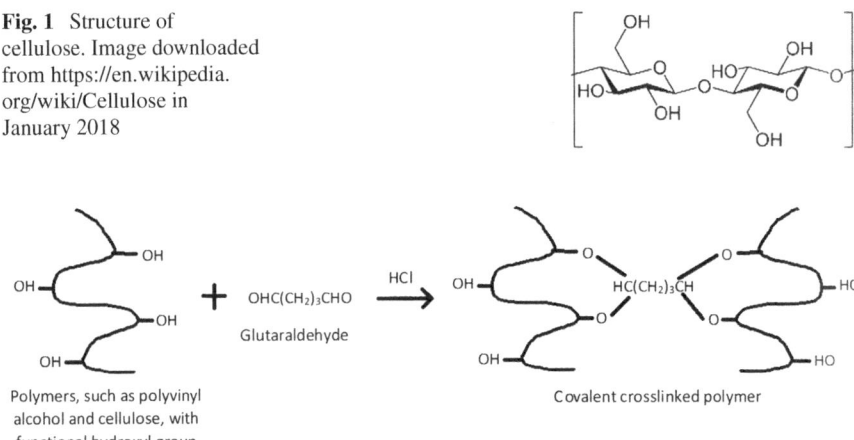

Fig. 1 Structure of cellulose. Image downloaded from https://en.wikipedia.org/wiki/Cellulose in January 2018

Polymers, such as polyvinyl alcohol and cellulose, with functional hydroxyl group

Covalent crosslinked polymer

Fig. 2 (Self-drawn) Mechanism of chemical crosslinking of polymers using glutaraldehyde

and the rest is inedible. Banana stem comprises of cellulose (Fig. 1), lignin and a small amount of ash [7] and has many hydroxyl functional groups on its surface thus making it a potential cheap renewable adsorbent for Na^+ from water [8, 9]. However, removal of Na^+ remains as a challenging task because of its high solubility in water [10]. In this study, banana stem waste was bleached first to extract the cellulose nanofibers and removing hemicelluloses, lignin and other functional groups, such as carbonyls and carboxylic acids that make it an interesting adsorbent [11].

Polymers have been experimented with for various desalination methods and good salt rejection was observed [12–14]. In particular, polymer hydrogels have been extensively developed into draw agents for forward osmosis desalination [15, 16] and as adsorbents for metal pollutants [17, 18]. The highly electronegative atoms, such as nitrogen and oxygen, present in polymers can bind to the metal ions via electron pair sharing to form a complex, therefore removing the metal ions from the solution [18]. In the current work, cellulose-based polymer hydrogels are synthesized to increase the adsorption capacity of the extracted cellulose. The hydrogels are synthesized by chemically crosslinking the polymers using glutaraldehyde [19] (Fig. 2). The hydrogen bonds within cellulose and the polymers were broken up to form acetal bonds with glutaraldehyde [19].

1.2 Objective

The aim of this project is to synthesize cellulose-based polymer hydrogels, via chemical crosslinking using glutaraldehyde, to increase the adsorption capacity of the extracted cellulose.

1.3 Scope of Work

The adsorbents were fully characterized by Scanning Electron Microscopy (SEM) and Fourier-Transform Infrared Spectroscopy (FTIR) while Na^+ concentration of samples was determined using Inductively Coupled Plasma-Optical Emission Spectrometer (ICP-OES), model Perkin Elmer Optima 5300DV. Results of batch studies depict that modified bio-waste could serve as viable second for desalination process.

2 Methodology

2.1 Bleaching Banana Stem Waste

Banana stem waste was first mashed into smaller pieces to allow larger surface area for the bleaching agent to act upon. NaOCl was used as the bleaching agent, along with HCl in water. When mixed with a material which can be oxidized, NaOCl gives out nascent oxygen which then oxidizes any impurities. After about 18 h, the bleached banana stem (BBS) would appear white. Then, BBS was filtered out and rinsed with ultra-pure (UP) water until Na^+ concentration of the solution is <0.10 ppm, determined from ICP-OES. The dried samples were then characterized using SEM and FTIR (Fig. 3).

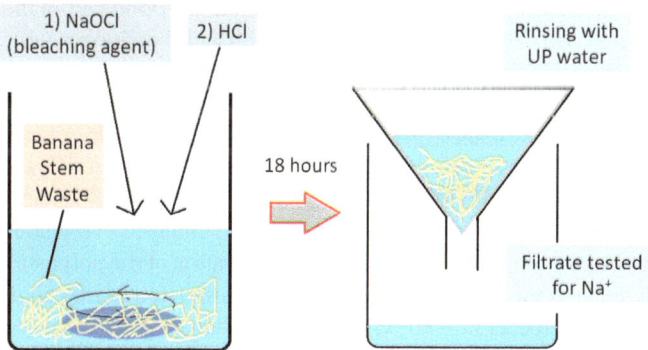

Fig. 3 (Self-drawn) Diagram depicting the process of bleaching banana stem waste

Fig. 4 (Self-drawn)
Diagram depicting the
process of dialysis of
polymers

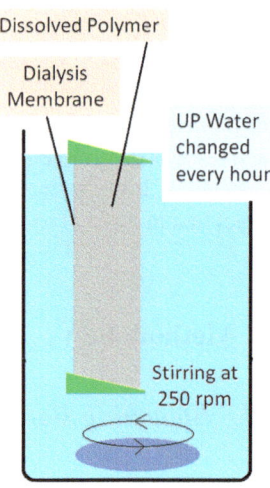

2.2 Dialysis of Polymers

This procedure was done to remove as much Na⁺ as possible from the impure solid polymers before use in the synthesis of the hydrogels. This decreases the number of washes needed to reduce Na^+ concentration in the gels to <0.10 ppm (Fig. 4).

First, 10 g of solid polymer was dissolved in 100 ml of UP water (heated at 200 °C and stirred at 500 rpm). The necessary length of the dialysis membrane (around the height of the container used for the dialysis) was cut out. One end of the membrane was then folded two to three times and sealed with a clip to make a tube, which was filled with the polymer solution until about two-thirds full. The tube was not fully filled as water would enter the tube during the dialysis process. After all air pockets in the tube were removed, the other end was also folded two to three times and sealed with a clip. The tube was then submerged in a container filled with UP water. One end of the tube was tied to the retort stand to ensure that it remains suspended in the water. The water was continuously stirred at 250 rpm, and changed every hour to increase the efficiency of the dialysis. Once the Na^+ concentration of the polymer solution was <0.10 ppm, the polymer solutions in the tubes were transferred into 50 ml centrifuge tubes and left in a freezer at −80 °C to freeze-dry before lyophilization.

2.3 Synthesis of Hydrogels

Different combination of polymers, including bleached banana stem (BBS), polyvinyl alcohol (PVA), polyacrylic acid (PAA) and chitosan, were experimented with to form a hydrogel. The protocol is as follows (Fig. 5):

Fig. 5 (Self-drawn) Diagram depicting the process of hydrogel synthesis

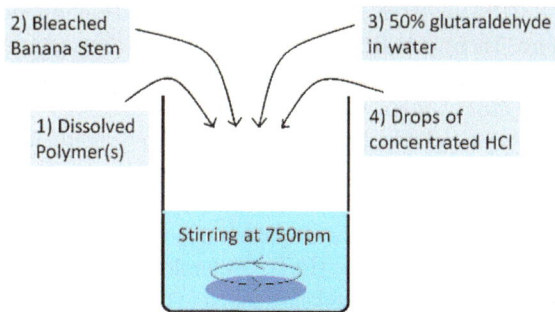

1. Lyophilized polymer was dissolved in UP water (heated when necessary). Note that chitosan powder was used as it is insoluble in water and was dissolved in an acidic solution instead of UP water.
2. All polymer solutions that were required for the particular combination were mixed and allowed to stir until the solution was homogenous.
3. BBS, if needed, was then added into the solution and allowed to stir until the mixture was homogenous.
4. A few drops of 50% glutaraldehyde in water was added dropwise into the mixture for cross-linking.
5. A few drops of concentrated hydrochloric acid were then added as a catalyst.
6. The mixture was left to evaporate while being stirred at 750 rpm until a gel formed.
7. UP water was added to the gel and replaced every 3 h until Na^+ concentration in the gel is <0.10 ppm, determined from ICP-OES. This is to ensure that as much Na^+ from the BBS is removed as possible before each batch adsorption studies (Table 1).

Table 1 Amount of polymers used to synthesize the hydrogels

Hydrogel in varying concentration	Amount of PVA used (grams)	Amount of PAA added (grams)	Amount of chitosan added (grams)	Amount of BBS used (grams)
PVA	5	0	0	0
BBS-PVA 1:10		0	0	0.5
BBS-PVA 1:5	5	0	0	1
BBS-PVA 1:2	0.5	0	0	0.25
BBS-PVA 1:1	5	0	0	5
BBS-PVA 2:1	2	0	0	4
BBS-PVA 5:1	1	0	0	5
BBS-PVA 10:1	5	0	0	50
PVA-PAA 1:1	5	5	0	0
Chitosan	0	0	5	0

BBS only hydrogel could not be formed but instead remains as a thick white mixture. Therefore, BBS in its fibrous form was used as a control instead of a hydrogel. Other polymers were tried as well and were concluded to be unsuitable as adsorbents. The observations were as follows:

Polyvinyl Alcohol (PA): A pale yellow and relatively soft hydrogel, compared to PVA hydrogel, was formed. When added into water in a 5 ml centrifuge tube and left to shake at 150mot for a few minutes, the hydrogel began breaking off into smaller pieces. Therefore, it was not suitable as an adsorbent.

Polyethyleneimine (PEI): Glutaraldehyde reacts too quickly with PEI, resulting in patches of red hydrogel suspended in the solution while glutaraldehyde was added in dropwise. Glutaraldehyde was mixed with a small amount of water to dilute it before attempting the procedure again. While no red patches were formed immediately upon adding the diluted glutaraldehyde, a red and paste-like mixture was formed. A hydrogel could not be obtained therefore PEI was not suitable as an adsorbent.

Polyethylene Glycol (PEG): The PEG solution remained as a liquid even after adding a large amount of glutaraldehyde. Since a hydrogel could not be obtained, PEG was not suitable as an adsorbent.

Starch: The gel formed was relatively soft and, breaks off easily into smaller when added into water in a 5 ml centrifuge tube and left to shake at 150mot for a few minutes. Therefore, it is not suitable as an adsorbent.

2.4   Batch Adsorption Studies

The adsorption of NaCl onto the synthesized cellulose-based polymer hydrogels was studied by a series of batch adsorption experiments. The efficiency of the synthetic adsorbents for the removal of sodium chloride ions from aqueous solutions has been determined at the different adsorbent dosage (100, 300, 500 mg) in water (10 ml), fixed Na^+ concentration (10 and 2500 ppm), agitation speed (150 rpm) and temperature (25 °C). The solutions after adsorption for a duration of 24 h were filtered and the Na^+ concentrations quantified using ICP-OES.

3   Results and Discussion

3.1   Characterization

Raw and bleached banana stem were characterized via various techniques such as SEM and FTIR. After bleaching treatment, the banana stem is significantly smoother with less residues on it as shown in the SEM photographs (Fig. 6). The fibrous structure obtained is representative of the vast cellulosic network present in banana

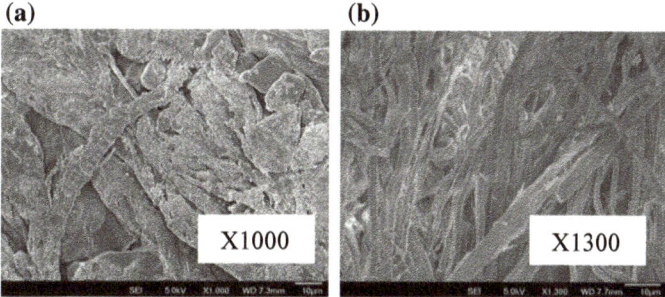

Fig. 6 **a** SEM of raw banana stem **b** SEM of bleached banana stem

stem. The SEM image has shown that the bleaching agent has removed the non-cellulosic matter, which are dissolved and washed out with water during the treatment.

Difference between the FTIR spectra of raw and bleached banana stem can be observed below (Fig. 7). Different absorption bands characteristic for cellulose were observed in raw banana stem, which became less pronounced after bleaching. The broad band from 3000 to 3700 cm^{-1} is characteristic of –OH stretching. The sharp peak at around 2900 cm^{-1} is due to C–H symmetrical stretching and that at around 2800 cm^{-1} is due to aliphatic C–H stretching. The small peak at around 1750 cm^{-1} is assigned to C=O stretching vibration which disappeared in the BBS sample. The sharp peak at about 1600 cm^{-1} is representative of –OH bending of absorbed water. The peak at about 1400 cm^{-1} which disappeared in the BBS sample is representative of –HCH and –OCH in-plane bending vibration. The broad absorption band at 800–1200 cm^{-1} may be due to C–C vibration within the cellulose backbone [20, 21]. As cellulose does not contain any carbonyl groups, it can be concluded that most of the non-cellulosic organic matter has been washed away by bleach.

Photographs of a piece of the hydrogels and cellulose were taken (Fig. 8). It was observed that the hydrogel becomes whiter than transparent, and also more solid, as the BBS:PVA ratio increases.

Fig. 7 FTIR of raw banana stem and BBS

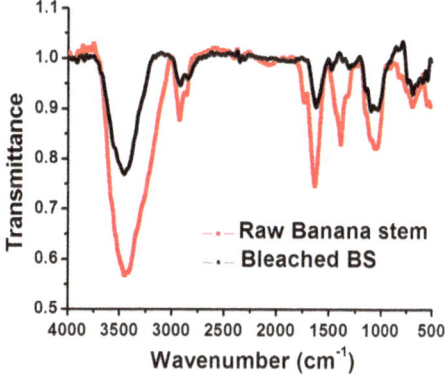

Fig. 8 a PVA hydrogel
b BBS-PVA 1:10 hydrogel
c BBS-PVA 1:5 hydrogel
d BBS-PVA 1:2 hydrogel
e BBS-PVA 1:1 hydrogel
f BBS-PVA 2:1 hydrogel
g BBS-PVA 5:1 hydrogel
h BBS-PVA 10:1 hydrogel
i BBS

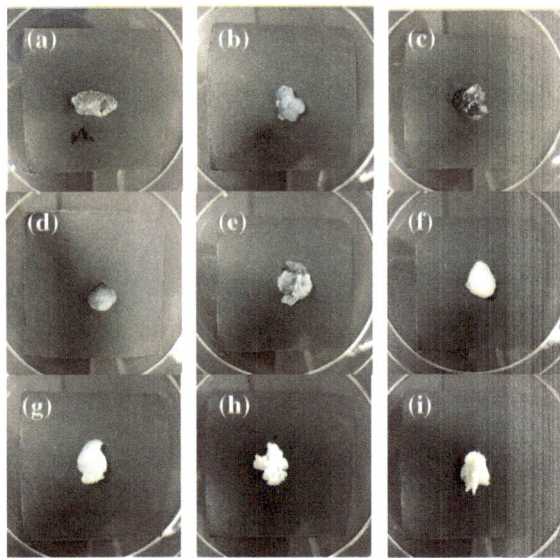

3.2 Batch Adsorption Studies

Batch adsorption studies were conducted with initial Na^+ concentration of 10 ppm (or 0.001%). The average adsorption capacity of BBS was 0.1027 mg/g which is not shown in the graph because the measured adsorption capacity of BBS fibers cannot be used for comparison with the hydrogels directly. This is because the hydrogels comprise water which makes the concentration of adsorbent different. To overcome this problem, we had lyophilized the synthesized hydrogels so that we could use its dry weight for a more accurate experiment, but the lyophilized hydrogel had an increased Na^+ concentration which made them unsuitable for use. Another solution we attempted was to synthesize a BBS only hydrogel, which we failed to obtain as cellulose is insoluble in water. We are still looking into other experimental methods to quantify the adsorption capacity of BBS, on top of the method of using varying ratios of BBS and another polymer to estimate its adsorption capacity which we will discuss next.

The synthesized hydrogels shown in Fig. 9 do not show a general trend in their adsorption capacity with the increasing BBS:PVA ratio. Note that for every cellulose monomer, which has a molecular weight of 324.3 g/mol, there are five oxygen atoms while for every PVA monomer, which has a molecular weight of 44.05 g/mol, there is one oxygen atom. This means that for the same mass of BBS and PVA (we used synthetic PVA that was about 90% hydrolyzed), there would be around the same number oxygen atoms. Assuming that all the oxygen molecules in both BBS and PVA have equal attraction force for Na^+ ions, we can approximate the adsorption capacity to be that of PVA. This also explains the results we obtained in Fig. 9 (Figs. 10 and 11).

Fig. 9 Comparison of adsorption capacities of PVA and BBS-PVA hydrogels in a 10 ppm NaCl solution

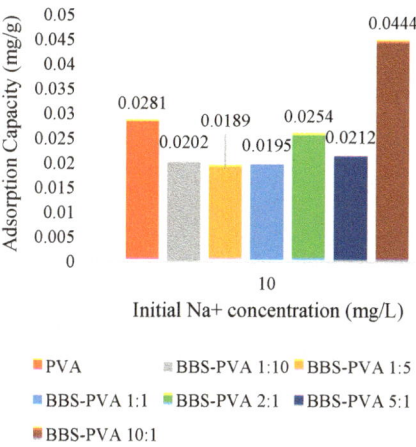

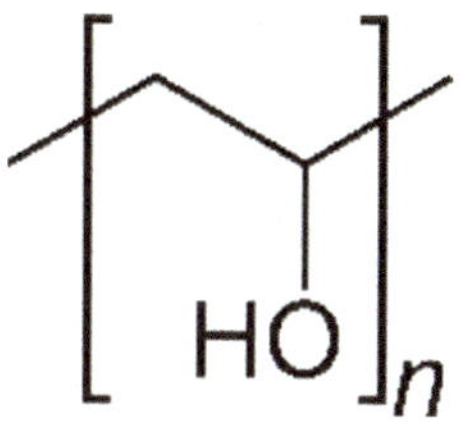

Fig. 10 Structure of Polyvinyl Alcohol (PVA). Image downloaded from https://en.wikipedia.org/wiki/Polyvinyl_alcohol in May 2018

Fig. 11 Comparison of adsorption capacities of PVA, PPA and chitosan hydrogels in a 2500 ppm NaCl solution

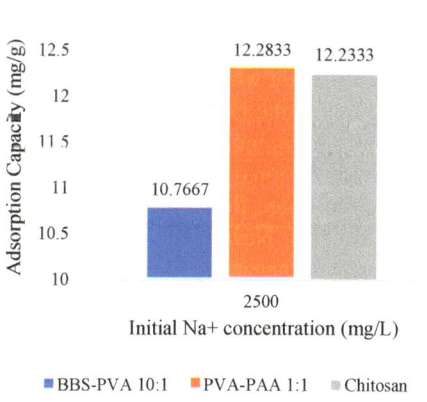

Batch adsorption studies were conducted with initial Na^+ concentration of 2500 ppm (or 0.25%). The increase in Na^+ concentration used was done to observe more significant differences between the adsorption capacities of the hydrogels and to minimize the possibility that any differences is due to chance.

PVA-PAA 1:1 hydrogel was synthesized instead of a PAA only hydrogel because PAA alone would remain as a viscous liquid rather than a gel. The addition of the highly viscous PVA solution could help solidify the PAA solution into a gel (Fig. 12).

Fig. 12 Comparison of estimated adsorption capacities of hydrogels in a 10 ppm NaCl solution

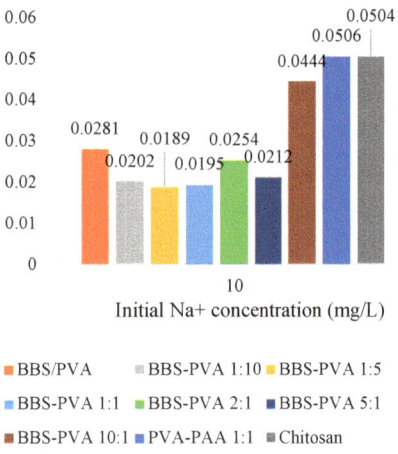

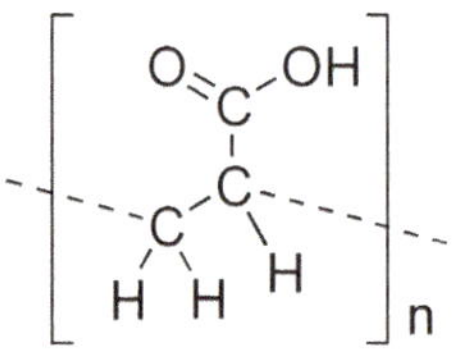

- BBS/PVA
- BBS-PVA 1:10
- BBS-PVA 1:5
- BBS-PVA 1:1
- BBS-PVA 2:1
- BBS-PVA 5:1
- BBS-PVA 10:1
- PVA-PAA 1:1
- Chitosan

Fig. 13 Structure of polyacrylic acid (PAA). Image downloaded from https://en.wikipedia.org/wiki/Polyacrylic_acid in May 2018

By scaling, we can make a rough estimation of the differences between all the gels. The data shows that PVA-PAA 1:1 hydrogel has an average adsorption capacity of 12.2833 mg/g which is only slightly greater than that of all PVA hydrogels. This may be due to the fact that PAA has twice the number of oxygen atoms of PVA but its molecular weight (72.06 g/mol) is also about twice as much, therefore having similar adsorption capacities (Fig. 13).

We also noted that chitosan has a similar adsorption capacity to all the other tried polymers. Nevertheless, the need to use acid to dissolve the chitosan powder may result in negative side effects such as degradation of BBS when we attempt to synthesize BBS-chitosan hydrogels. Hence, chitosan is highly unsuitable as an adsorbent.

The mechanism of adsorption is based on the interaction of electronegative atoms on the cellulose fiber and salt in water, as shown below cartoon (Fig. 14) and appears to be consistent with the data we have obtained.

4 Conclusion

Desalination via adsorption is a promising field of research but remains a challenging task due to the high solubility of Na^+ in water. In this study, we experimented using cellulose-based polymer hydrogels as renewable adsorbents and concluded that BBS,

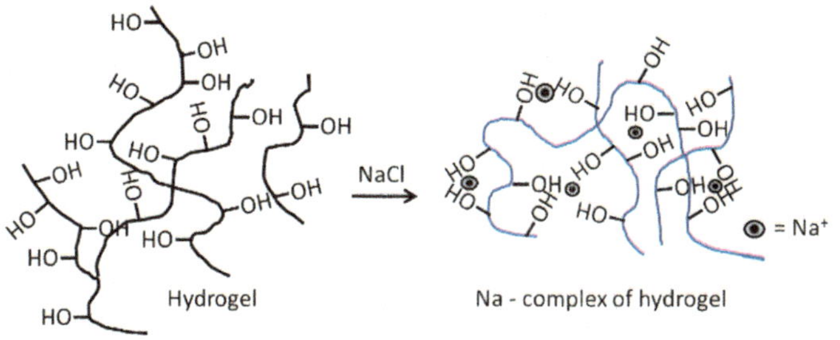

Fig. 14 (Self-drawn) Cartoonistic representation of the sodium complexation of hydrogels

PVA, PAA and chitosan show the equal potential as adsorbents of salt due to the similar number of electronegative atoms present in the polymer hydrogels which is in line with our hypothesized mechanism of sodium complexation in hydrogels. From our experiments, we can also conclude that PVA is the most suitable polymer to form a hydrogel with BBS as it is the most stable and can achieve similar adsorption capacities as the other tried synthetic polymers.

5 Future Work

We can look into increasing the concentration of the polymers or cellulose extracted from other bio-wastes for optimization. We can also be more selective in choosing the polymers to work with by looking at the structures now that we are more certain of the properties of the hydrogels. We can also try using other methods to synthesize the hydrogels. The hydrogels will be used for batch adsorption studies before proceeding with kinetic studies to decide whether cellulose in hydrogel form has any advantage over cellulose in solid form.

Acknowledgements I wish to express my sincere gratitude to Professor Suresh for providing me with the opportunity to do my first research project at Department of Chemistry, NUS. I also sincerely thank Dr. Teh Yun Ling for her guidance and encouragement throughout the duration of the project. Lastly, I wish to express my gratitude to all the students and staff working in the same laboratory who helped me in carrying out this project work despite their busy schedules, motivated me to continue working after even after continuous failures and treated me as part of them.

References

1. Slesarenko, V. V. (2001). Heat pumps as a source of heat energy for desalination of seawater. *Desalination, 139*(1), 405–410.

2. Liu, F., Chung, S., Oh, G., & Seo, T. S. (2012). Three-dimensional graphene oxide nanostructure for fast and efficient water-soluble dye removal. *ACS Applied Materials & Interfaces, 4,* 922–927.
3. Dou, X. Q., Li, P., Zhang, D., & Feng, C. L. (2012). C2-symmetric benzene-based hydrogels with unique layered structures for controllable organic dye adsorption. *Soft Matter, 8,* 3231–3238.
4. Zhu, X. D., Liu, Y. C., Zhou, C., Zhang, S. C., & Chen, J. M. (2014). Novel and high-performance magnetic carbon composite prepared from waste hydrochar for dye removal. *ACS Sustainable Chemistry & Engineering, 2,* 969–977.
5. Liu, L., et al. (2015). Adsorption removal of dyes from single and binary solutions using a cellulose-based bioadsorbent. *ACS Sustainable Chemistry & Engineering, 3,* 432–442.
6. Mallampati, R., & Valiyaveettil, S. (2013). Apple Peels-A Versatile Biomass for Water Purification? *ACS Applied Materials & Interfaces, 5,* 4443–4449.
7. Reddy, N., & Yang, Y. (2015). Fibers from Banana Pseudo-Stems. *Innovative Biofibers from Renewable Resources* (pp. 25–27). Berlin: Springer.
8. Wang, L.-Y., & Wang, M.-J. (2016). Removal of heavy metal ions by poly (vinyl alcohol) and carboxymethyl cellulose composite hydrogels prepared by a Freeze-Thaw Method. *ACS Sustainable Chemistry & Engineering, 4,* 2830–2837.
9. Mahanta, N., Leong, W., & Valiyaveettil, S. (2012). Isolation and characterization of cellulose-based nanofibers for nanoparticle extraction from an aqueous environment. *Journal of Materials Chemistry, 22*(5), 1985–1993.
10. Harvie, C. E., Moller, N., & Weare, J. H. (1984). The prediction of mineral solubilities in natural waters: The Na-K-Mg-Ca-Cl-SCVOH-HCCb-Cos-CCb-HaO system to high ionic strengths at 25 °C. *Geochimica et Cosmochimica Acta, 48,* 723–751.
11. Xiao, S., Gao, R., Gao, L., & Li, J. (2016). Poly (vinyl alcohol) films reinforced with nanofibrillated cellulose (NFC) isolated from corn husk by high intensity ultrasonication. *Carbohydrate Polymers, 136,* 1027–1034.
12. Kamachi, M., Kurihara, M., & Stille, J. K. (1972). Synthesis of block polymers for desalination membranes. Preparation of block copolymers of 2-vinylpyridine and methacrylic acid or acrylic acid. *Macromolecules., 5*(2), 161–167.
13. Liu, Y., Wang, W., & Wang, A. (2010). Adsorption of lead ions from aqueous solution by using carboxymethyl cellulose-g-poly (acrylic acid)/attapulgite hydrogel composites. *Desalination, 259*(1–3), 258–264.
14. Ahmed, E. (2015). Hydrogel: Preparation, characterization, and applications: A review. *Journal of Advanced Research, 6*(2), 105–121.
15. Li, D., et al. (2011). Stimuli-responsive polymer hydrogels as a new class of draw agent for forward osmosis desalination. *Chemical Communications., 47*(6), 1710–1712.
16. Li, D., et al. (2013). Forward osmosis desalination using polymer hydrogels as a draw agent: Influence of draw agent, feed solution and membrane on process performance. *Water research, 47*(1), 209–215.
17. Zheng, Y., Hua, S., & Wang, A. (2010). Adsorption behavior of Cu 2 + from aqueous solutions onto starch-g-poly (acrylic acid)/sodium humate hydrogels. *Desalination, 263*(1), 170–175.
18. Jin, L., & Bai, R. (2002). Mechanisms of lead adsorption on chitosan/PVA hydrogel beads. *Langmuir, 18*(25), 9765–9770.
19. Kumar, G. T., et al. (2012). Modified chitosan hydrogels as drug delivery and tissue engineering systems: Present status and applications. *Acta Pharmaceutica Sinica B, 2*(5), 439–449.
20. Fan, M., Dai, D., & Huang, B. (2012). *Fourier transform infrared spectroscopy for natural fibres.* InTech: Fourier Transform-Materials Analysis.
21. Hao, L., Wang, P., & Valiyaveettil, S. (2017). Successive extraction of As (V), Cu (II) and P (V) ions from water using spent coffee powder as renewable bioadsorbents. *Scientific Reports, 7,* 42881.

Dye-Sensitised Solar Cells Using Flora Extracts

Hui Yu Cherie Lee, Xin Yi Ariel Ho, Di er Kaylee Therese Poh and William Phua

Abstract Dye-sensitized solar cells (DSSC) are simple and low-cost solar cells that can efficiently generate electricity from visible light, serving as environmentally friendly sources of energy. However, recent studies have shown that DSSCs have not been effective in generating a significant amount of voltage for usage as compared to the traditional silicon solar cells. Hence, this research aims to improve the efficacy of DSSCs for future uses with the addition of 1% agarose gel with 4% $NaCl$ and natural sensitisers from flower or leaf extracts on TiO_2-based cells. With a hypothesis that the spectrum of light absorbed from passing through the dye would affect the amount of voltage generated by solar cells, natural dyes extracted from various flowers and leaves were then tested for the range of light absorbed by a spectrophotometer. A thin layer of agarose gel that reduces internal resistance within the solar cells was also used to obtain a higher yield of voltage as compared to previous researches. With successful results of a highest 0.58 V from a single solar cell with the dye extract from the Wall Daisy flower (Erigeron Karvinskianus), the sustainability of these DSSCs in terms of efficiency were also considered to achieve a better dye-sensitised solar cell for future use.

Keywords Low-cost · Environmentally friendly · Flora extracts · Agarose gel · Dye-sensitised solar cells

1 Introduction

In an increasingly polluted environment, it is necessary to find alternative sources of energy that are environmentally friendly and will not become a cause of pollution. Such a source is dye-sensitized solar cells (DSSC), which are simple and low-cost solar cells that can efficiently generate electricity from visible light, serving as environmentally friendly sources of energy (Fig. 1).

Dye sensitized solar cells (DSSC) mainly consist of three components: dye sensitized mesoporous TiO_2 nano crystallites, a counter electrode and an electrolyte

H. Y. C. Lee · X. Y. A. Ho · D. K. T. Poh · W. Phua (✉)
National Junior College, Singapore 288913, Singapore
e-mail: williamphua3@gmail.com

© Springer Nature Singapore Pte Ltd. 2019
H. Guo et al. (eds.), *IRC-SET 2018*,
https://doi.org/10.1007/978-981-32-9828-6_18

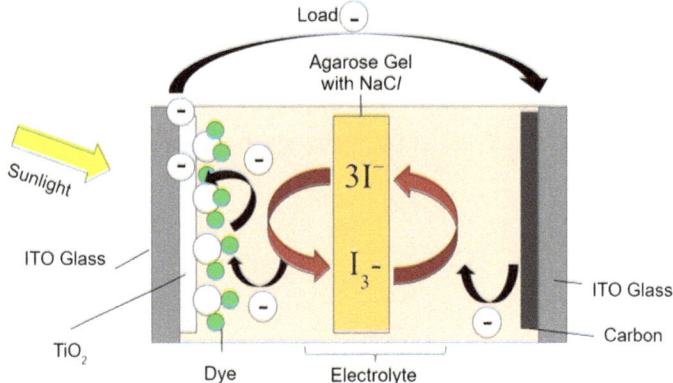

Fig. 1 Working principle of dye-sensitised solar cells

containing an iodide (I^-) or tri-iodide (I_3^-) redox couple. With TiO_2 as a semicon-ductor, this layer is formed between the dye-sensitised anode and the electrolyte. When light is shone, dye molecules get excited and inject electrons into the TiO_2 conduction band, oxidising the dye molecules. The injected electrons diffuse in the TiO_2 conduction band and reach the outer circuit. Oxidized dyes are rapidly regenerated by I^- ions, resulting in the formation of I_3^- ions. The I_3^- ions are then reduced by the electrons reaching the counter electrode through the external circuit. To produce electricity, the photo-injected electrons do not recombine with acceptors, but are transferred to an external circuit instead [1, 2].

As DSSC is a great alternative to the conventional silicon solar cells, many recent studies have made use of DSSC to generate electricity. However, these studies have shown that DSSC have not been effective in generating a significant voltage potential, with a peak conversion efficiency of 11.1%. [3]. This percentage is significantly lower than the traditional silicon solar cells, which have a peak conversion efficiency of 26.6% [4].

Hence, this research aims to improve the efficacy of DSSCs for future uses with the addition of 1% agarose gel with 4% NaCl and natural sensitisers from leaf and flower extracts on TiO_2-based cells, so that DSSCs can be widely used around the world. The agarose gel is used to separate the iodine from the layer of TiO_2 to obtain better results. Where liquid NaCl is solidified within the agarose gel, the thin layer of agar can reduce energy loss due to electron transfer between the layer of TiO_2 and iodine. Agarose gel is also easily accessible and environmentally friendly as it is an extract from seaweed and is biodegradable. The natural dyes used in this research study are extracted from Red Flame Ivy (Hemigraphis Alternata), Moses in the Cradle (Tradescantia Spathacea), Wall Daisy (Erigeron Karvinskianus) and Red Ti (Cordyline Terminalis). It is hypothesised that the spectrum of light absorbed after it passes through the flora extract would affect the voltage generated by dye-sensitized solar cells.

2   Methods

2.1   Dye Extraction

Natural dyes were extracted from Red Flame Ivy, Wall Daisy, Red Ti and Moses in a Cradle by cutting them up into small pieces and soaking them in 30 ml acetone overnight. The mixture was then filtered to remove the leaves or petals and obtain the liquid dye. The range of light that the liquid dye for each type of leaf or flower absorbs was also tested for. This was done by taking a small sample of the liquid dye for each type of leaf or flower and placing it in a spectrophotometer. The graphs generated from the spectrophotometer were then analysed to find out the range of light absorbed by each liquid dye.

2.2   Preparation of TiO_2 Paste

The TiO_2 paste was made by adding 6 g of TiO_2 to 9 cm^3 of acetic acid (4pH) to 1 cm^3 of water, before letting it sit for 15 min. A thin layer of TiO_2 paste was then applied on the conductive side of the ITO glass plate and left to dry. Once dried, the plate was placed in the furnace for a total of an hour, 30 min to reach 450 °C, 15 min at 450 °C and the last 15 min to cool down. When the glass plate was removed from the furnace, it was soaked in the natural dye overnight.

2.3   Preparation of Carbon Coated Glass Plates

The carbon coated plate was made by coating the conductive side of another ITO glass plate with graphite pencil lead. It was then held over a flame from a Bunsen burner to form a thin layer of soot.

2.4   Preparation of 1% Agarose Gel with 4% NaCl

1.0 g of agarose gel powder and 2 cm^3 of 4% NaCl was added to 100 ml of 1X TAE buffer. The mixture was then placed in a microwave, and heated for one minute intervals. Once all the agarose gel powder has completely dissolved and the mixture is clear, pour a thin layer over agar plate and allow to solidify (Fig. 2).

2.5   Assembly of Dye-Sensitised Solar Cells

The TiO_2 coated ITO glass plate was placed at the bottom while the carbon coated ITO glass plate was placed above it, with both conductive sides facing inwards. Iodine was then dripped into the solar cell through the seams at the sides. Binder clips were then used to hold the entire structure together. Two protruding sides on

Dye-Sensitised Solar Cell

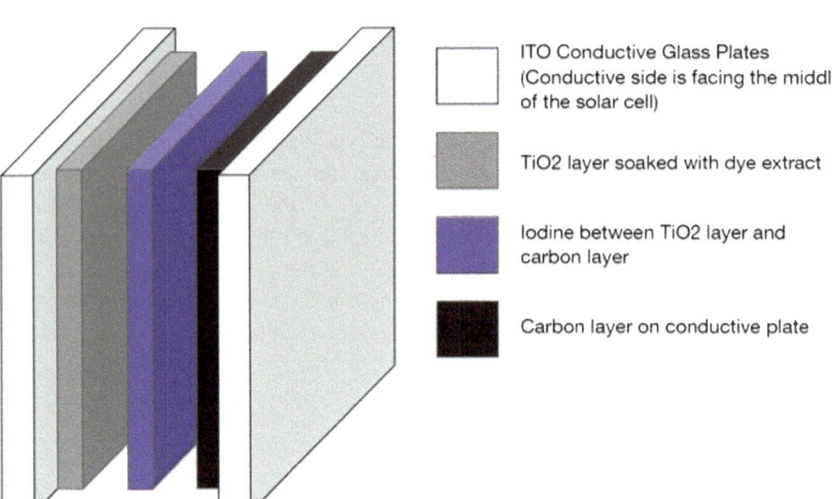

ITO Conductive Glass Plates
(Conductive side is facing the middle
of the solar cell)

TiO2 layer soaked with dye extract

Iodine between TiO2 layer and
carbon layer

Carbon layer on conductive plate

Fig. 2 Structure of Dye-Sensitised solar cells

the DSSCs would be made in order for connection to the digital multimeter used for testing of voltage.

A thin layer of agarose gel was placed above the TiO_2 side of the conductive ITO glass plate. Iodine was then dripped onto the agarose gel. Lastly, the carbon plate was placed above the iodine layer and binder clips were used to hold the entire structure together (Fig. 3).

2.6 Test for Voltage

After complete assembly of both types of DSSCs, they are immediately tested for the voltage yielded under direct sunlight. A digital multimeter would be connected to the two protruding conductive sides of the DSSCs, with the negative lead to the TiO_2 glass and the positive lead to the other glass. With one side facing the direction of the sunlight, the voltage (V) shown is recorded.

3 Results and Discussion

3.1 Analysis of Experimental Results

With reference to both Figs. 4 and 5, the natural dye extracted from the flower, Wall Daisy, could produce the highest voltage amongst the four dyes used. Comparing

Dye-Sensitised Solar Cell with Agarose Gel

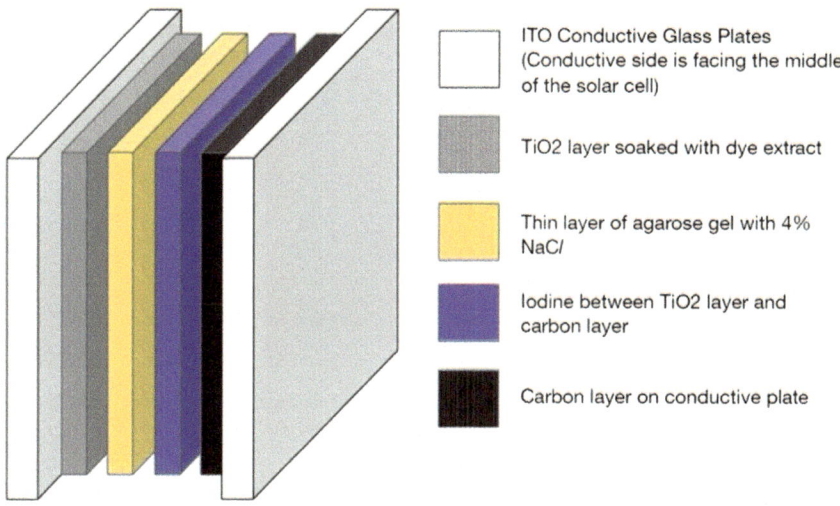

ITO Conductive Glass Plates
(Conductive side is facing the middle
of the solar cell)

TiO2 layer soaked with dye extract

Thin layer of agarose gel with 4%
NaC*l*

Iodine between TiO2 layer and
carbon layer

Carbon layer on conductive plate

Fig. 3 Structure of Dye-Sensitised solar cells with agarose gel

Fig. 4 Measurements of
voltage produced by solar
cells without agarose gel

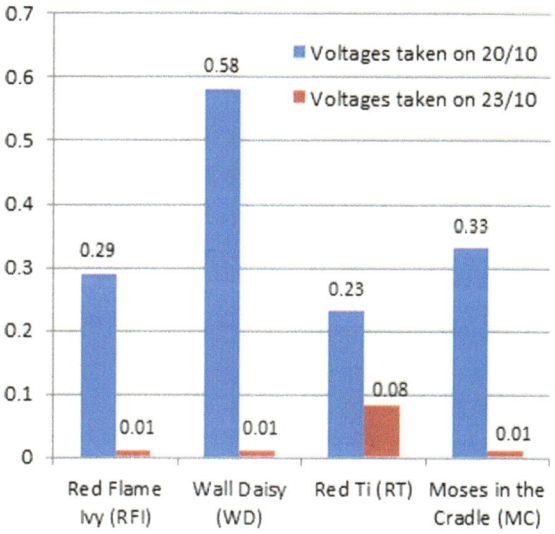

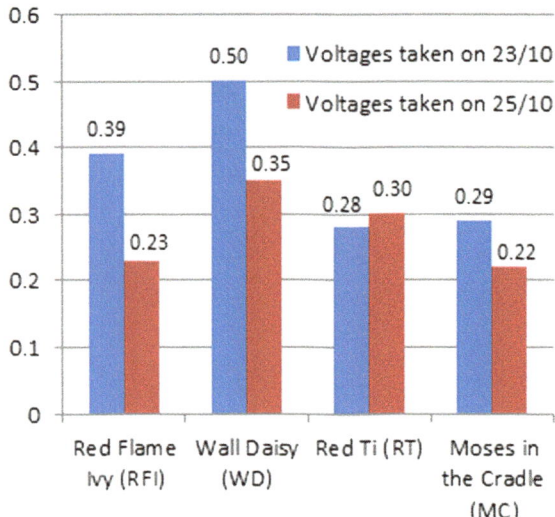

Fig. 5 Measurements of voltage produced by solar cells with agarose gel

only the dyes extracted from the leaves, Red Flame Ivy and Moses in the Cradle had shown a higher yield than Red Ti, ranging from 0.29 to 0.39 V.

It was also observed that the solar cells made degraded over time. Decreasing in voltage produced when tested after about two days from when the solar cells were made, these cells were unable to sustain their original states of efficacy. Regardless of the presence of the agarose gel, all results showed this trend. However, it could be seen in the two graphs that the addition of the agarose gel could sustain the solar cells much better than those in the absence of the gel. Comparing the decrease in voltage for both types of solar cells, it was obvious that solar cells with agarose gel did not reduce in the production of voltage as much as the basic structure.

It could be deduced that the addition of the 1% agarose gel with 4% NaCl aided in the electron transfer between the redox mediator and the layer of TiO_2 with attached dye particles within the improved solar cells. Acting as an additional electrolyte, it was hypothesised that the aqueous Na^+ and Cl^- ions would be able to increase the voltage produced by solar cells. However, according to results shown in the Figs. 4 and 5, the effectiveness of 1% agarose gel with 4% NaCl has not been proven as there is no significant increase in the voltage generated by improved solar cells. It was yet observed that the addition of agarose gel was able to sustain the voltage produced by the solar cells with only a slight decrease of about 0.15 V in Fig. 5 as compared to a complete degradation of the cells in Fig. 4. This trend could be explained where the agarose gel helped to prevent the excessive loss of liquid electrolytes and iodine solution, giving the cell a higher durability as shown.

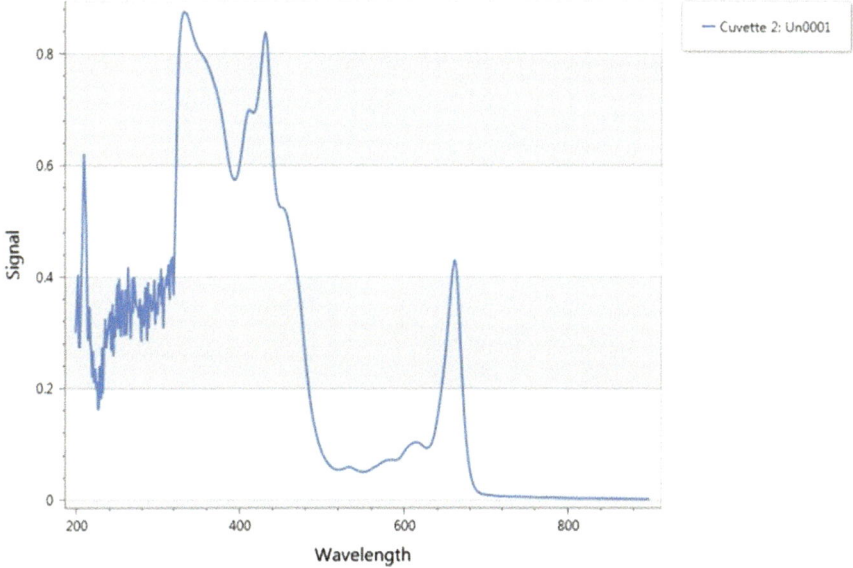

Fig. 6 Spectrum graph of Moses in the Cradle dye extract

3.2 *Spectrum Graph Analysis*

By comparing the spectrum graphs of the various natural dyes, it was shown that the light range from ultraviolet to visible light rays was absorbed by all the dyes. It could also be seen that the range of light absorbed by the dye affected the voltage produced. Since the range of the light spectrum shows the number of photons absorbed by each dyes, it is deduced that the larger the range of peaks across the light spectrum, the more the number of photons absorbed by the dye. Comparing the higher peaks on the graphs in Figs. 6, 7, 8 and 9 within the range of 300–400 nm, dye extracts from Moses in a Cradle and Red Flame Ivy leaves had shown a higher yield of voltage from Figs. 4 and 5. Where the range of light absorbed by both dye extracts are similar, it can be seen that the similar high voltages produced was caused by the current conversion of excited dye molecules due to the similar number of photons.

4 Conclusion

In this study, assembling DSSC and generating voltage using DSSC was successful. It was shown from results that the addition of a thin layer of agarose gel could possibly increase the voltage produced by most solar cells since two out of the four solar cells produced a considerable amount of voltage as compared to basic cells. This could be an addition to future dye-sensitised solar cells and increase the 11.1% efficiency after

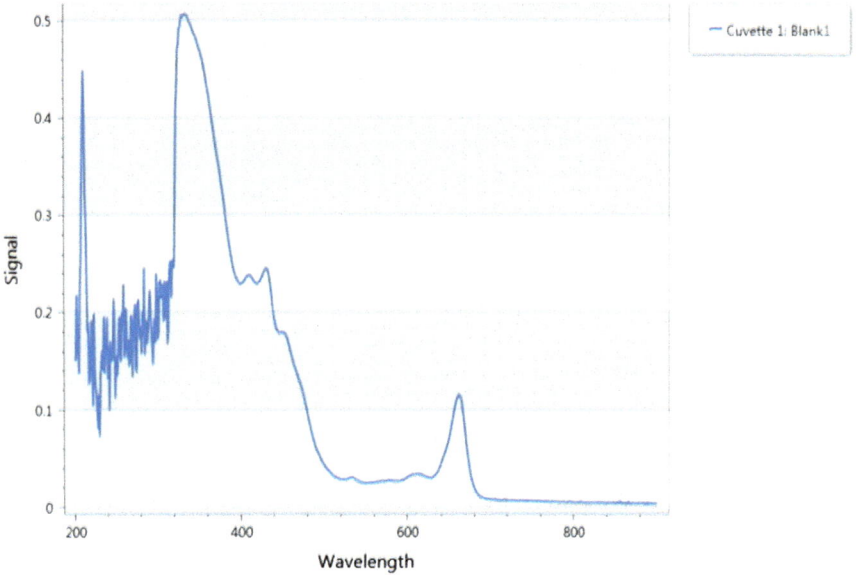

Fig. 7 Spectrum graph of Red Ti dye extract

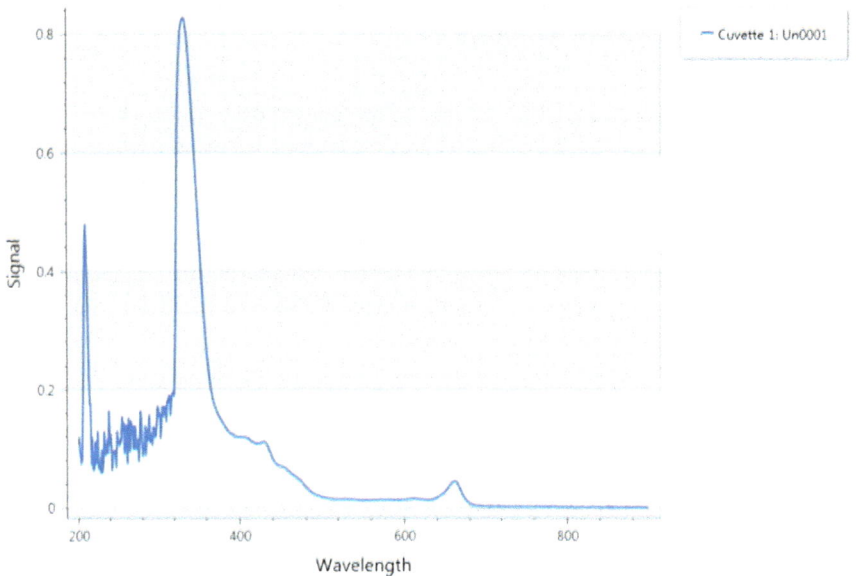

Fig. 8 Spectrum graph of Red Flame Ivy dye extract

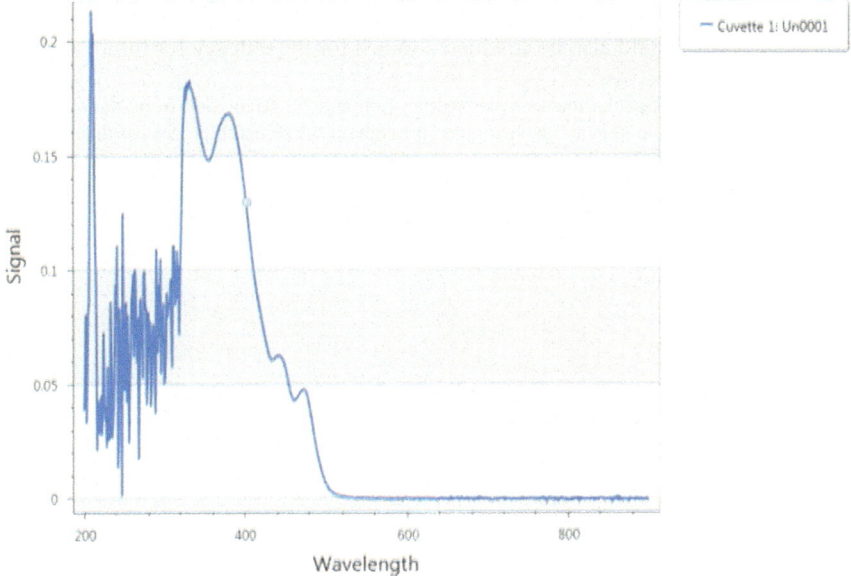

Fig. 9 Spectrum graph of Wall Daisy dye extract

further research. From the results of the research, it was also found that the DSSCs created degraded over time and produced even less voltage than when it was first made. This trend is shown to be similar to those in researches done by Yuvapragasam et al. [5] and Calogero et al. [6].

5   Future Work

In the future, experiments should be conducted to study the effect of the agarose gel on the voltage generated by solar cells. A solar panel could also be created by combining many solar cells together so as to observe the amount of voltage produced. As this research has not succeeded in generating enough voltage, hence to increase the voltage that each individual solar cell produces, future research should seal the sides of the solar cell properly with a good sealant to prevent any air from entering and oxidising the dye. More types of flora dyes could also be tested. Only specific chosen parts of the plants should be used to extract the natural flora dye, for example the petals or stem. Solar cells are used to produce clean energy, however materials such as titanium dioxide can be hard to find in certain places, hence materials used to make the solar cells should be more easily accessible to everyone. As the flora dye made in this research deteriorated over time and affected the efficiency of the solar cells, there is a need to find a way to preserve the flora dye such that it will not deteriorate over time. The application of successful dye-sensitised solar cells in powering loads should also be conducted to test for the efficacy for future uses.

Acknowledgements Special thanks to Dr Adrian Loh and Mr Allan Goh from National Junior College for the valuable help and advice given throughout the research process of this project. It is also greatly appreciated with gratitude towards National Junior College for the provision of the opportunity to engage in scientific research.

References

1. Mathew, A., Anand, V., Rao, G. M., & Munichandraiah, N. (2013). Effect of iodine concentration on the photovoltaic properties of dye sensitized solar cells for various I 2/LiI ratios. *Electrochimica Acta, 87,* 92–96.
2. Nazeeruddin, Md. K., Baranoff, E., & Grätzel, M. (2011). Dye-sensitised solar cells: A brief overview. *Institute of Chemical Sciences and Engineering. School of Basic Sciences. Swiss Federal Institute of Technology.*
3. Qin, Y., & Peng, Q. (2012). Ruthenium sensitizers and their applications in dye-sensitized solar cells. *International Journal of Photoenergy, 2012.*
4. Yoshikawa, K., Kawasaki, H., Yoshida, W., Irie, T., Konishi, K., Nakano, K., … & Yamamoto, K. (2017). Silicon heterojunction solar cell with interdigitated back contacts for a photoconversion efficiency over 26%. *Nature Energy, 2,* 17032.
5. Yuvapragasam, A., Muthukumarasamy, N., Agilan, S., Velauthapillai, D., Senthil, T.S., & Sundaram, S. (2015). Natural dye sensitized TiO2 nanorods assembly of broccoli shape based solar cells. *Journal of Photochemistry and Photobiology B: Biology.*
6. Calogero, G., Bartolotta, A., Marco, G. D., Carlob, A. D., & Bonaccorso, F. (2015) Vegetable-based dye-sensitized solar cells. *Royal Society of Chemistry.*

Genes Associated with Disease-Free Survival Prognosis of Renal Cancers

A Computational Screening for Potential Biomarkers and Targets for Gene Therapy

Gideon Tay Yee Chuen, Timothy Lim Tyen Siang, Shannon Lee Xin Ying, Lee Wai Yeow and Caroline G. Lee

Abstract Renal Cell Carcinoma (RCC) incidence has consistently been on the rise in recent years. There are 4 main types of RCC, namely Bladder Urothelial Carcinoma (BLCA), Kidney Chromophobe (KICH), Kidney Renal Clear Cell Carcinoma (KIRC), and Kidney Renal Papillary Cell Carcinoma (KIRP). The aim of this investigation is to identify genes in the tumors across the various renal cancers that can best distinguish patients with good versus those with poor disease-free survival (DFS), and determine pertinent cancer-related pathways that genes associated with DFS reside in. We hypothesized that genes significantly associated with DFS are associated with pathways that can be targeted for gene therapy and be identified as potential biomarkers for RCC. Genes in the tumors of RCC patients significantly associated with DFS were identified from The Cancer Genome Atlas (TCGA) database using Kaplan-Meier analyses. Genes with high expression that are associated with poor survival of patients might serve as potential biomarkers and/or targets for gene therapy across renal associated cancers with the exception of BLCA.

Keywords Renal Cell Carcinoma (RCC) · Bladder Urothelial Carcinoma (BLCA) · Kidney Chromophobe (KICH) · Kidney Renal Clear Cell Carcinoma (KIRC) · Kidney Renal Papillary Cell Carcinoma (KIRP) · Gene therapy · Biomarkers · Disease-free survival · Prognosis

G. T. Y. Chuen · T. L. T. Siang · S. L. X. Ying
Anglo-Chinese School (Independent), Singapore, Singapore

L. W. Yeow · C. G. Lee (✉)
Department of Biochemistry, Graduate School for Integrative Sciences and Engineering,
National University Singapore, Singapore, Singapore
e-mail: bchleec@nus.edu.sg

C. G. Lee
Division of Medical Sciences, National Cancer Centre Singapore, Singapore, Singapore

Cancer and Stem Cell Biology Programme, Duke-NUS Graduate Medical School,
Singapore, Singapore

© Springer Nature Singapore Pte Ltd. 2019
H. Guo et al. (eds.), *IRC-SET 2018*,
https://doi.org/10.1007/978-981-32-9828-6_19

1 Introduction

Renal Cell Carcinoma (RCC), a form of kidney cancer, originates from cells in the renal cortex [1]. 59% of Renal Cell Carcinoma (RCC) cases occur in developed countries, with statistics showing rates of diagnosis of RCC being three times lower in developing regions compared to developed regions. This could be due to obesity being a significant cause of renal cancer, as reflected in various studies [2]. With worldwide obesity rates nearly tripling since 1975, and almost 39% of adults aged 18 years and above being overweight, RCC incidence is naturally expected to increase with time [3]. In the last decade, there has been an increased risk of renal cancer in many countries, in particular, Korea, China, Hong Kong, Singapore and Japan [4]. Due to the growing significance and risk of RCC, it was selected as the focus of this study. Here, four subsets of RCC documented in The Cancer Genome Atlas (TCGA), a collaboration between the National Cancer Institute (NCI) and the National Human Genome Research Institute (NHGRI), were studied. They are Bladder Urothelial Carcinoma (BLCA), Kidney Chromophobe (KICH), Kidney Renal Clear Cell Carcinoma (KIRC), and Kidney Renal Papillary Cell Carcinoma (KIRP).

In this investigation, we identify genes whose expression are significantly higher in the tumors of RCC patients with good disease-free survival (DFS) compared to those with poorer DFS, are involved in pertinent cancer-related pathways, and have functions that may affect cancer relapse. In this study, DFS has been defined as the time between the point of diagnosis and the first sign of relapse of RCC [5]. DFS was studied instead of Overall Survival given the high chance of cancer recurrence following nephrectomy [6].

It is hypothesized that genes significantly associated with DFS are associated with pathways that can be targeted for gene therapy or can serve as potential biomarkers for RCC.

2 Methodology

2.1 Overview

The flow of our methodology has been presented in Fig. 1. Based on data from TCGA database, genes with the most significant difference in relapse time ($p < 0.004$) between high and low expression of the genes were taken for each of the 4 subsets of RCC. RCC patients in the data set were arranged in order of their expression of a given gene, with the bottom quartile taken as the high expression population and the top quartile taken as the low expression population.

Kaplan-Meier plots were plotted for each of these genes, and a rigorous selection procedure was employed to allow for the screening of potential biomarkers and targets for gene therapy, which will be further explained below.

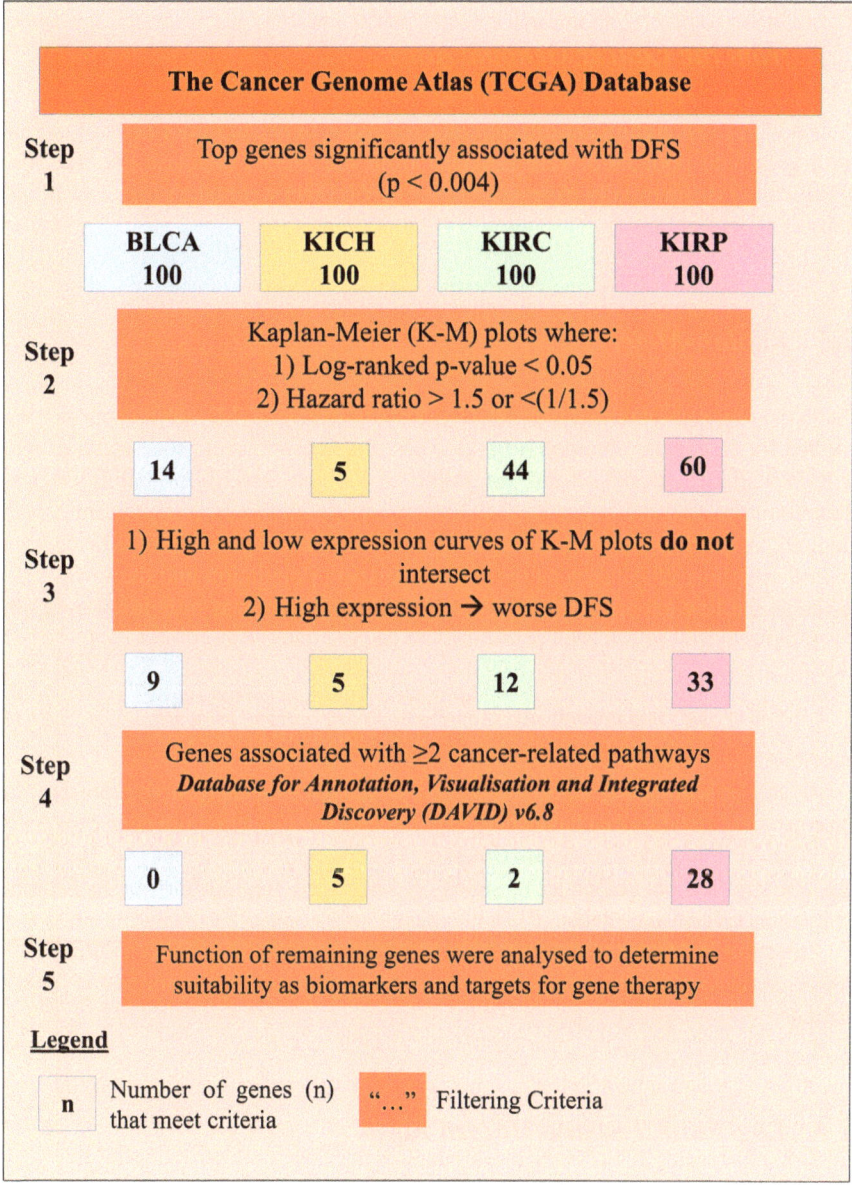

Fig. 1 Overview of methodology reflecting the number of genes selected for further downstream analysis at each step of our project. The orange boxes include the criteria used to select the genes while the colour-coded boxes reflect number of genes for each RCC subtype that meet criteria following the step taken

2.2 Selecting Differentially Expressed Genes from the Cancer Genome Atlas

Genes with most significant differential expression between tumor cells of patients with good and poor DFS were obtained from the TCGA database [7]. The most significantly differentially expressed genes ($p < 0.004$) were taken for each of the 4 subsets of RCC.

2.3 Kaplan-Meier Plots

From the raw data obtained from TCGA, Kaplan-Meier (K-M) survival curves were plotted for each gene obtained (Fig. 2). These curves represent processed data from a K-M statistical analysis, and are graphs of Percent Survival/probability against Time/months, and are hence used to estimate population survival over time. RCC patients in the data set were arranged in order of their expression of a given gene, with the highest 25% taken as the high expression population and the lowest 25% taken as the low expression population. Each population was monitored over time for relapse, allowing survival probability to be calculated according to (1):

$$\text{survival probability} = \frac{\text{number of patients yet to relapse}}{\text{total number of patients}}. \tag{1}$$

As can be seen in Fig. 2, each plot has two curves, and each curve represents the proportion of patients that have yet to experience a relapse with high gene expression (red line) or low gene expression (blue line) for the respective genes [8]. The curves are not smooth, but rather are a series of downward steps occurring each time a patient experiences a relapse [9].

From the K-M plots, we were able to determine whether higher expression or lower expression of the genes in the tumors were associated with better DFS of patients.

2.4 Log-Rank Test and Hazard Ratio

Each K-M plot also includes the log-rank p-value obtained from a log-rank test of the data. The log-rank test calculates the chi-square value for each event time (i.e. the time taken for RCC relapse to occur) for each curve and sums the results, which was then added to derive the final chi-square value to compare the two curves of each plot.

If the final p-value was less than 0.05, the survival times of the two plots were taken as significantly different from each other, indicating that level of gene expression of

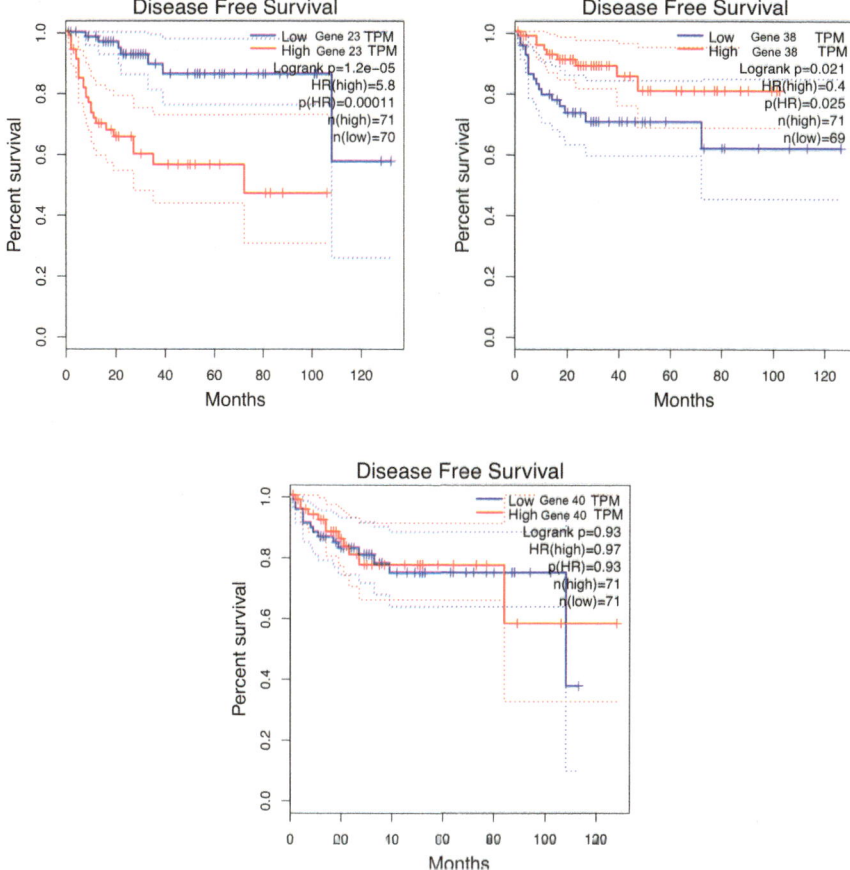

Fig. 2 Kaplan-Meier survival plots showing the difference in DFS between KIRP patients with high gene expression (red line) and low gene expression (blue line) of Genes 23, 38 and 40

the specific gene was more likely to be associated with the DFS of the patients. As such, genes with log-ranked p-values above 0.05 were removed from the list of genes for each cancer.

Each plot also has a hazard ratio, derived from the Cox proportional hazards model, which is a regression method for survival data [10]. It indicates the relative likelihood of the hazardous outcome occurring (i.e. relapse of RCC) in the group with patients experiencing high levels of gene expression compared to patients with low levels of gene expression. A hazard ratio of greater than or less than 1 indicates that DFS was better in one of the groups [11].

However, unlike the log rank test, there is no standard value for hazard ratios upon which the difference between the two groups would be considered significant. As such, to filter out genes not significantly associated with DFS in patients, the gene list was narrowed down to those with hazard ratios greater than 1.5 or less than 1/1.5.

Hence, only genes whose increased expression led to a 1.5 times increased likelihood of experiencing no DFS (hazard ratio greater than 1.5) or DFS (hazard ratio less than 1/1.5) were considered.

2.5 Criterion for Kaplan-Meier Plots

From the remaining genes (log-ranked p-value < 0.05, hazard ratio > 1.5 or < 1/1.5), those with high expression and low expression curves crossing in their respective Kaplan-Meier plots were not considered for future stages. This is because in such a case, the differential expression of the gene is unlikely to predict the DFS status. Furthermore, the curves do not indicate that there is a consistent effect of gene expression on DFS of the respective type of RCC, and thus the genes were rejected. An example of such a plot would be that of Gene 40 in Fig. 2.

Additionally, only genes whose high expression were found to lead to poorer DFS were further analyzed. Such genes are preferred as it is easier to downregulate gene expression rather than attempting to upregulate it in gene therapy, and products of genes with high expression would be more easily detected, making them also better potential biomarkers. This is reflected in the K-M plot as the high expression curve being below the low expression curve. An example of such a plot would be that of Gene 23 in Fig. 2.

Genes whose low expression were more likely to lead to poorer DFS were thus eliminated. This is reflected in the K-M plot as the low expression curve being below the high expression curve. An example of such a plot would be that of Gene 38 in Fig. 2.

Hence, using the examples given in Fig. 2, Gene 40 is rejected due to intersecting curves and gene 38 is rejected as high gene expression (red curve) led to better DFS compared to low gene expression (blue curve), while Gene 23 will be further analysed as it met both criteria set for Kaplan-Meier plots (Fig. 2).

2.6 Database for Annotation, Visualisation and Integrated Discovery (DAVID) v6.8

DAVID v6.8 is a web-based functional annotation tool that includes an integrated annotation knowledgebase, and provides various bioinformatics tools to analyze pathways enriched by specific gene sets [12, 13]. Using DAVID v6.8, genes whose expression can significantly predict the DFS status of patients with hazard ratios greater than 1.5 or less than −1/1.5 were analyzed, and then grouped according to their functional pathways. For each cancer, genes that were found in two or more pathways were identified as genes with great potential to be useful as targets for gene therapy or biomarkers for better prognosis.

3 Results and Discussion

3.1 Overview

In the discussion below, gene names have been substituted to alternative names due to intellectual property matters.

After being grouped into their functional pathways using the DAVID program, the open source software platform Cytoscape, in conjunction with the Agilent Literature Search Software, was used to illustrate the interaction network between the various genes in each group for easier visualization [14–17]. However, not all of the genes in this study were recognized by Agilent Literature Search Software due to the limitations of the databases linked, though they were still taken into account in later stages of the methodology, just not reflected in the respective pathway diagrams.

Aside from BLCA, significant genes in the other three RCC types were found to be associated with DFS. The resulting genes obtained are of interest as they can serve as potential biomarkers or as targets for gene therapy.

Biomarkers are used to better guide therapy and enable more accurate prognosis of renal cancers. Biomarkers are defined as "any substance, structure, or process that can be measured in the body or its products and influence or predict the incidence of outcome or disease" by the World Health Organization (WHO) [18]. Biomarkers are helpful in early prognosis, and treatment of various diseases, and is important for cancer. For example, the BRCA1 germline mutation has been used in estimating the risk of developing breast and ovarian cancer while the prostate specific antigen biomarker has been used in the screening for prostate cancer [19, 20]. However, there has yet to be a satisfactory biomarker (i.e. associated with survival of patients and easily measured) introduced for the prognosis of the different subsets of renal cancers, thereby adding to the significance of our work.

At the same time, the resulting genes obtained can also be targets of gene therapy. For instance, genes whose high expression are associated with poor DFS rates and are involved in pathways associated with the hallmarks of cancer can be targeted by therapy to downregulate their expression to prevent relapse in RCC patients. Given the rising prevalence of renal cancers, the possible useful applications of our results are thus laudable.

3.2 Kidney Chromophobe

Five genes, Genes 1, 2, 3, 4 and 5, fulfilled all given criteria (Fig. 3).

Given the high hazard ratios of these genes ranging from 4.9 to 10, their high expression is shown to negatively impact DFS of KICH patients. Furthermore, these genes all code for secretory proteins found in blood plasma, making them good candidates as KICH plasma biomarkers.

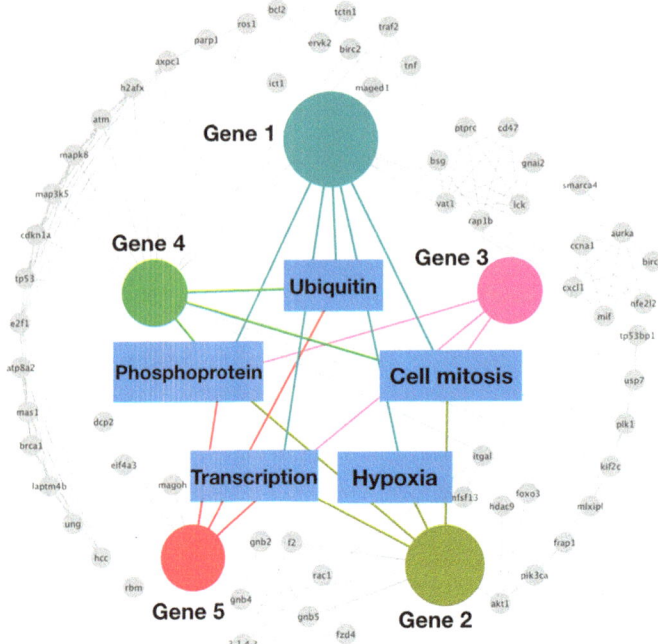

Fig. 3 Network of genes and pathways where higher expression of these genes is associated with poorer DFS of KICH patients. Blue rectangles represent pathways, while the circles represent genes. Non-grey circles represent the genes significantly associated with DFS while grey circles represent genes associated with the said genes. The larger the circle, the greater the number of pathways the gene is found in

Gene 1 and Gene 5, with high hazard ratios of 10 and 9.9, along with low log ranked p-value of 0.0077 and 0.011 respectively. These two genes are involved in cancer-related pathways: they affect ubiquitin function, and are involved in transcription and thus could regulate the expression of other cancer-related genes. The latter is the main function of Gene 5, which codes for a protein that is involved in mRNA splicing. The two genes also code for phosphoproteins, and excessive phosphorylation of proteins has been known to be associated with the emergence of cancers [21].

Furthermore, Gene 1 is also involved in cell mitosis and thus affects tumor growth, as well as cellular response to hypoxia, a pathway that is important in circumventing tumor hypoxia common in solid tumors. Specifically, Gene 1 functions to encode for cyclins involved in transition through cell cycle checkpoints, and upregulation of Gene 1 would thus result in increased proliferative signaling in the cell, which is a hallmark of cancer [22].

Given the high association of these two genes with DFS, and their apparent involvement in cancer-related pathways, they can potentially be used as targets for gene therapy. As the high expression of these two genes are strongly associated with poorer DFS, it may be possible to employ gene therapy or other approaches to reduce the expression of these genes to better control the disease.

3.3 Kidney Renal Clear Cell Carcinoma

44 genes had significant p-values and hazard ratios. Out of these 44 genes, 12 had Kaplan-Meier plots without any intersecting survival curves, and when upregulated, were associated with poorer DFS. Out of the 12 genes, eight were recognized by Agilent Literature Search, and are reflected above (Fig. 4). As can be seen, only two genes were found to be associated with two or more pathways: Gene 6 and Gene 7.

Gene 6 is involved in the transcription and sprouting angiogenesis pathways. Sprouting angiogenesis is the mechanism of blood vessel growth, which includes growth toward tumor cells. This process provides oxygen and nutrients to enable further tumor growth, and the blood vessel proximity to the tumor facilitates metastasis [23]. Metastasis, a hallmark of cancer, will greatly increase chances of cancer relapse due to difficulty in eliminating all secondary tumor and circulating tumor

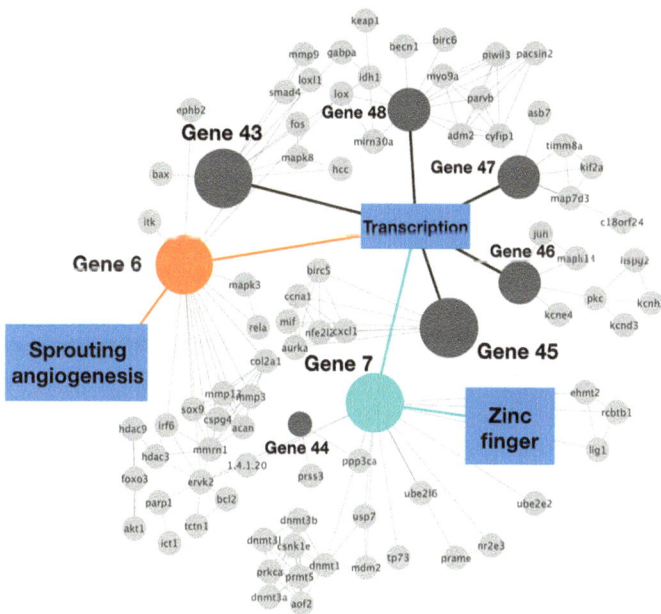

Fig. 4 Network of genes and pathways where higher expression of these genes is associated with poorer DFS of KIRC patients. Blue rectangles represent pathways, while the circles represent genes. Non-pale-grey circles represent the genes significantly associated with DFS while pale-grey circles represent genes associated with the said genes. Colourful circles represent genes that are associated with two or more pathways. The larger the circle, the greater the number of pathways the gene is found in

cells during treatment [24]. Promoting metastasis could be one of the reasons why high expression of Gene 6 is associated with poorer DFS.

Meanwhile, Gene 7 is involved in transcription and affects zinc finger function. Since zinc fingers are the largest transcription family in humans and are involved in RCC progression, this suggests Gene 7 affects cancer progression through affecting gene expression in cancer cells [25].

The hazard ratios of Gene 6 and Gene 7 are 2.7 and 2.2, while their log ranked p-value is 0.00027 and 0.0021 respectively.

Hence, they are clearly associated with DFS, and their function further suggests their association with relapse of KIRC. Since products of expression of Gene 6 and Gene 7 are not secreted, they can only function as biomarkers if there are circulating tumor cells which are extractable from the patient's blood. Additionally, these two genes could potentially be targeted for gene therapy, as their high expression is associated with the relapse of KIRC.

3.4 Kidney Renal Papillary Cell Carcinoma

60 genes had significant p-values and hazard ratios. Out of these 60 genes, 33 had Kaplan-Meier plots without any intersecting survival curves, and when upregulated, were associated with poorer DFS.

23 of the 33 remaining significant genes were recognized by Agilent Literature Search (Fig. 5). However, there were 28 genes that were associated with two or more pathways. Of these 28, there were 3 genes, Gene 8, Gene 9 and Gene 10, with hazard ratios more than or equal to 10, and log rank p-values of less than 10^{-7}. This signifies an extremely strong association with DFS.

All three of these genes are involved in transcription and cell mitosis pathways. Gene 8 encodes for a protein belonging to a dual specificity protein phosphatase family which regulates the cell cycle. Gene 9 codes for a component of the essential kinetochore-associated NDC80 complex, which is required for chromosome segregation, spindle checkpoint activity and kinetochore stability. Gene 10 codes for histone H3-like nucleosomal protein that is specifically found in centromeric nucleosomes.

Upregulation of these three genes will thus increase the proliferative rate of cells, which is a hallmark of cancer. The functions and pathways these 3 genes are involved in are intrinsically involved in the cell cycle, which in turn could affect cancer progression when upregulated and thus rates of cancer relapse. Hence, Genes 8, 9 and 10 present themselves as potential targets for gene therapy. Additionally, genes 9 and 10 codes for secretory proteins found in blood plasma, making them good candidates for biomarkers.

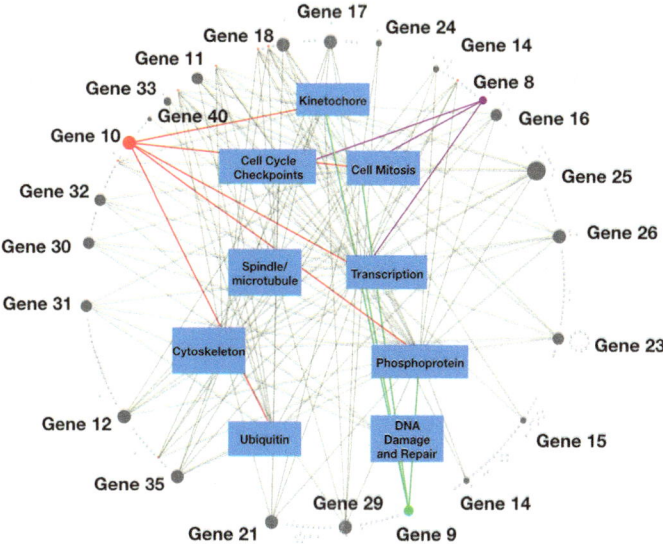

Fig. 5 Network of genes and pathways where higher expression of these genes is associated with poorer DFS of KIRP patients. Blue rectangles represent pathways, while the circles represent genes. Non-pale-grey circles represent the genes significantly associated with DFS while pale-grey circles represent genes associated with the said genes. Colourful circles represent genes that are associated with two or more pathways. The larger the circle, the greater the number of pathways the gene is found in

3.5 *Bladder Urothelial Carcinoma*

As can be seen in Fig. 6, only Gene 36 and 37 were found to be associated with one pathway, related to the centrosome.

As such, when genes that were found in less than two pathways were removed from the gene list, there were no genes left associated with the DFS of BLCA patients. Therefore, in this investigation, none of the significant genes of BLCA patients were associated with DFS of the patients, and thus no potential biomarkers or targets for gene therapy can be inferred for BLCA.

4 Conclusion

Out of the four cancers studied, three cancers (KICH, KIRC and KIRP) had genes that met all the set criteria of having suitable K-M plots, and were associated with two or more cancer-related pathways. The expression levels of these genes were found to be significantly associated with DFS, with high expression leading to poor DFS, and the genes are associated with two or more cancer-related pathways.

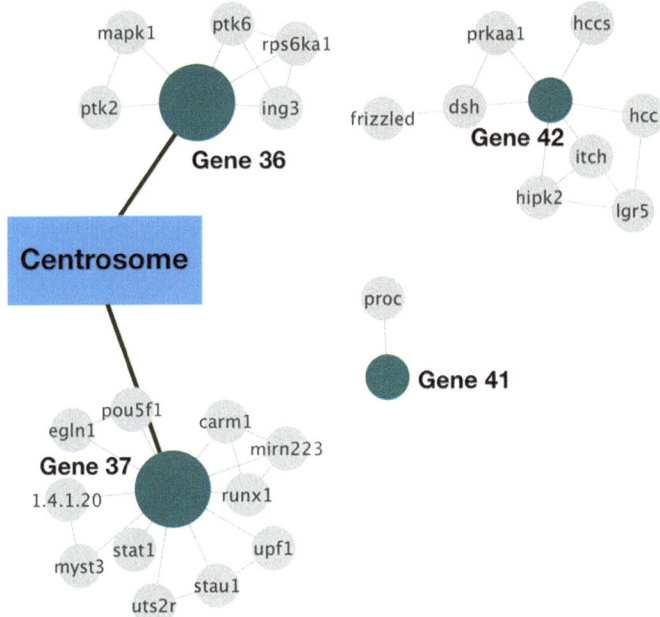

Fig. 6 Network of genes and pathways where higher expression of these genes is associated with poorer DFS of BLCA patients. Blue rectangles represent pathways, while the circles represent genes. Non-pale-grey circles represent the genes significantly associated with DFS while pale-grey circles represent genes associated with the said genes. The larger the circle, the greater the number of pathways the gene is found in

These genes obtained could be considered as potential biomarkers and/or targets for gene therapy. This would ensure better prognosis and more effective treatments for patients of the respective cancers.

In summary, genes identified as potential biomarkers and targets for gene therapy are Genes 1–5 for KICH, Genes 6 and 7 for KIRC, and Genes 8–10 for KIRP. Potential biomarkers of particular interest would be the protein products of Genes 1–5 for KICH and Genes 9 and 10 for KIRP, as these genes' protein products are secreted, making them more easily detectable given a blood sample. The results have been summarized in Table 1, with all the noteworthy genes having been included.

As can be seen in Table 1, most of the genes that have been identified are involved in the pathways of transcription, cell cycle and metastasis. It is also interesting to note that the genes significant to DFS of KIRC patients differ vastly from those significant to DFS of the other two cancers, in terms of the pathways that the genes are associated with. This could be due to different developmental pathways of the cancer.

Table 1 Summary table

Cluster	Pathway	Gene Number KICH					KIRC		KIRP			Total in Pathway	Total in Cluster
		1	2	3	4	5	6	7	8	9	10		
Transcription	Transcription											9	21
	Phosphoprotein											7	
	Ubiquitin											4	
	Zinc Finger											1	
Cell cycle	Mitosis											7	10
	Kinetochore											2	
	Cell Cycle Checkpoints											1	
Metastasis	Hypoxia											2	3
	Angiogenesis											1	

Acknowledgements The authors thank their mentors, A/P Caroline Lee Guat Lay and Mr. Lee Wai Yeow for their constant guidance and support.

References

1. Motzer, R. J., Bander, N. H., & Nanus, D. M. (1996). Renal-cell carcinoma. *New England Journal of Medicine, 335*(12), 865–875. https://doi.org/10.1056/nejm199609193351207.
2. World Cancer Research Fund International. *Kidney cancer statistics*. Retrieved from http://www.wcrf.org/int/cancer-facts-figures/data-specific-cancers/kidney-cancer-statistics.
3. World Health Organisation. (2017). *Obesity and overweight*. Retrieved from http://www.who.int/mediacentre/factsheets/fs311/en/.
4. Chow, W., Dong, L. M., & Devesa, S. S. (2010). Epidemiology and risk factors for kidney cancer. *Nature Reviews Urology, 7*(5), 245–257. https://doi.org/10.1038/nrurol.2010.46.
5. Denkert, C., et al. (2003). Elevated expression of cyclooxygenase-2 is a negative prognostic factor for disease free survival and overall survival in patients with breast carcinoma. *Cancer, 97*(12), 2978–2987. https://doi.org/10.1002/cncr.11437.
6. Chin, A. I., et al. (2006). Surveillance strategies for renal cell carcinoma patients following nephrectomy. *Reviews in Urology, 8*(1), 1–7.
7. Weinstein, J. N., et al. (2013). The Cancer Genome Atlas pan-cancer analysis project. *Nature Genetics, 45*(10), 1113–1120.
8. Chakraborty, R., Rao, C. R., & Sen, P. K. (2012). *Bioinformatics in human health and heredity* (Vol. 28). Newnes Books.
9. Barraclough, H., Simms, L., & Govindan, R. (2011). Biostatistics primer: What a clinician ought to know: Hazard ratios. *Journal of Thoracic Oncology, 6*(6), 978–982. https://doi.org/10.1097/jto.0b013e31821b10ab.
10. Rich, J. T., et al. (2010). A practical guide to understanding Kaplan-Meier curves. *Otolaryngology—Head and Neck Surgery, 143*(3), 331–336. https://doi.org/10.1016/j.otohns.2010.05.007.
11. Spruance, S. L., Reid, J. E., Grace, M., & Samore, M. (2004). Hazard ratio in clinical trials. *Antimicrobial Agents and Chemotherapy, 48*(8), 2787–2792. https://doi.org/10.1128/AAC.48.8.2787-2792.2004.
12. Huang, D. W., Sherman, B. T., & Lempicki, R. A. (2009). Systematic and integrative analysis of large gene lists using DAVID bioinformatics resources. *Nature Protocols, 4*(1), 44–57. https://doi.org/10.1038/nprot.2008.211.

13. Huang, D. W., Sherman, B. T., & Lempicki, R. A. (2008). Bioinformatics enrichment tools: Paths toward the comprehensive functional analysis of large gene lists. *Nucleic Acids Research, 37*(1), 1–13. https://doi.org/10.1093/nar/gkn923.
14. Shannon, P., et al. (2003). Cytoscape: A software environment for integrated models of biomolecular interaction networks. *Genome Research, 13*(11), 2498–2504. https://doi.org/10.1101/gr.1239303.
15. Vailaya, A., et al. (2004). An architecture for biological information extraction and representation. *Bioinformatics, 21*(4), 430–438. https://doi.org/10.1093/bioinformatics/bti187.
16. King, J. Y., et al. (2005). Pathway analysis of coronary atherosclerosis. *Physiological Genomics, 23*(1), 103–118. https://doi.org/10.1152/physiolgenomics.00101.2005.
17. Ashley, E. A., et al. (2006). Network analysis of human in-stent restenosis. *Circulation, 114*(24), 2644–2654. https://doi.org/10.1161/circulationaha.106.637025.
18. World-Health-Organisation. (2001). *Biomarkers in risk assessment: Validity and validation*. Retrieved from http://www.inchem.org/documents/ehc/ehc/ehc222.htm.
19. Easton, D. F., et al. (1995). Breast and ovarian cancer incidence in BRCA1-mutation carriers. *American Journal of Human Genetics, 56*(1), 265–271.
20. Lin, K., Lipsitz, R., Miller, T., & Janakiraman, S. (2008). Benefits and harms of prostate-specific antigen screening for prostate cancer: An evidence update for the U.S. Preventive Services Task Force. *Annals of Internal Medicine, 149*(3), 192–199.
21. Singh, V., et al. (2017). Phosphorylation: Implications in cancer. *The Protein Journal, 36*(1), 1–6. https://doi.org/10.1007/s10930-017-9696-z.
22. Hanahan, D., & Weinberg, R. (2011). Hallmarks of cancer: The next generation. *Cell, 144*(5), 646–674. https://doi.org/10.1016/j.cell.2011.02.013.
23. Hillen, F., & Griffioen, A. W. (2007). Tumour vascularization: Sprouting angiogenesis and beyond. *Cancer and Metastasis Reviews, 26*(3–4), 489–502. https://doi.org/10.1007/s10555-007-9094-7.
24. Flemming, A. (2015). Cancer stem cells: Targeting the root of cancer relapse. *Nature Reviews Drug Discovery, 14*(165). https://doi.org/10.1038/nrd4560.
25. Jen, J., & Wang, Y.-C. (2016). Zinc finger proteins in cancer progression. *Journal of Biomedical Science, 23*(53). https://doi.org/10.1186/s12929-016-0269-9.

Regeneration of α-Cellulose Nanofibers Derived from Vegetable Pulp Waste to Intercept Airborne Microparticulates

Jing Wesley Leong, Jie Hui Deon Lee, Wan Yu Jacqueline Tan, Kaede Saito and Haruna Watanabe

Abstract One of the most pertinent challenges we face is to increase protection to harmful and carcinogenic airborne particulate matter smaller than 2.5 μm, which can accumulate in the human respiratory system and cause irreversible impacts to human health. Currently existing N95 masks are relatively expensive and insufficient in supply during periods of severe air pollution. Furthermore, it is not accessible to people worldwide due to its cost. Nanofiber masks made from electrospun Cellulose Acetate are regarded as a promising filtration material due to its effectiveness in sieving out microparticulate matter as compared to existing solutions like the N95 mask. However, electrospinning technology used to generate such masks is often expensive and is of a large scale. This project proposes a novel way in regenerating Cellulose Nanofibers (CN) while reducing by-product food waste from vegetables. Fibril aggregation was carried out through dissolving leftover carrot pulp from juicing, employing the Kraft Process and Acetylation. Scanning Electron Microscopy was applied to demonstrate the intricacy of the CN polymers as compared to other respiratory mask materials. The Nanofiber polymer sheet was tested for its filtration capabilities and its efficiency was evaluated against the existing N95. The Cellulose Nanofibers regenerated from leftover vegetable waste are highly effective in reducing exposure to microparticulates and display remarkable potential as an alternative to N95.

Keywords Microparticulate · Cellulose · Nanofiber · Regeneration · Paper-making

J. W. Leong (✉) · J. H. D. Lee · W. Y. J. Tan
National Junior College, 37 Hillcrest Road, Singapore 288913, Singapore
e-mail: leongw717@gmail.com

K. Saito · H. Watanabe
Ritsumeikan Keisho Senior High School, 640-1 Nishinopporo, Ebetsu,
Hokkaido 069-0832, Japan

J. H. D. Lee
Nanyang Technological University, 50 Nanyang Avenue, Singapore 639798, Singapore

© Springer Nature Singapore Pte Ltd. 2019
H. Guo et al. (eds.), *IRC-SET 2018*,
https://doi.org/10.1007/978-981-32-9828-6_20

235

1 Introduction

Causing approximately 6.5 million premature deaths globally in 2016, air pollution is the leading cause of death worldwide—and almost all (94%) of these deaths occur in low- and middle-income countries due to insufficient protection. Airborne microparticulates (PM) are the deadliest form of air pollution, as these aerosols ranging from 0.1 to 2.5 μm in size can penetrate deep into the lungs and bloodstream. They are unable to be naturally expelled by the human body [1], eventually leading to devastating and irreversible damage health such as increasing the risk of lung cancer. Hence, these are designated as carcinogens.

Currently, N95 respirators are sought after as an effective form of filtration against Particulate Matter originating from industrialization, such as slash-and-burn agricultural methods and industrial smoke. Despite its proven usefulness in reducing exposure to pollution, the masks are not readily available to many around the world due to its high costs. The lack of access to effective aerosol respirators caused many people to be unable to receive enough protection. The 2013 Singapore Haze saga was a case whereby the Pollution Standards Index (PSI) hit 400, causing N95 masks to be sold out across the island. This is even though the government had a stockpile of 9 million N95 masks. When the Pollution Standards Index is high, demand for N95 masks often overbears its supply. In China, due to industrial factories producing carbon emissions on a regular basis, air pollution is a common sight. Locals tend to use common household materials which do not have filtration properties to prevent themselves from inhaling the Particulate Matter (PM).

On the other hand, food wastage has continued to be a pertinent challenge to overcome as waste by-products from the food industry are not maximized to their full potential. At the same time, cellulose exists as the most abundant polymer on Earth [2] and it can be extracted from plant-based waste that are high in cellulosic content, such as those in carrots. Carrots can easily be found in all parts of the world, even in less developed countries such as China. Furthermore, the alpha cellulose in carrots are around 7000–15,000 μm units per polymer, which makes it a highly intricate structure. In recent years, vegetables with Cellulose Nanofibers (CN) such as carrots are of interest and have been substituted as alternatives to carbon fibres. However, no research has been done to apply these versatile yet abundant fibres as polymer sheets. Hence, given carrots' intricate structure and its abundance, we have decided to use leftover vegetable pulp as a novel material in creating filtration polymers to combat pollution.

2 Hypothesis

We hypothesize that the carrot-derived Cellulose Nanofiber is equally as effective as existing N95 respiratory masks in filtration efficiency and show high potential in being able to reduce waste while minimizing human exposure to air pollution.

Table 1 Comparison between various classes of cellulose

	Polyose (β-, γ-)	Cellulose (α-)
Units per polymer	500–3000	7000–15,000
Structure	Random, amphorous	Crystalline, linear
	Highly branched	Unbranched
Strength	Little strength	Strong

3 Methodology

3.1 Selection of Fibre Composition

Out of the 3 classes of cellulose, only α-cellulose is desired as it is the most stable with the highest degree of polymerization and has a high predicted stiffness of 130 GPa [3].

β-cellulose and γ-cellulose, collectively referred to as *polyose* or *hemicellulose*, and are branched (Table 1).

3.2 Dissolving Pulp

Various methods have been employed to utilize extracted cellulose from plant fibres. Other of such methods is the Viscose process, which dissolves pulp using intermediate chemicals such as the highly toxic C_2S and produces pollution to the environment as well as harms factory workers during the production [4]. In contrast, the newly discovered Lyocell process used to make materials such as Tencel is pollution-free but has a high cost as it involves dry wet jet spinning and is unsustainable to create low-cost polymers. Hence, the papermaking method of dissolving pulp was chosen as an in-between solution to address both concerns. The process of dissolving pulp requires high chemical purity and with distinctly low polyose content, as the chemically similar polyose can interfere with fibril aggregation [5]. Cellulose pulp with high purity is typically extracted using two industrial methods, the Sulphite and Kraft processes.

During the Sulphite process (SP), sulphur is burnt with oxygen to create SO_2 which is then absorbed in water, releasing acidic sulphurous acid (H_2SO_3). However, during the burning process, SO_2 is unavoidably oxidized to form SO_3 that gives undesirable H_2SO_4 when dissolved in water, which promotes hydrolysis of cellulose without contributing to delignification. Undesirably, does not fully degrade lignin to the extent which the Kraft Process does [6] and hence allows for more useful by-products such as lignosulfonates used to make concrete and vanilla flavouring. However, it is undesirable for make cellulosic fibres as the pH 1.5 acidic conditions hydrolyse part of the α-cellulose, making them not as strong as those regenerated using the Kraft Process.

Fig. 1 Chemical illustration of delignification

Dissolving pulp allows for derivatization into a homogenous solution, making the pulp chemically accessible and removes remaining undesirable fibrous structure. This allows for high brightness and uniform molecular weight distribution.

3.3 Fibril Aggregation via Kraft Process

While polyose is easily hydrolysed in dilute acids or bases, α-cellulose is resistant to hydrolysis. Hence, the carrot fibres were dissolved in NaOH to separate α-cellulose from polyose for further treatment. Lignin and polyose degrade to give fragments that are soluble in the strongly basic liquid, by breaking the bonds between them (Fig. 1).

Carrot pulps obtained after juicing were pre-treated by washing with distilled water as per laboratory practice. The pulp was then dissolved for derivatization into a chemically accessible homogenous solution. It was then treated with an alkali of 17.5 wt% NaOH solution at 80 °C for 2 h under mechanical stirring to purify cellulose by removing other constituents and to achieve uniform molecular weight distribution. The polyose which was soluble in NaOH was then disposed of, leaving the residue of α-cellulose with high brightness. It was then filtered and washed with distilled water to eliminate the alkali, which was essential not to affect pH, as exact values are required during later steps.

Other works, such as that described by Siqueira et al. [7] had used 2 wt% NaOH as an alkali to decrease hemicellulose content for assay purposes and not achieve complete elimination, which was undesirable in this research. The bonds between lignin and hemicellulose were broken and degraded, giving fragments of polyose that were soluble in the strongly basic NaOH (Image 1).

3.4 Acetylation, Hypochlorite Bleaching and Drying

A subsequent bleaching treatment was carried out to remove non-cellulosic components such as lignin, tannin and β-carotene. With the use of a pH meter, 1 M NaOH

Image 1 Recovery of
α-cellulose from carrot waste

was added to 1 M CH$_3$COOH until a pH value of 4.5 was achieved. Then, the amount of solution was measured and an equal amount of NaClO (1.7 wt% in water) was added. The bleaching treatment was performed at 80 °C for 2 h under a magnetic stirrer for a shorter completion time. Once again, the material was filtered and washed with distilled water to achieve a neutral pH and to purify the solid recoveries The cellulose gel was thinly laid out into a sheet and dried in a drying oven at 50 °C for 2 h until completely dry.

The NaOH in the set-up also contributes to saponification, removing acetyl groups from the surface to leave them with a cellulose coating which reduces the tendency of fibres to acquire static charges. Furthermore, the NaOH treatment allows for a stable re-dispersed suspension. This translates in possibilities for future, further recycling of the re-dispersed fibres and its stability in a solvent when dissolved (Image 2).

4　Results and Analysis

4.1　Characterisation via Scanning Electron Microscopy

Scanning Electron Microscopy (SEM) was applied to analyse the arrays of fibre networks in the various polymer surfaces. As the diameter of Nanofibers, as well as pore size gap was of the magnitude of about 10 nm, SEM was critical in allowing

Image 2 Dried cellulose
nanofiber sheet

estimation of the pore size, as well as to allow visualisation of the spatial arrangement
and distortion of the fibre networks.

The main filtration mechanism employed was the Interception mechanism, which
works by filter fibre diameters smaller and more intricate than that of target particles
and hence carry out a size-based filtration (Fig. 2).

The gaps between Nanofibers should be smaller than 2.5 μm to perform filtration
by size. The Nanofiber regenerated from spent carrot has a pore gap size of smaller
than 1 μm, compared to about 20 μm (Surgical/Clinical Mask) and 10 μm (N95
Mask) (Images 3, 4, 5, 6, 7 and 8).

Fig. 2 Visual representation
of interception filtration
mechanism

Image 3 Clinical mask at **400×** magnification under SEM

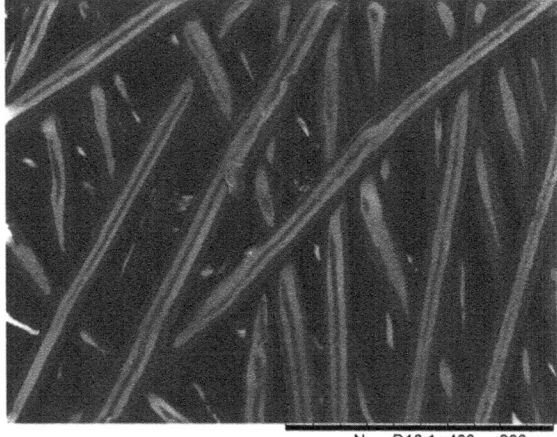

Image 4 Clinical mask at **2500×** magnification under SEM

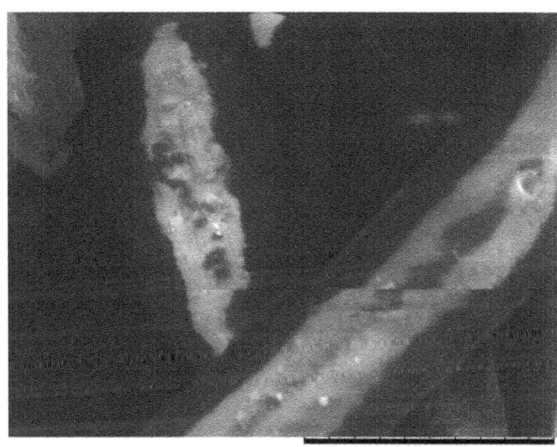

4.2 Efficiency Test via Aerosol Microparticulate Counter Test

To compare the efficiency of the mask, a filtration test was conducted with the aid of an Aerosol Particulate Counter. The experiment was carried out in an indoor laboratory setting with temperature of 25.0 °C in Ebetsu, Hokkaido, Japan (Image 9 and Fig. 3).

It was used to assess the permeability of the regenerated Cellulose Nanofiber polymer to microparticulates, hence testing for its filtration efficiency as a respirator. The experiment was conducted in an indoor laboratory under room temperature. Control readings were first taken by measuring the amount of existing microparticulates in the room. Then, the two samples of CN and N95 were separately placed to cover the particulate counter to allow for a decrease in aerosol concentration readings. The

Image 5 N95 mask at **400×** magnification under SEM

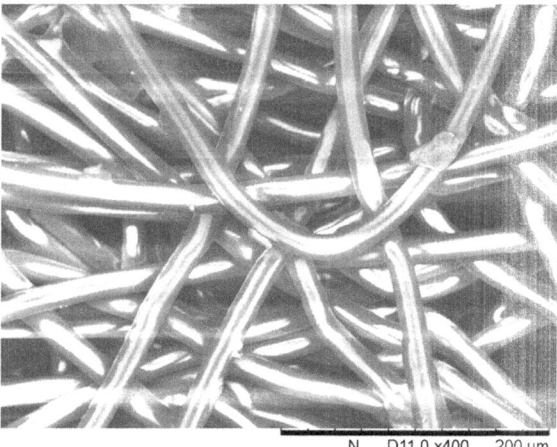

Image 6 N95 mask at **4000×** magnification under SEM

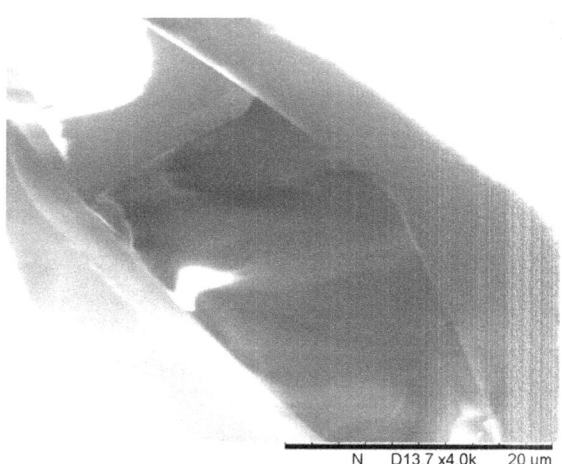

decrease in number of particulates that had permeated was attributed to the filtering surface. This method had allowed for accurate comparison of the effectiveness of both polymers as it had reduced the possibility of human error and allowed for consistency of results. The device was placed in a fixed position to prevent drastic changes in air composition and degree of microparticulate concentration. Repeated readings were collected to ensure consistency and reliability of data collected to attribute effective filtration due to the regenerated Cellulose Nanofiber masks, and not due to inaccuracy during experimentation.

From Fig. 4, it can be inferred that most particulate matter were between the size of 0.3 and 0.5 μm. Hence, an effective mask should have pore sizes smaller than 1 μm in size for effective filtration by size to prevent permeation by extremely fine aerosols, leading to filtration inability. However, this does include filtration resulting from

Image 7 Cellulose nanofiber at **400×** magnification under SEM

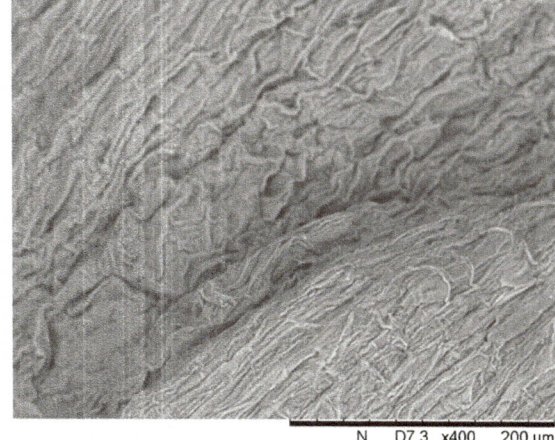

Image 8 Cellulose nanofiber at **6000×** magnification under SEM

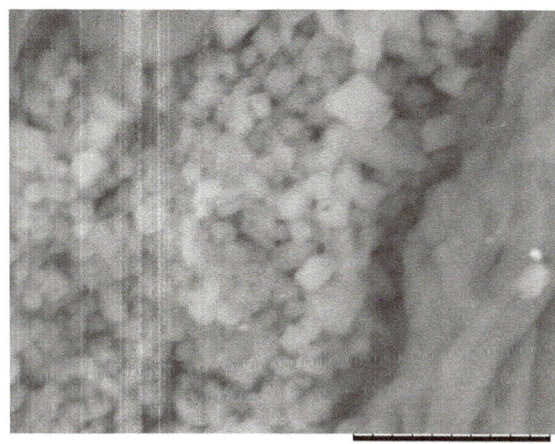

Image 9 Aerosol particle counter set-up

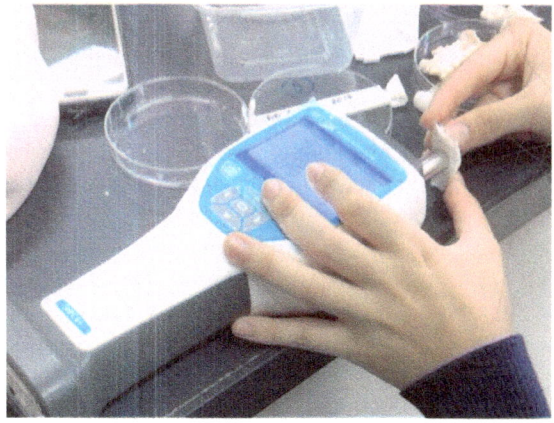

Fig. 3 Visual representation of particle counter

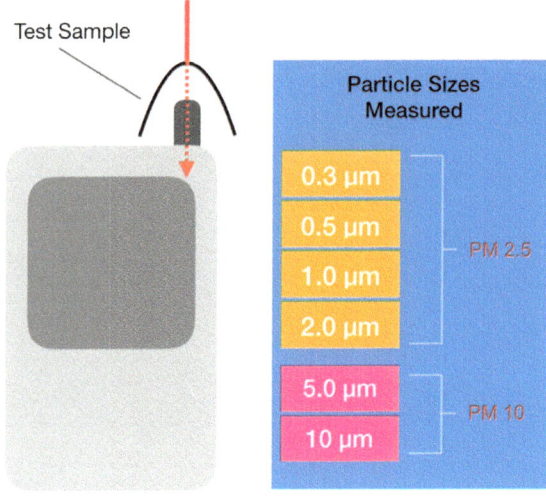

Fig. 4 Graph of average particle size abundant in air

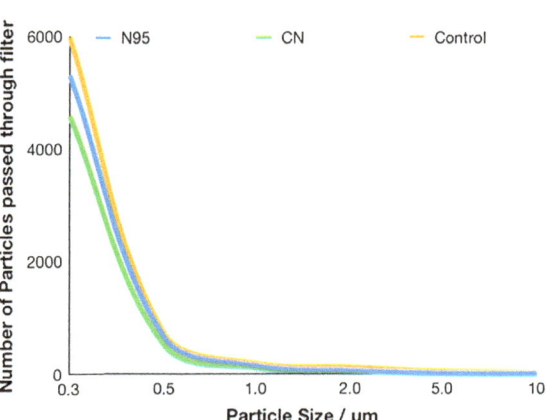

other reasons such as inertial impaction and particle attraction due to electrostatic charges.

From Fig. 5, it was shown that the Cellulose Nanofiber mask was over 200% as effective in filtering out microparticulate between the sizes of 0.3 and 2.0 μm, proving to be effective in reducing exposure to PM2.5 particulate matter.

For particles about 5.0 μm in size, the Cellulose Nanofiber polymers were less effective than the N95 masks in filtration efficiency. This was probably due to inconsistent heat that was applied to the sample during the drying of the regenerated Cellulose, and the methods used to lay out the fibres into a thin sheet, resulting in small pore holes in the fibre polymer that had allowed 5.0 μm microparticulates to permeate through.

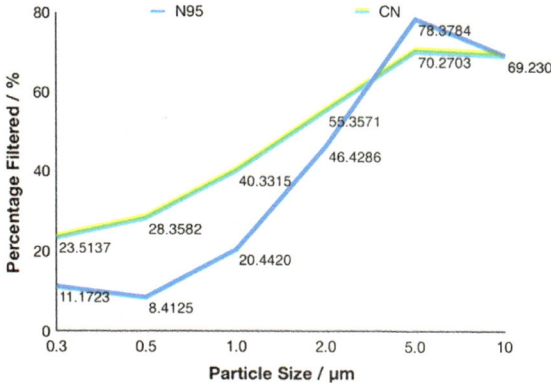

Fig. 5 Graph of percentage of particles filtered against particle size

Also, the percentage of 10.0 μm sized particles that were filtered was the same: this is most likely because such particulates exist in very small quantities in the atmosphere. Nevertheless, this is not of relevance or concern as the human body is able to naturally expel particulates that are larger 2.5 μm in size, and hence the mask is not required to achieve high levels of protection against (Tables 2, 3 and 4).

Table 2 Specific data from microparticulate counter test (normal air, control set-up)

Particulate size/μm	Normal air (control)				
	1	2	3	4	Average
0.3	5370	6300	6089	6191	5987.5
0.5	767	799	658	724	737
1.0	204	194	141	184	180.75
2.0	124	122	80	120	111
5.0	42	42	16	49	37.25
10.0	13	18	5	17	13.25

Table 3 Specific data from microparticulate counter test (N95 mask, variable 1)

Particulate size/μm	N95 mask				
	1	2	3	4	Average
0.3	5300	5269	5007	5700	5319
0.5	660	682	661	697	675
1.0	180	135	120	141	144
2.0	61	62	56	61	60
5.0	9	8	7	8	8
10.0	5	4	3	4	4

Table 4 Specific data from microparticulate counter test (regenerated cellulose fibre, variable 2)

Particulate size/μm	Cellulose nanofibers derived from leftover carrot pulp				
	1	2	3	4	Average
0.3	4639	4570	3973	5139	4580.25
0.5	499	524	487	601	527.75
1.0	108	120	97	114	109.75
2.0	50	55	47	48	50
5.0	15	11	8	10	11
10.0	4	5	2	5	4

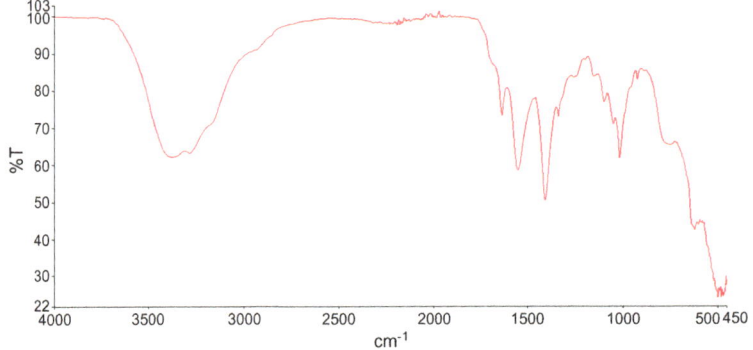

Fig. 6 Transmittance values (arb.) of CN fibre at varying wavelengths [8, 9]

4.3 *Characterisation via Fourier-Transform Infrared Spectroscopy*

Fourier-Transform Infrared Spectroscopy allows for characterisation of fibres to ensure that desirable properties and qualities are achieved (Fig. 6 and Table 5).

5 Conclusion

Carrot-derived Cellulose Nanofiber polymers are 200% effective as compared to present N95 masks in terms of filtration efficiency. In addition, waste carrot is an abundant source of starting material and can be easily found in all parts of the world, making it readily available to everyone. Producing carrot-derived Cellulose Nanofiber polymers can also reduce large amounts of food waste and these Cellulose Nanofiber polymers are biodegradable.

Table 5 FT-IR data relevant for cellulose nanofibers

Wavelengths peaks at: (cm^{-1})	Significance
Lower peaks, 3300–3335	Due to stretching vibrations of OH, indicating partial acetylation
2900–2800	Stretching of C–H groups of cellulose
Absence, 1840–1760	Free of unreacted acetic anhydride (CH$_3$CO)$_2$O, undesirable by-product
Absence, 1700	No carboxylic group [COOH] in acetic acid by-product
1700–1745 and 1235–1240	Acetylated fibres C=O stretching of carbonyl in ester bonds Vibration peaks due to C–O stretching of acetyl groups
1320–1330	Bending vibration of C–H, C–O aromatic rings in polysaccharides
1020–1030	Stretching of C–O and O–H

6 Future Work

One aspect to improving the Cellulose Nanofiber polymers is to test its filtration efficiency in varying climates, where there are different types of Particulate Matter (PM) present. Unlike N95 masks which are made of charged polypropylene fibres, carrot-based masks are non-charged and can possibly be more versatile in its application in various climates globally—from being able to filter smog in winter climates to forest fire haze in summer climates. Being able to work in different climates would greatly benefit different groups of people living in all parts of the world as these carrot-derived Cellulose Nanofiber polymers would be able to have a 200% filtration efficiency against any type of Particulate Matter (PM) at different climates.

Also, commercial masks provide vague instructions of when it is necessary to replace for a new mask, which may be wasteful during use in relatively less-polluted environments. It would be of benefit if the Cellulose Nanofiber polymers are able to change colour visibly when its performance falls below a certain threshold, so consumers are aware of having to dispose and recycle the mask. This can be tested with the use of UV-vis Spectroscopy.

Lastly, research can be done on improving the mask to be applicable and effective against various types of microparticulate globally, from pollen dispersal to dust from Aeolian sand. On top of this, it is possible to investigate the effectiveness of Cellulose Nanofiber polymers in the medical field, such as filtering bacteria, viruses and other airborne pathogens. Different types of vegetable waste can be tested for their relative effectiveness as well. We anticipate Cellulose Nanofibers as an alternative, sustainable solution to reducing exposure to air pollution, and eventually improving overall respiratory health worldwide.

Acknowledgements The authors would like to thank our research mentors for their constant care and guidance throughout the research journey, as well as the Affiliate Schools of the Ritsumeikan Trust, and National Junior College for the opportunity to take up this collaborative research.

References

1. Dockery, D. (1994). Acute respiratory effects of particulate air pollution. *Annual Review of Public Health, 15*(1), 107–132.
2. Klemm, D., Heublein, B., Fink, H., & Bohn, A. (2005). Cellulose: Fascinating biopolymer and sustainable raw material. *ChemInform, 36*(36).
3. Yao, J., Bastiaansen, C., & Peijs, T. (2014). High strength and high modulus electrospun nanofibers. *Fibers, 2*(2), 158–186.
4. Markowitz, G. (2017). *Review of Paul David Blancs fake silk: The lethal history of viscose rayon* (p. 309). New Haven: Yale University Press. *American Journal of Industrial Medicine, 60*(4), 408–409.
5. Duchesne, I., Hult, E., Molin, U., Daniel, G., Iversen, T., & Lennholm, H. (2001). The influence of hemicellulose on fibril aggregation of kraft pulp fibres as revealed by FE-SEM and CP/MAS 13C-NMR. *Cellulose, 8*(2), 103–111.
6. Pérez, J., Muñoz-Dorado, J., Rubia, T. D., & Martínez, J. (2002). Biodegradation and biological treatments of cellulose, hemicellulose and lignin: An overview. *International Microbiology, 5*(2), 53–63.
7. Siqueira, G., Oksman, K., Tadokoro, S. K., & Mathew, A. P. (2016). Re-dispersible carrot nanofibers with high mechanical properties and reinforcing capacity for use in composite materials. *Composites Science and Technology, 123,* 49–56.
8. Jonoobi, M., Harun, J., Mathew, A. P., Hussein, M. Z., & Oksman, K. (2009). Preparation of cellulose nanofibers with hydrophobic surface characteristics. *Cellulose, 17*(2), 299–307.
9. Alemdar, A., & Sain, M. (2008). Isolation and characterization of nanofibers from agricultural residues—Wheat straw and soy hulls. *Bioresource Technology, 99*(6), 1664–1671.

Biofabrication of Organotypic Full-Thickness Skin Constructs

Abby Chelsea Lee, Yihua Loo and Andrew Wan

Abstract Organotypic skin constructs have gained significant research and commercial interest in the face of EU bans on animal-tested cosmetic products. In this project, a simplified, full-thickness organotypic skin construct was prepared using fibroblasts encapsulated in a synthetic peptide hydrogel matrix, over which keratinocytes were allowed to proliferate, differentiate and stratify. The self-assembling peptide was synthesized using solid phase chemistry. The peptide was then used to prepare hydrogels of varying concentrations and the formulation was optimized based on gelation kinetics and mechanical strength. The long-term biocompatibility of this matrix with dermal fibroblasts was also evaluated. Finally, the in vitro skin constructs were characterized using histology and electron microscopy. Potential primers to evaluate gene expression of epithelial biomarkers were also identified. In conclusion, the peptide hydrogel is an appropriate matrix for culturing organotypic skin constructs due to its stability and low cytotoxicity. Building on this model, more elaborate systems can be cultured with the addition of more cell types. These biological constructs can potentially be used to screen therapeutic candidates, as well as to evaluate the effects of compounds on skin tissue viability, permeability and cellular gene expression.

Keywords Organotypic skin construct · Ultra-small peptides · Peptide hydrogel

1 Background and Purpose of Research

Three-dimensional (3D), multi-cellular, tissue mimetic models are powerful experimental platforms. They enable the in vitro study of mammalian tissue development, the modelling of human disease and the evaluation of novel therapeutics [1]. Organotypic 3D cultures are more biologically relevant and allow better understanding of in vivo physiology and function as integral biological processes are altered when

A. C. Lee (✉)
Raffles Institution, Singapore, Singapore
e-mail: abbiclee@gmail.com

Y. Loo · A. Wan
Institute of Bioengineering and Nanotechnology, Singapore, Singapore

© Springer Nature Singapore Pte Ltd. 2019
H. Guo et al. (eds.), *IRC-SET 2018*,
https://doi.org/10.1007/978-981-32-9828-6_21

cells are cultured in 2D versus 3D. These include immune system activation, defense response, cell adhesion and tissue development [2, 3].

Organotypic skin constructs, in particular, have gained significant research and commercial interest in the face of EU bans on animal-tested cosmetic products [4]. Human cadaver skin and excised animal skin have been traditionally used as topical and transdermal permeation models [5]. However, while human cadaver skin replicates in vivo permeation performance to some extent, there is a high sample to sample variation. Animal skin, though easily procured, is inherently different in physiology and gene expression from human skin. A simple in vitro culture of full-thickness skin can be constructed using dermal fibroblasts encapsulated in a 3D gel matrix, over which keratinocytes are seeded. Various biomaterials which mimic the extracellular matrix (ECM) have demonstrated applicability as matrices [6, 7]. Collagen is currently the most commonly used matrix for building organotypic full-thickness skin. It is a good scaffold material since it contains naturally-occurring ligands that facilitate cell attachment and proliferation [8]. However, batch-to-batch variations reduce the degree of experimental control; while potential risks of immunogenicity and pathogen transmission hinder future applications as tissue transplants [8]. Experimentally, one of the major challenges of using collagen hydrogels is remodelling by dermal fibroblasts, resulting in significant and uneven shrinkage of the tissue construct after 3 weeks of culture [9]. Hence, we propose the use of self-assembling, synthetic peptide hydrogels as a substitute supporting matrix. Short peptides are versatile building blocks for fabricating supramolecular structures. A specific motif enables ultrasmall peptides with 3–6 amino acids to self-assemble to helical supramolecular fibres [10]. This amphiphilic peptide motif consists of a tail of aliphatic nonpolar amino acids with decreasing hydrophobicity and a hydrophilic head group [10]. The peptides undergo a structural transition pathway from random coils to α-helical intermediates to β-turn structures with increasing concentration. The β-turn fibers further condense into meshed 3D nanofibrous networks, which closely resemble the ECM [10]. Due to their strong amphiphilic nature, the peptide networks entrap water, forming 3D nanofibrous hydrogels. In this study, we use a hexamer peptide, IK6, which demonstrates salt-enhanced gelation. They are highly suitable due to their outstanding biocompatibility, mild gelation conditions [10], and excellent mechanical properties. Moreover, they have been investigated as bioink candidates to facilitate bioprinting of 3D tissue constructs [11].

2   Engineering Goal

The objective of this project is to synthesize, optimize and characterize the formulation of peptide hydrogels for building full-thickness organotypic skin constructs.

3   Methods and Materials

3.1   Solid-Phase Peptide Synthesis

The ultrashort peptides were synthesized manually, using solid phase peptide chemistry (Appendix) [12]. Briefly, L-lysine residues conjugated to polystyrene beads via a Rink-amide linkage (GL Biochem, China) were swelled in dimethylformamide (JT Baker, USA), de-protected and each succeeding amino acid residue added sequentially. After the terminal amino acid had been added, the N-terminus amine group was acetylated using acetic anhydride (Sigma Aldrich, USA). The peptides were then cleaved using trifluoroacetic acid (Acros Organics, USA), precipitated and washed with diethyl ether (JT Baker, USA). After vacuum drying, the crude products were dissolved in dimethylsulfoxide (Kanto Chemical Co. Inc., Japan) and purified using reverse-phase high-performance liquid chromatography mass spectrometry. The purified peptide solution was subsequently lyophilized to obtain a dry powder.

3.2   Preparation of Peptide Hydrogels

Hydrogel samples were prepared in polydimethysiloxane moulds to obtain 8 mm diameter discs, approximately 1 mm thick. Peptide powder was first dissolved in milliQ water. 10% volume of 10× phosphate-buffered saline (PBS) was subsequently added to stimulate gelation.

3.3   Rheological Analysis

Dynamic strain and oscillatory frequency sweep experiments were carried out using the ARES-G2 Rheometer (TA Instruments, USA) with 8 mm titanium parallel plate geometry. The storage (G') and loss (G'') moduli were recorded in response to varying strain (γ) and angular frequency (ω). The readings of 3 samples were averaged for each condition.

3.4   HDF and HDK Culture

Human dermal fibroblasts (HDFs) were obtained from Lonza (Basel, Switzerland), cultured in Dulbecco's minimum essential media (DMEM) supplemented with 10% fetal bovine serum and 1% penicillin-streptomycin. Human dermal keratinocytes (HDKs) were also purchased from Lonza (Basel, Switzerland), cultured in Keratinocyte Serum Free Medium (Life Technologies, USA).

3.5 Encapsulation of HDF and Cytotoxicity Assay

HDFs cultured were either cultured on tissue culture plates (24-well) or encapsulated in 6 mg/mL peptide hydrogel, and cultured for 1 or 3 weeks. Cytotoxicity was evaluated using the LIVE/DEAD assay (Thermo Fisher Scientific, USA). Constructs were incubated at 37 °C for 5 min before being viewed underneath a fluorescence microscope (Olympus IX71 Research Inverted microscope).

3.6 Organotypic Skin Co-cultures

HDFs were combined with the IK6 hydrogel and NaOH and seeded onto a 24-well transwells (Corning, USA) at 4×10^6 cells/mL (Fig. 1). After a week, the HDKs seeded onto the constructs. The constructs were maintained in 50% DMEM media and 50% Keratinocyte SFM media for another week. After which, they were switched to air-liquid interphase culture in stratification media for another 2 weeks [13].

3.7 Histology

Samples were fixed in 4% paraformaldehyde. Paraffin embedding was carried out and 5 μm sections were stained with hematoxylin and eosin (H&E) for routine histological evaluation under an Olympus IX71 microscope.

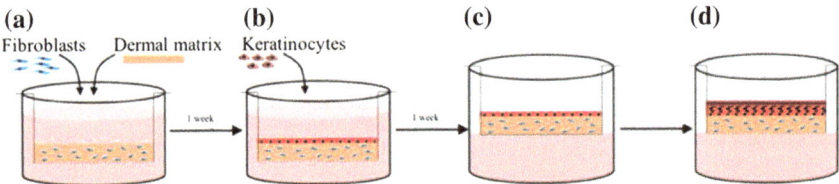

Fig. 1 Schematic representation of preparation of 3D organotypic skin constructs. **a** Peptide hydrogels encapsulating human dermal fibroblasts are pipetted into transwells and allowed to gel at 37 °C for 30 min. The hydrogels were subsequently submerged in fibroblast media for a week. **b** Keratinocytes are seeded onto the hydrogel surface and the construct is maintained under submerged conditions in mixed media for another week. **c** Air-liquid interface culture wherein the construct is exposed to stratification media on the basolateral surface. This results in keratinocyte differentiation into a stratified epidermis

3.8 Field Emission Scanning Electron Microscopy

Organotypic skin constructs were dehydrated by sequential immersion in ethanol solutions of increasing concentration, followed by crucial point drying using an Autosamdri-815 Series A Critical Point Dryer. The dried samples were then sputter-coated with platinum in a JEOL JFC 1600 High Resolution Sputter Coater. The coated sample was then examined with a JEOL JSM-7400F FESEM system using an accelerating voltage of 5 kV.

3.9 Qualitative Gene Expression Analysis Using Real Time PCR

The constructs were dissolved in 500 μL of TRIzol (Life Technologies, USA). The mRNA was extracted using chloroform extraction, followed by isopropanol-ethanol precipitation. The concentration of mRNA content was quantified via absorbance reading using NanoDrop 2000 Spectrophotometer (Thermo Fisher Scientific, USA). 500 ng mRNA was subsequently used to prepare cDNA via reverse transcription using Super Script (Thermo Fisher Scientific, USA). 4 μL of template cDNA, 5 μL SYBR Green (Kapa Biosystems, USA) and 0.2 μL ROX dye were combined to make the template mix. 9 μL of the template mix and 1 μL of diluted primer (col3A1, CRABP2, KLK5, CDSN2 and GADPH) were added to the walls of a 96 well plate. The plate was sealed with thermosensitive film and incubated in the 7500 Fast Real-time PCR System (Applied Biosystems, USA).

4 Results and Discussion

4.1 Peptide Synthesis

The peptide was synthesized by solid-phase peptide synthesis and subsequently puri-fied using high performance liquid chromatography. The collection of the desired product is triggered by the detection of the 641 mass peak in the spectrometer. With reference to the chromatograph (Fig. 2a) and mass spectra of the collected fragment (Fig. 2b), there is little contamination of the crude product by deletion, addition or substitution peptide sequences. The collected product (Fig. 2c) is very pure.

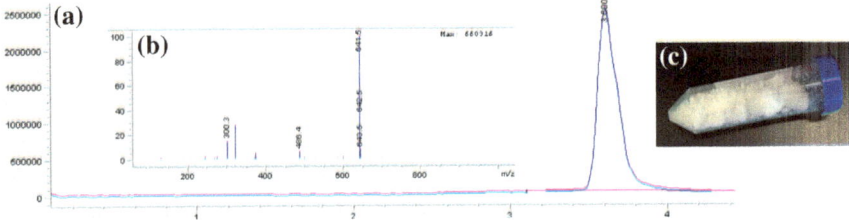

Fig. 2 Purification of the peptide product. The **a** chromatograph and **b** mass spectra of the collected fraction reflect the purity of the product. The desired product has a molecular weight of 640. It is detected in the mass spectrometer as a singly charged species as denoted by the 641 peak. The 642 and 643 peaks contain deuterium isotopes, while the 321 peak is indicative of the double charged species. **c** The purified, lyophilized peptide product is a fluffy white powder

4.2 Gelation Behavior and Mechanical Properties

Rigid hydrogels with storage moduli (G′) in the range of 10 kPa were obtained, using PBS as the buffer. From the frequency sweep study (Fig. 3a), as frequency increases, the gel remains fairly stable. Increasing peptide concentration increases both storage and loss (G″) moduli. Variation between samples is very small (n = 3). From the amplitude sweep study (Fig. 3b), G′ and G″ drops sharply when the strain increases past a certain point, indicating that the hydrogel's microstructure is breaking down. In both studies, G′ remains higher than G″, demonstrating that the samples remain as gels throughout the measurement. When peptide concentration increases from 6 to 7.5 mg/mL, G′ increases from 10^4 to 1.5×10^4 Pa. When peptide concentration increases from 7.5 to 9 mg/mL, G′ increases from 1.5×10^4 to 6×10^4 Pa.

The 6 mg/mL hydrogel formulation was chosen for preparation of tissue constructs as it sufficiently stable (above 10^4 Pa). The lower concentration hydrogel conserves

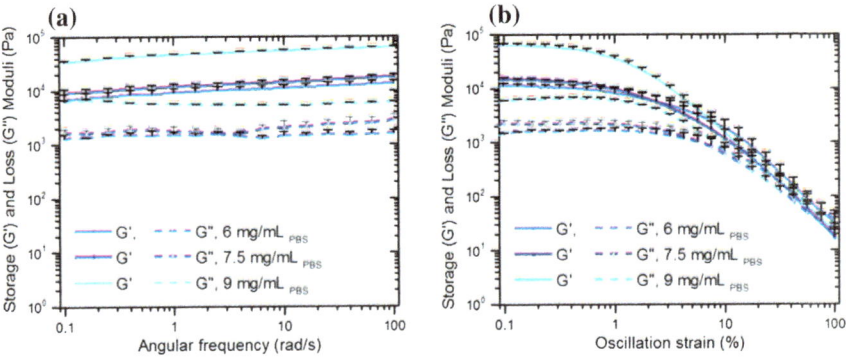

Fig. 3 The peptide hydrogels demonstrate good mechanical strength, as reflected by high storage moduli in the range of 10 kPa. **a** Frequency and **b** amplitude sweep studies show that hydrogel rigidity increases with peptide concentration

HDFs on tissue culture
(a)   plate (control)

HDFs encapsulated in
peptide hydrogel

(b)

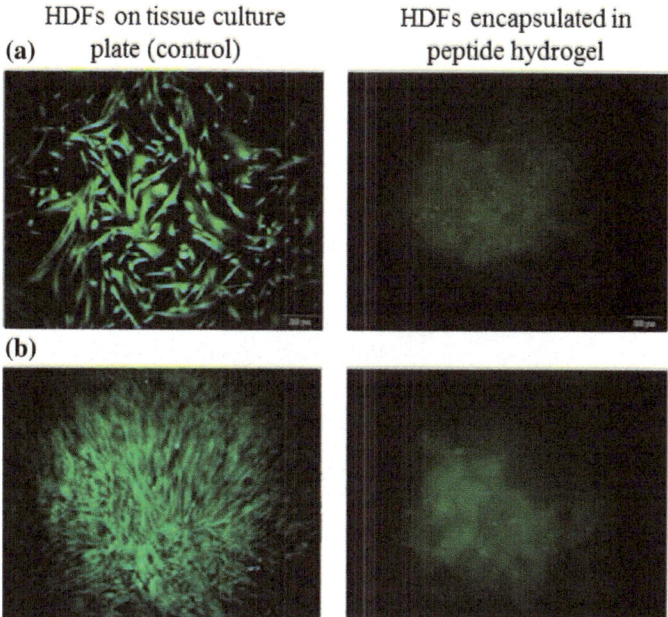

Fig. 4 Encapsulated fibroblasts maintain their viability after **a** one and **b** three weeks of culture, as evident from calcein (green) staining

material and reduces cost, resulting in significant savings when producing tissue constructs on a large scale. Moreover, the gelation time of the 6 mg/mL hydrogel was about 30 min, which is comparable to that observed in literature. During this window, the gel is very soft and fluid. It can be mixed to distribute the cells uniformly, and consistently dispensed into the transwells before solidification.

4.3 HDF Encapsulation

The cell viability of the encapsulated HDFs is high following encapsulation at the one- and three-week time points (Fig. 4), demonstrating that the hydrogel has low cytotoxicity. These results suggest that the peptide hydrogel can support the long-term culture of fibroblasts in organotypic skin constructs.

4.4 Organotypic Skin Culture

The gel retained its shape and size after 3 weeks of culture; no shrinkage was observed (Fig. 5a). The surface of the gel at the air-liquid interphase was smooth.

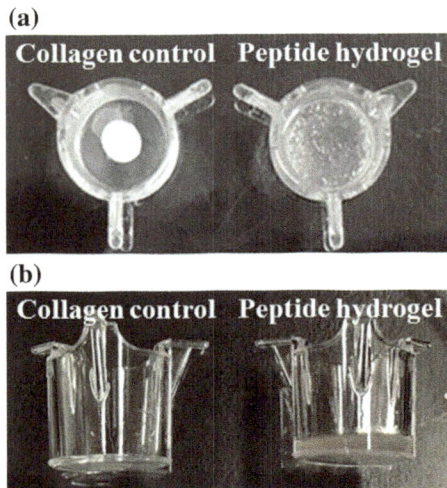

Fig. 5 Organotypic skin constructs after culturing for 3 weeks. Comparison of the collagen control (left) and peptide hydrogels (right) reflect the dramatic shrinkage of the former while the latter remains largely unchanged in both the **a** top view and **b** side view

From the histology images (Fig. 6), the stratified layer of the organotypic skin construct can be observed. Similar to the keratinized stratified squamous epithelium of human skin, the upper most layer of the construct consists of tightly packed, flattened keratinocytes, layered on top of more cuboidal cells near the hydrogel.

The surface topology of the organotypic skin constructs was smooth (Fig. 7a). From the SEM image of the underside, the hydrogel is visibly present and the nanofibrous microarchitecture is preserved even after 3 weeks of culture (Fig. 7b).

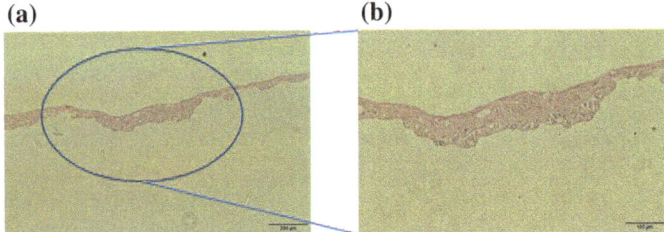

Fig. 6 Histology of the tissue constructs after 2 weeks of stratification culture. **a** At low magnification (4×), it is observed that the keratinocytes formed a continuous layer on the hydrogel. **b** At higher (10×) magnification, the organization of differentiated keratinocytes into a pluristratified epidermis was seen

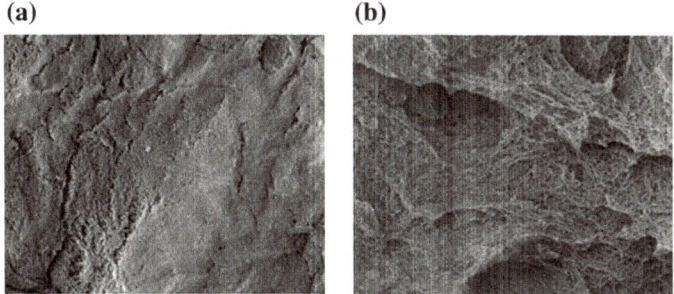

Fig. 7 Field emission scanning microscopy images of the **a** apical and **b** basolateral surfaces of the tissue constructs. **a** The stratified keratinocytes have a smooth, continuous surface. **b** The basolateral surface reflects the nanofibrous microarchitecture of the peptide hydrogel

4.5 Gene Expression

After exploring different primer sequences, the following primers are found to be appropriate for detecting gene expression in these organotypic skin constructs. Col3A1 codes for type 3 collagen, while CRABP2 codes for the Cellular Retinoic Acid Binding Protein, both of which are likely expressed by the fibroblasts. CDSN2 codes for corneodesmosin, which is involved in corneocyte maturation, and KLK5 codes for Kallikrein-related peptidase 5, which is suggested to regulate desquamation, which are likely expressed by the keratinocytes (Fig. 8).

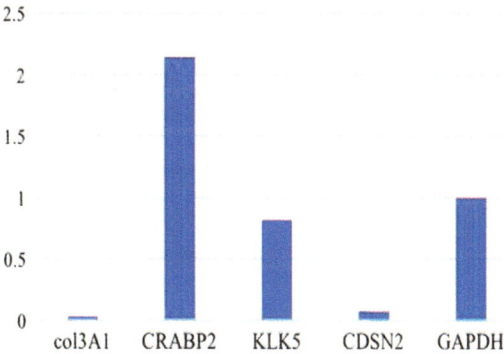

Fig. 8 Relative expression of genes

5 Conclusions, Challenges Encountered and Recommendations for Future Work

During the course of this project, the specific aims to synthesize a suitable peptide candidate, optimize the formulation of peptide hydrogels and characterise a simple organotypic skin construct were achieved. These biological constructs can subsequently be used to screen therapeutic candidates, as well as to evaluate the effects of compounds on skin tissue viability, permeability and cellular gene expression.

Due to the limited time frame of this project, the evaluation of a model test compound—retinoic acid, was limited in scope. The results are presented in Appendix. Briefly, two concentrations of retinoic acid (RA) (0.05 and 0.25%) were applied to the constructs. The preliminary data suggests that KLK5 and CDSN2 were upregulated with increasing concentration of RA. However, the small sample set and limited data points did not allow reliable conclusions to be drawn. One of the major challenges encountered was designing the dosages for the in vitro model to correlate well with published studies on human samples. Majority of the published literature utilize human patient participants who applied the test compound as a cream to parts of their skin, which was subsequently either biopsied or evaluated non-invasively. To eliminate potentially confusing results from possible permeation enhancers or toxic additives in the cream, attempts were made to dissolve RA in dimethylsulfoxide and dilute in PBS for application to the in vitro constructs. Due to differences in application medium, as well as differences in surface area of application, there were discrepancies between our results and literature published date. The RA applied in our study is likely to be of a higher concentration than what is applied to skin in vivo per unit area. Therefore, in the future, determining concentrations of RA more similar to that of what is applied topically can be explored in order for more reliable comparisons between organotypic skin constructs and in vivo studies can be made. Additionally, a wider variety of model compounds, such as salicylic acid or benzoic acid, across a wide range of concentrations should be tested, to validate the model for evaluating novel therapeutic candidates.

In this project, a simplified organotypic skin construct was prepared using fibroblasts encapsulated in a synthetic peptide hydrogel matrix, over which keratinocytes were allowed to proliferate, differentiate and stratify. Building on this model, more elaborate systems can be cultured with the addition of more cell types such as immune cells and adipocytes into the dermal compartment, as well as melanocytes into the epidermis. For example, the addition of normal versus malignant melanocytes can be used to study the mechanism of epidermal pigmentation and melanoma progression [14].

In conclusion, IK6 is an appropriate matrix for culturing organotypic skin constructs due to its stability and low cytotoxicity. With great advancements in biomedical engineering in the last decade, it can be hoped that such constructs will be reliable and accurate solution to the complex problems of animal testing, including inherent differences in skin physiology and gene expression between human and animal samples.

Acknowledgements This work was supported by the Youth Research Program (YRP) at the Institute of Bioengineering and Nanotechnology (Biomedical Research Council, Agency for Science, Technology and Research, Singapore).

Appendix

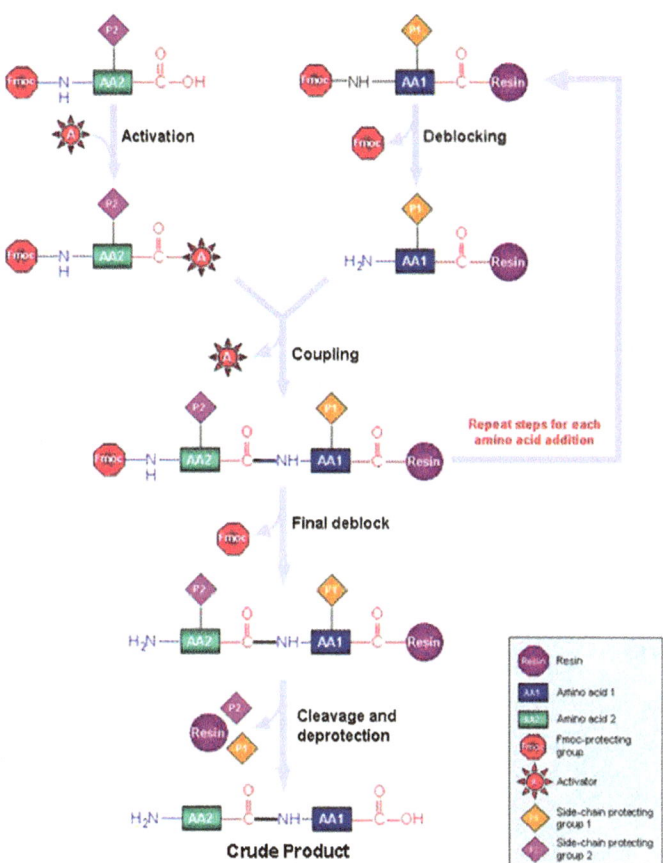

Solid phase peptide synthesis scheme. *Source* https://www.sigmaaldrich.com/life-science/custom-oligos/custom-peptides/learning-center/solid-phase-synthesis.html

Primers used		
Gene	Forward (5'-3')	Reverse (5'-3')
col3A1	5'-TCTTGGTCAGTCCTATGCGGATA-3'	5'-CATCGCAGAGAACGGATCCT -3'
CRABP2	5'- CAAGACCTCGTGGACCAGAGA-3'	5'- ACCCTGGTGCACACAACGT-3'
KLK5	5'- CCGGTGACAAAGCAGGTAG-3	5'-GAGCCATTGCAGACCACA-3'
CDSN2	5'-ATGATGGCACTGCTGCTG-3'	5'- AAGGTGCCAATGCTCTTAGC-3'
GAPDH	5'-TCCACTGGCGTCTTCACC-3'	5'-GGCAGAGATGATGACCCTTT-3'

Primers used in real time PCR

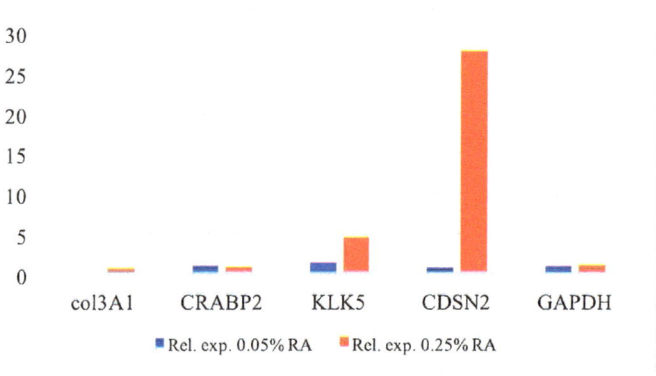

Relative expression of genes under 0.05 and 0.25% retinoic acid

References

1. Shamir, E., & Ewald, A. (2014). Three-dimensional organotypic culture: Experimental models of mammalian biology and disease. *Nature Reviews Molecular Cell Biology, 15*(10), 647–664. https://doi.org/10.1038/nrm3873.
2. Weigelt, B., Lo, A. T., Park, C. C., Gray, J. W., & Bissell, M. J. (2010). HER2 signaling pathway activation and response of breast cancer cells to HER2-targeting agents is dependent strongly on the 3D microenvironment. *Breast Cancer Research and Treatment, 122,* 35–43. https://doi.org/10.1007/s10549-009-0502-2.
3. Kenny, P. A., Lee, G. Y., Myers, C. A., Neve, R. M., Semeiks, J. R., Spellman, P. T., et al. (2007). The morphologies of breast cancer cell lines in three-dimensional assays correlate with their profiles of gene expression. *Molecular Oncology, 1,* 84–96. https://doi.org/10.1016/j.molonc.2007.02.004.
4. *EUR-Lex—32009R1223—EN—EUR-Lex.* (2017). *Eur-lex.europa.eu.* Retrieved December 28, 2017, from http://eur-lex.europa.eu/legal-content/EN/TXT/?uri=CELEX:32009R1223.
5. Panchagnula, R., Stemmer, K., & Ritschel, W. A. (1997). Animal models for transdermal drug delivery. Retrieved from https://www.ncbi.nlm.nih.gov/pubmed/9379782.

6. Andersen, T., Auk-Emblem, P., & Dornish, M. (2015). 3D cell culture in alginate hydrogels. *Microarrays, 4*(2), 133–161. https://doi.org/10.3390/microarrays4020133.

7. Lee, K., & Mooney, D. (2001). Hydrogels for tissue engineering. *Chemical Reviews, 101*(7), 1869–1880. https://doi.org/10.1021/cr000108x.

8. Rosso, F., Giordano, A., Barbarisi, M., & Barbarisi, A. (2004). From cell-ECM interactions to tissue engineering. *Journal of Cellular Physiology, 199*(2), 174–180. https://doi.org/10.1002/jcp.10471.

9. Timpson, P., Mcghee, E., Erami, Z., Nobis, M., Quinn, J., Edward, M., et al. (2011). Organotypic collagen I assay: A malleable platform to assess cell behaviour in a 3-dimensional context. *Journal of Visualized Experiments,* (56). http://dx.doi.org/10.3791/3089.

10. Hauser, C., Deng, R., Mishra, A., Loo, Y., Khoe, U., Zhuang, F., et al. (2011). Natural tri- to hexapeptides self-assemble in water to amyloid β-type fiber aggregates by unexpected α-helical intermediate structures. *Proceedings of the National Academy of Sciences, 108*(4), 1361–1366. https://doi.org/10.1073/pnas.1014796108.

11. Loo, Y., Lakshmanan, A., Ni, M., Toh, L., Wang, S., & Hauser, C. (2015). Peptide bioink: Self-assembling nanofibrous scaffolds for three-dimensional organotypic cultures. *Nano Letters, 15*(10), 6919–6925. https://doi.org/10.1021/acs.nanolett.5b02859.

12. Kirin, S., Noor, F., Metzler-Nolte, N., & Mier, W. (2017). Manual solid–phase peptide synthesis of metallocene–peptide bioconjugates.

13. Sriram, G., Bigliardi, P., & Bigliardi-Qi, M. (2015). Fibroblast heterogeneity and its implications for engineering organotypic skin models in vitro. *European Journal of Cell Biology, 94*(11), 483–512. https://doi.org/10.1016/j.ejcb.2015.08.00.

14. Eves, P., Haycock, J., Layton, C., Wagner, M., Kemp, H., Szabo, M., et al. (2003). Anti-inflammatory and anti-invasive effects of α-melanocyte-stimulating hormone in human melanoma cells. *British Journal of Cancer, 89*(10), 2004–2015. https://doi.org/10.1038/sj.bjc.6601349.

Resolving the Diagnostic Odyssey of a Patient with an Undefined Neuromuscular Disorder Using Massively Parallel Sequencing Approaches

Yu Yiliu, Ong Hui Juan, Swati Tomar, Grace Tan Li Xuan, Raman Sethi, Tay Kiat Hong and Lai Poh San

Abstract We investigated the effectiveness of DNA-based next-generation sequencing (NGS) targeting different genomic region and coverage from five platforms in resolving the diagnostic odyssey of a patient with an unidentified neuromuscular disorder. There were advantages and limitations associated with the different platform approaches. On average, over 22,000 rare protein effecting single nucleotide variants (SNVs), 1955 structural variants (SVs), and 229 copy number variants (CNVs) were identified and analyzed. Seven candidate SNVs fulfilled filtration criteria but were likely to be non-causative due to their classification or disease phenotype. Assessment of an intronic event through IGV and PCR suggested a region of structural rearrangement in DMD gene, which mismatched to two other genes on chromosome X hinting towards a possible novel mechanism of gene inactivation. Our results show that NGS platforms detect candidate variants but some disease mechanisms may remain undetected and points for need for caution when applying NGS for diagnostic purposes.

Keywords Next generation sequencing · Targeted sequencing · Whole exome sequencing · Whole genome sequencing · Neuromuscular disorder

1 Introduction

A diagnostic odyssey is a journey of searching for an accurate diagnosis of a patient, typically involving multiple clinical evaluations and laboratory tests for effective disease management [1]. During this period, patients may seek advice from multiple

Y. Yiliu · O. H. Juan (✉)
Raffles Institution, Singapore, Singapore
e-mail: atrainwreckofthoughts@gmail.com

S. Tomar · G. T. L. Xuan · R. Sethi · T. K. Hong · L. P. San
Department of Paediatrics, YYL School of Medicine, National University of Singapore, Singapore, Singapore

© Springer Nature Singapore Pte Ltd. 2019
H. Guo et al. (eds.), *IRC-SET 2018*,
https://doi.org/10.1007/978-981-32-9828-6_22

263

specialists, where many receive at least one incorrect diagnosis [2]. Hence, a diagnostic odyssey is a period of financial and emotional burden for the affected family [3].

In recent years, there has been an increase in the use of next-generation sequencing (NGS) to resolve diagnostic odysseys due to its high throughput platform and cost effectiveness [4]. There are three common types of DNA-based NGS platforms used clinically-targeted/exome panel sequencing, whole exome sequencing (WES) and whole genome sequencing (WGS). However, the diagnostic yield is still low and ranges from 8 to 45% [5–7], which is in turn linked to uneven coverage, incomplete capture [8] and discovery of variants of unknown significance (VUS) [9]. The choice of NGS platform also influences the diagnostic success rate, as they are categorized by varying read depth, average genomic region covered and types of variants detected [10–12]. Variations in read depth can affect the number and accuracy of detected variants, even when using same sequencing platforms [13]. Higher sequencing depth is recommended for accurate results for current NGS platforms which are primarily short-read sequencing (100–150 bp paired end) with a higher error rate (0.1–1% depending on sequencing platform) [14].

Neuromuscular disorders (NMDs) are a group of diseases which affect the peripheral nervous system and muscles [15]. Major challenges associated with diagnosis of NMDs include genetic heterogeneity [16, 17], overlapping phenotypes [18], and large repertoire of genes associated with muscular disorders [15]. In this project, we aimed to determine the candidate gene or genetic mechanism underlying an undiagnosed case of neuromuscular disorder (NMD) via five different NGS platforms encompassing targeted (to investigate specific genes closely associated with NMD), whole exome (to investigate all coding genes of human genome) and whole genome (to investigate the non-coding regions of human genome) sequencing. We also wanted to compare the effectiveness of the different sequencing platforms.

2   Hypothesis

We hypothesized that the use of multiple DNA based NGS platforms targeting different genomic region and coverage can increase the effectiveness of detecting disease associated candidate gene(s).

3   Methods

The anonymized patient's data was provided to us. The patient had no known family history and displayed following phenotypes: elevated serum creatine phosphokinase, muscular hypotonia, muscular dystrophy, flexion contracture, childhood onset, and autistic behaviour. Based on the above phenotypes, a panel of 109 genes was shortlisted for targeted sequencing (Platform T) which included top candidates associated with muscular dystrophy like DMD, FPRK, LAMA2 and LMNA among others. The overall workflow of this project was categorized into three stages: variant identification, variant filtration and variant prioritization. Variant identification involved the use of three different kinds of NGS to sequence genomic DNA from the patient. WES platform was called Platform E. WGS was conducted through three different sequencing coverages, namely Platform G (40x), Platform N (60x) and Platform A (90x). Trios analysis was performed through data generated from Platform G, as DNA samples from father, mother and the proband were sequenced (Appendix 2). The identified variants were categorized based on sequence ontology and size as follows: Single nucleotide variations, insertions and deletions (SNV-indels), structural variations (SVs) and copy number variations (CNVs).

3.1 Variant Filtration

SNV-indels (Fig. 1) were detected from all five sequencing platforms. Variants affecting the coding regions of the genes—exonic/splice site, were selected as more likely candidates to alter gene function. For splice site variants, they were considered to be deleterious if they had a dbscSNV_ADA score and dbscSNV_RF score of ≥ 0.6 [19] and only essential splice site variants (± 2 bp) were considered to have a significant

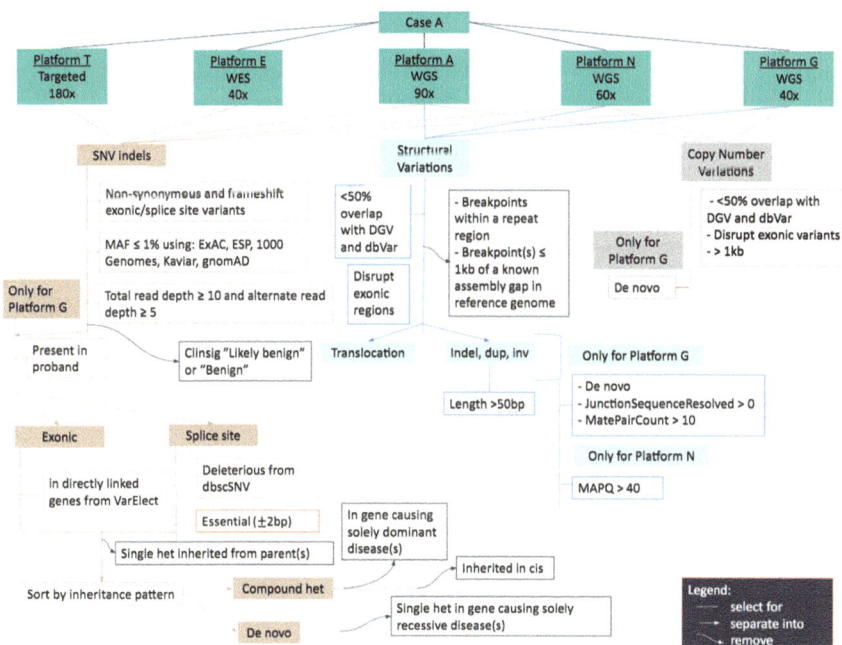

Fig. 1 Inhouse variant analysis pipeline

impact on the gene product. The dbscSNV scores predict splicing impact and a larger score indicates more likely splice altering. We further filtered out synonymous and non-frameshift variants. All variants were annotated against five common population databases—ExAC [20], ESP [21], 1000 Genomes [22], Kaviar [23] and gnomAD [20]. Only variants with a minor allele frequency (MAF) of ≤1% were categorized as rare variants. An exception was made for the filtration of Platform G whereby the MAF filtration criteria was relaxed to 5% MAF as no results could be obtained from the former criterion. Furthermore, a total read depth ≥ 10 and an alternate read depth ≥5 was considered as a cutoff for true positive variant.

Gene inheritance patterns were obtained from Online Mendelian inheritance In Man (OMIM) [24]. All single heterozygous variants inherited from either or both parents were removed, for they could not sustain a dominant disease model as the parents were healthy, nor a recessive model as a single variant would be incapable of disease manifestation. The remaining variants were then categorized into de novo, compound heterozygous and X-linked (inherited in heterozygous state from mother only) models. For compound heterozygous variants, genes that were associated with only dominant disease models and variants that were inherited from same parent, i.e. inherited in cis, were ruled out.

SVs and CNVs (Fig. 1) were detected by WGS platforms only (Platforms A, N and G). SVs were categorized into deletions, duplications (or insertions), inversions and translocations. We used BEDTools [25] to compare variants against common variations database—DGV [26] and dbVar [27] and variants with >50% overlaps across normal known variants were removed. Variants were further filtered out based on (i) any breakpoint located within 1 kb of a known assembly gap region in the reference genome, or with both breakpoints within repeat region, (ii) <50 bp and (iii) disrupted non coding regions. Platform G was filtered only for de novo SV and CNV variants as it contained trios data. High quality SVs were selected based on JunctionSequenceResolved >0 (for physical connection between the left and right genomic sequences of a breakpoint junction) and MatePairCount >10 (for sufficient DNB support for the junction) [28]. For Platform N, mapping quality (MAPQ) >40 was considered. Only variants involving exonic regions of genes or entire copy of gene were considered for CNVs.

3.2   Variant Prioritisation

Genes of selected variants were prioritized using VarElect [29] and only those with a direct association to the given phenotype were selected as they were more likely to be causative. Final SNV-indel candidates were classified by incorporating American College of Medical Genetics and Genomics (ACMG) guidelines [30]. The top SVs and CNVs were visually assessed using IGV (Integrative Genomics Viewer) [31].

4 Results and Discussion

4.1 Sequence Variations

When considering only rare ($\leq$1% MAF) protein effecting variants, 15,185 variants were identified from Platform A (~90x), 2501 from Platform N (~60x), 2611 from Platform G (~40x), 1694 from Platform E (~60x) and 36 variants from Platform T (~140x) (Appendix 3). Of these, only 7 SNV-indels in five genes fulfilled all the filtration criteria (Table 1). TTN is directly associated with neuromuscular diseases such as limb-girdle muscular dystrophy (LGMD) type 2J and FRG1 is associated with facioscapulohumeral muscular dystrophy (FSHD) while MYOM3 overexpression is a biomarker for Duchenne muscular dystrophy and LGMD phenotype (32). LDB3 causes cardiomyopathy, left ventricular noncompaction 3 and myofibrillar myopathy 4. This was ruled out, as our patient did not present any heart related phenotype and myofibrillar myopathy 4 has a late onset (from 44 to 73 years) and our patient is a child. Lastly, the missense mutation in FRG1 was removed as a candidate as the phenotype of FSHD included muscle weakness that first affects the facial muscles and upper extremities which was inconsistent with our patient's phenotype. Finally, there were 3 candidate variants remaining from the genes TTN and MYOM3. Both variants in TTN were classified as likely benign while the variant in MYOM3 was classified as a VUS.

4.2 Structural and Copy Number Variations

Final CNV events as obtained from Platforms A, N, and G were 203, 3, and 0 respectively. A total of 63, 1888, and 4 SVs were detected from each platform, respectively. Only two genes associated with neuromuscular disorders harboured copy number losses (LARGE (Chr 22), ATXN7 (Chr 3), one copy loss each), and none harboured any gains. Only one gene associated with neuromuscular disorders (FRG1) harboured an SV (translocation) from Platform A. Among the genes harbouring final SV variants from Platform N, 55 muscle panel genes harboured deletions, 25 harboured duplications, 33 harboured inversions, and 2 harboured translocations. All 33 muscle panel genes carrying final SVs from Platform G harboured inversions (Fig. 2). Overlapping SVs across the three WGS platforms were detected through Shell script (Appendix B). A total of 112 overlapping SVs effecting 575 genes were detected. Of which, one muscle-panel gene (GNB4) was a part of 40 Mb deletion at Chr3, and 5 muscle-panel genes (AKAP9, AP4M1, GATAD1, HSPB1, SGCE) were contained in a 38 Mb duplication at Chr7. However, these variants were unlikely to be true as the number of discordant reads as seen through IGV were too low (<15%). No overlapping CNVs were found.

Multiple large SVs (of size >3 mb) were found by Platforms N (108 SVs) and G (29 SVs), spanning 96 muscle panel genes. Of which, 2 deletion with insert sizes

Table 1 Final filtered SNV-indels

Gene	Details	Zygosity	Population frequency	ACMG classification
LDB3	Missense; LDB3:NM_001171610: exon10:c.T1348C:p.Y450H	het	esp6500: 0; ExAC: 0; 1000 genomes: 0	VUS (Potentially pathogenic)—PM2, PP2, PP3
LDB3	Missense; LDB3:NM_001171610: exon10:c.A1349C:p.Y450S	het	esp6500: 0; ExAC: 0; 1000 genomes: 0	VUS (Potentially pathogenic)—PM2, PP2, PP3
LDB3	Frameshift; LDB3:NM_001171610: exon10:c.1305dupC:p.T435fs	het	esp6500: 0; ExAC: 0; 1000 genomes: 0; Kaviar: 0; gnomAD (exome): 0; gnomAD (genome): 0	VUS (Potentially pathogenic)—PM2
MYOM3	Stopgain; MYOM3:NM_152372: exon5:c.G430T:p.E144X	het	esp6500: 0; ExAC: 0.0063; 1000 genomes: 0.00998403; Kaviar: 0.0052716; gnomAD (exome): 0.0055; gnomAD (genome): 0.0013	VUS
FRG1	Missense; FRG1:NM_004477: exon3:c.A196G:p.K66E	het	esp6500: 0; ExAC: 0.00003349; 1000 genomes: 0	Likely benign—BP1, BP4
TTN	Missense; TTN:NM_001267550: exon300:c.A59282G:p.N19761S	het	esp6500: 0; ExAC: 0.0006; 1000 genomes: 0.00079872; Kaviar: 0.0004851; gnomAD (exome): 0.0004; gnomAD (genome): 0.0001	Likely benign—BS1, BP1
TTN	Missense; TTN:NM_001267550: exon81:c.T23638A:p.S7880T	het	esp6500: 0; ExAC: 0; 1000 genomes: 0; Kaviar: 0; gnomAD (exome): 0.000004126; gnomAD (genome): 0	Likely benign—BP1, BP4

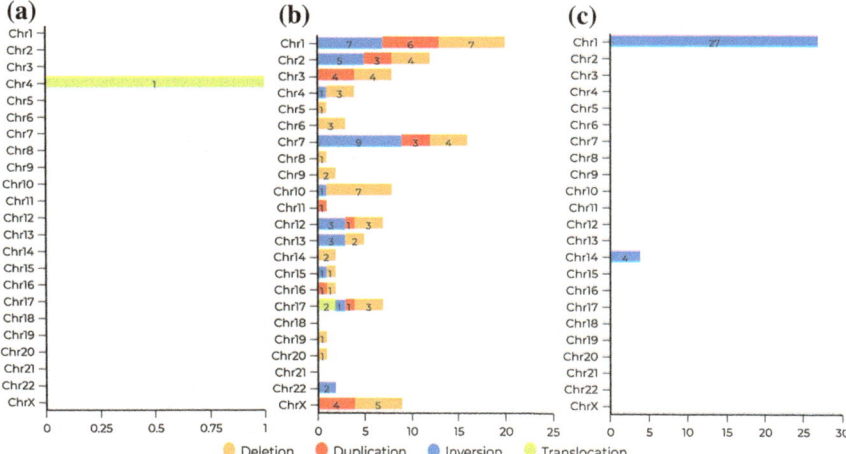

Fig. 2 Number of muscle panel genes affected by final SV variants in each chromosome. **a** Number of genes harbouring variants from Platform A. **b** Number of genes harbouring variants from Platform N. **c** Number of genes harbouring variants from Platform A. Annotation: Chr, chromosome

3Mbp and 18Mbp were observed in DMD by Platform N, which is associated with Duchenne muscular dystrophy. The deletions were matched to CFAP47 gene and GLRA2 gene respectively, each with one breakpoint residing in Intron 9 of DMD. The breakpoint matched to CFAP47 gene had approximately 51% (18/35) of discordant reads from Platform N and 37% (10/27) from Platform A. The breakpoint with reads mapping to GLRA2 gene had approximately 41% (12/29) of discordant reads from Platform N and 52% (15/29) from Platform A. A PCR was performed on normal and patient's DNA encompassing this region to verify whether raw DMD sequence could be amplified. Our PCR result showed that there was no amplification in patient's DNA vs the control DNA (Fig. 3). However, the region between the two breakpoints

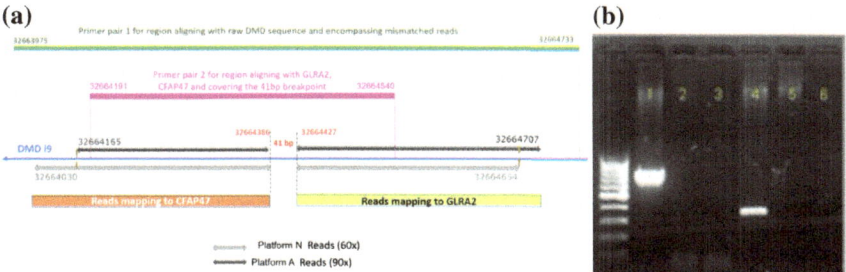

Fig. 3 **a** 2 primers were designed for PCR. **b** Lane 1: normal control for 759 bp amplicon; Lane 2: DMD sample; Lane 3: negative control for 759 bp amplicon; Lane 4: normal control for 350 bp amplicon; Lane 5: DMD sample; Lane 6: negative control for 350 bp amplicon. Test samples were not amplified, suggesting the presence of a deletion

were well covered in general and showed no significant signs of a deletion as viewed from IGV. Bam file visualization for both Platform A (90x) and Platform N (60x) both confirmed similar phenomena although the large deletion was reported in the NGS data from both platforms. Previous research based on a well-studied pedigree NA12878 reported SV identification tools DELLY (used by Platform N) and Manta (used by Platform A) to have high false discovery rates [32, 33]. While an abnormality is confirmed to be present in Intron 9 of DMD gene, in light of the inconclusive results from IGV visualisation, it is unclear whether this variant truly effects the entire DMD gene. If the causal variant is really the structural rearrangement as observed in intron 9 of DMD gene, it is rather complex and cannot be detected by the other testing technologies, and was hence missed in previous and current studies.

4.3 Discussion

The overlapping SNV-indels identified across five sequencing platforms included 13 stop-gain, 10 frameshift, 600 missense and 1139 splice site variants. However, we excluded final filtered variants in TTN that were classified under ACMG clinical guidelines as "Likely benign". Another candidate variant in MYOM3 has been previously reported in dbSNP (rs143187236) but not in ClinVar [34]. MYOM3 belongs to a family of structural proteins that localize to the M-band of the sarcomere in striated muscle and are involved in sarcomere stability as well as resistance during intense or sustained stretch. It was reported that MYOM3 protein were overexpressed in sera of 39 DMD patients but not in 38 controls [35]. However, since our proband harbours a stopgain mutation in this gene, which would most likely result in a truncated protein or result in nonsense-mediated mRNA decay, it is thus unlikely to have an association with the previously reported observation in DMD patients.

Platform N's CNV events were called using Control-FREEC [36] and it reported copy number losses in Chromosome X, with copy number of only "1". However, they were found to be false positives because the proband is a male with one copy of X chromosome.

The top candidate variants for SV and CNV were a copy number loss in LARGE gene and a translocation in FRG gene. Both were muscle panel genes and were directly related to the patient's identified phenotypes. They are probable true variants. Mutation in the LARGE gene is known to cause muscular dystrophy—dystroglycanopathy type A, 6 and type B, 6, both autosomal recessive. Phenotypes matched with the patient were "Elevated serum creatine phosphokinase", "Muscular dystrophy", "Childhood onset", "Muscular hypotonia" and "Flexion contracture". However, mental retardation (muscular dystrophy—dystroglycanopathy type B, 6), and brain and eye anomalies (muscular dystrophy—dystroglycanopathy type A, 6) were not present in the patient. FRG1 is associated with facioscapulohumeral muscular dystrophy (autosomal dominant). Phenotypes matched were "Elevated serum creatine phosphokinase", "Muscular dystrophy" and "Childhood onset". Despite this, the involvement of facial muscles, characteristic of facioscapulohumeral muscular dystrophy, did not coincide with the patient's phenotype. Thus, it was unlikely that these 2 variants were the cause of the patient's disease.

Our results showed different platforms yielded varying results in detecting candidates genes. This is related to different sequencing depths and approaches. Ability to use targeted (Platform T) and lower coverage (either WES or WGS) to detect causal variants would save cost. We noted that there are limitations and strengths for these platforms. Nonetheless, our results showed candidate genes could be detected for which further assessment is required which may rule them out. Multiple approaches can enhance effectiveness of determining candidate gene expression for challenging undiagnosed cases. In this project, we found many candidate variants but they were likely to be benign based on our analysis. It is possible that other disease mechanisms may be involved such as methylation, post-translational modification, etc., that cannot be detected by our DNA-based NGS approaches. Hence, this points towards caution when applying such technologies for diagnosis.

5 Conclusion

While the use of multiple platforms did provide us with a plethora of variants, most of the identified variants were filtered out based on our in-house pipeline and only a very small portion of them fulfilled all of our criteria. This observation highlights that even though multiple DNA-NGS platforms may yield large data for analysis, it may still be unable to find causative variants in some rare cases. This may be explained due to the presence of undetectable sequence alterations or require other means of NGS such as long read sequencing or whole transcriptome sequencing, where feasible. Lastly, the presence of a disease-causing intronic variants cannot be ruled out from our observation.

Acknowledgements We would like to thank our mentors: Professor Lai Poh San, Dr. Swati Tomar, Ms. Grace Tan, Mr. Raman and Mr. Tay Kiat Hong for their invaluable guidance and help throughout the entire course of the project.

References

1. Thevenon, J., Duffourd, Y., Masurel-Paulet, A., Lefebvre, M., Feillet, F., El Chehadeh-Djebbar, S., et al. (2016). Diagnostic odyssey in severe neurodevelopmental disorders: Toward clinical whole-exome sequencing as a first-line diagnostic test. *Clinical Genetics, 89*(6), 700–707. https://doi.org/10.1111/cge.12732.
2. Zurynski, Y., Deverell, M., Dalkeith, T., Johnson, S., Christodoulou, J., Leonard, H., et al. (2017). Australian children living with rare diseases: Experiences of diagnosis and perceived consequences of diagnostic delays. *Orphanet Journal of Rare Diseases, 12*(1), 68. https://doi.org/10.1186/s13023-017-0622-4.

3. Wong, S. H., McClaren, B. J., Archibald, A. D., Weeks, A., Langmaid, T., Ryan, M. M., et al. (2015). A mixed methods study of age at diagnosis and diagnostic odyssey for duchenne muscular dystrophy. *European Journal of Human Genetics, 23*(10), 1294–1300. https://doi.org/10.1038/ejhg.2014.301.

4. O'Donnell-Luria, A. H., & Miller, D. T. (2016). A clinician's perspective on clinical exome sequencing. *Human Genetics, 135*(6), 643–654. https://doi.org/10.1007/s00439-016-1662-x.

5. Gilissen, C., Hehir-Kwa, J. Y., Thung, D. T., van de Vorst, M., van Bon, B. W. M., Willemsen, M. H., et al. (2014). Genome sequencing identifies major causes of severe intellectual disability. *Nature, 511*(7509), 344–347. https://doi.org/10.1038/nature13394.

6. Savarese, M., Sarparanta, J., Vihola, A., Udd, B., & Hackman, P. (2016). Increasing role of titin mutations in neuromuscular disorders. *Journal of Neuromuscular Diseases, 3*(3), 293–308. https://doi.org/10.3233/JND-160158.

7. Dixon-Salazar, T. J., Silhavy, J. L., Udpa, N., Schroth, J., Schaffer, A. E., Olvera, J., et al. (2012). Exome sequencing can improve diagnosis and alter patient management. *Science Translational Medicine, 4*(138). https://doi.org/10.1126/scitranslmed.3003544.exome.

8. Zhang, X. (2014). Exome sequencing greatly expedites the progressive research of mendelian diseases. *Frontiers of Medicine in China, 8*(1), 42–57. https://doi.org/10.1007/s11684-014-0303-9.

9. Pal, L. R., Kundu, K., Yin, Y., & Moult, J. (2017). CAGI4 SickKids clinical genomes challenge: A pipeline for identifying pathogenic variants. *Human Mutation, 38*(9), 1169–1181. https://doi.org/10.1002/humu.23257.

10. Marian, A. J. (2014). Sequencing your genome: What does it mean? *Methodist DeBakey Cardiovascular Journal, 10*(1), 3–6. https://doi.org/10.14797/mdcj-10-1-3.

11. Liew, W. K. M., Ben-Omran, T., Darras, B. T., Prabhu, S. P., De Vivo, D. C., Vatta, M., et al. (2013). Clinical application of whole-exome sequencing. *JAMA Neurology, 70*(6), 788. https://doi.org/10.1001/jamaneurol.2013.247.

12. Belkadi, A., Bolze, A., Itan, Y., Cobat, A., Vincent, Q. B., Antipenko, A., et al. (2015). Whole-genome sequencing is more powerful than whole-exome sequencing for detecting exome variants. *Proceedings of the National Academy of Sciences, 112*(17), 5473–5478. https://doi.org/10.1073/pnas.1418631112.

13. Fumagalli, M. (2013). Assessing the effect of sequencing depth and sample size in population genetics inferences. *PLoS One, 8*(11), e79667. https://doi.org/10.1371/journal.pone.0079667.

14. Desai, A., Marwah, V. S., Yadav, A., Jha, V., Dhaygude, K., Bangar, U., et al. (2013). Identification of optimum sequencing depth especially for de novo genome assembly of small genomes using next generation sequencing data. *PLoS One, 8*(4), e60204. https://doi.org/10.1371/journal.pone.0060204.

15. Laing, N. G. (2012). Genetics of neuromuscular disorders. *Neuromuscular Disorders of Infancy, Childhood, and Adolescence: A Clinician's Approach, 49,* 17–31. https://doi.org/10.1016/B978-0-12-417044-5.00002-0.

16. Timmerman, V., Strickland, A. V., & Züchner, S. (2014). Genetics of Charcot-Marie-Tooth (CMT) disease within the frame of the human genome project success. *Genes, 5*(1), 13–32. https://doi.org/10.3390/genes5010013.

17. Nigro, V., & Savarese, M. (2014). Genetic basis of limb-girdle muscular dystrophies: The 2014 update. *Acta Myologica, 33*(1), 1–12.

18. Bönnemann, C. G., Wang, C. H., Quijano-Roy, S., Deconinck, N., Bertini, E., Ferreiro, A., et al. (2014). Diagnostic approach to the congenital muscular dystrophies. *Neuromuscular Disorders, 24*(4), 289–311. https://doi.org/10.1016/j.nmd.2013.12.011.diagnostic.

19. Li, Q., & Wang, K. (2017). InterVar: Clinical interpretation of genetic variants by the 2015 ACMG-AMP guidelines. *American Journal of Human Genetics, 100*(2), 267–280. https://doi.org/10.1016/j.ajhg.2017.01.004.

20. Lek, M., Karczewski, K. J., Minikel, E. V., Samocha, K. E., Banks, E., Fennell, T., et al. (2016). Analysis of protein-coding genetic variation in 60,706 humans. *Nature, 536*(7616), 285–291. https://doi.org/10.1038/nature19057.

21. NHLBI grand opportunity exome sequencing project (ESP). (2017). Accessed December 26. https://esp.gs.washington.edu/drupal/.
22. Auton, A., Abecasis, G. R., Altshuler, D. M., Durbin, R. M., Abecasis, G. R., Bentley, D. R., et al. (2015). A global reference for human genetic variation. *Nature, 526*(7571), 68–74. https://doi.org/10.1038/nature15393.
23. Glusman, G., Caballero, J., Mauldin, D. E., Hood, L., & Roach, J. C. (2011). Kaviar: An accessible system for testing SNV novelty. *Bioinformatics, 27*(22), 3216–3217. https://doi.org/10.1093/bioinformatics/btr540.
24. Hamosh, A., Scott, A. F., Amberger, J. S., Bocchini, C. A., & McKusick, V. A. (2004). Online Mendelian Inheritance in Man (OMIM), a knowledgebase of human genes and genetic disorders. *Nucleic Acids Research, 33*(Database issue), D514–D517. https://doi.org/10.1093/nar/gki033.
25. Quinlan, A. R. (2014). BEDTools: The Swiss-army tool for genome feature analysis. *Current Protocols in Bioinformatics, 47*, 11.12.1–11.12.34. https://doi.org/10.1002/0471250953.bi1112s47.
26. MacDonald, J. R., Ziman, R., Yuen, R. K. C., Feuk, L., & Scherer, S. W. (2014). The database of genomic variants: A curated collection of structural variation in the human genome. *Nucleic Acids Research, 42*(Database issue), D986–D992. https://doi.org/10.1093/nar/gkt958.
27. Lappalainen, I., Lopez, J., Skipper, L., Hefferon, T., Spalding, J. D., Garner, J., et al. (2013). DbVar and DGVa: Public archives for genomic structural variation. *Nucleic Acids Research, 41*(Database issue), D936–D941. https://doi.org/10.1093/nar/gks1213.
28. Genomics Incorporated, Complete. (2013). *Standard sequencing service data file formats.* http://www.completegenomics.com/documents/DataFileFormats_Standard_Pipeline_2.5.pdf.
29. Stelzer, G., Plaschkes, I., Oz-Levi, D., Alkelai, A., Olender, T., Zimmerman, S., et al. (2016). VarElect: The phenotype-based variation prioritizer of the GeneCards Suite. *BMC Genomics, 17*(Suppl 2), 444. https://doi.org/10.1186/s12864-016-2722-2.
30. Rehm, H. L., Bale, S. J., Bayrak-Toydemir, P., Berg, J. S., Brown, K. K., Deignan, J. L., et al. (2013). ACMG clinical laboratory standards for next-generation sequencing. *Genetics in Medicine, 15*(9), 733–747. https://doi.org/10.1038/gim.2013.92.
31. Thorvaldsdóttir, H., Robinson, J. T., & Mesirov, J. P. (2013). Integrative Genomics Viewer (IGV): High-performance genomics data visualization and exploration. *Briefings in Bioinformatics, 14*(2), 178–192. https://doi.org/10.1093/bib/bbs017.
32. Chen, X., Schulz-Trieglaff, O., Shaw, R., Barnes, B., Schlesinger, F., Cox, A. J., et al. (2015). Manta: Rapid detection of structural variants and indels for clinical sequencing applications. *bioRxiv, 32*, 24232. https://doi.org/10.1101/024232.
33. Kronenberg, Z. N., Osborne, E. J., Cone, K. R., Kennedy, B. J., Domyan, E. T., Shapiro, M. D., et al. (2015). Wham: Identifying structural variants of biological consequence. *PLoS Computational Biology, 11*(12), 1–19. https://doi.org/10.1371/journal.pcbi.1004572.
34. Reference SNP (refSNP) Cluster Report: rs143187236. (2018). Accessed January 3. https://www.ncbi.nlm.nih.gov/projects/SNP/snp_ref.cgi?rs=143187236.
35. Rouillon, J., Poupiot, J., Zocevic, A., Amor, F., Leger, T., Garcia, C., et al. (2015). Serum proteomic profiling reveals fragments of MYOM3 as potential biomarkers for monitoring the outcome of therapeutic interventions in muscular dystrophies. https://helda.helsinki.fi/handle/10138/225083.
36. Boeva, V., Popova, T., Bleakley, K., Chiche, P., Cappo, J., Schleiermacher, G., et al. (2012). Control-FREEC: A tool for assessing copy number and allelic content using next-generation sequencing data. *Bioinformatics (Oxford, England), 28*(3), 423–425. https://doi.org/10.1093/bioinformatics/btr670.

Eccentric Mass Designs of Membrane-Type Acoustic Metamaterials to Improve Acoustic Performance

Ruochen Xu and Zhenbo Lu

Abstract Membrane-type acoustic metamaterials (MAMs) are engineered to be both extremely compact and lightweight with periodic structure. These materials reveal dramatically better sound insulation especially at lower frequencies than the conventional systems of similar dimensions and weight. It promises for the applications of lightweight construction and vehicular cabin to insulate sound at a specific frequency range. In this experiment, MAMs with eccentric masses were designed, and these prototypes were used for investigating their acoustic performances for seeking the optimised distribution of eccentric masses that can drastically increase the frequency band of sound absorption, which occurs especially in the low frequency range. Through simulations, their acoustic performances were quantified and graphed, and their surface displacement diagrams and transmission loss graphs were generated and analysed. The results showed that the MAMs with optimised eccentric mass distribution are able to drastically widen the frequency band of sound absorption in the low-frequency range, with the effectiveness varying slightly for different designs.

Keywords Metamaterials · Membrane-type acoustic metamaterials · Sound insulation · Eccentric masses · Acoustics

1 Introduction

Metamaterials are man-made materials engineered to possess properties that are otherwise undiscovered in nature [1]. Their unique properties stem from their newly designed structures, enabling them to manipulate electromagnetic, and even sound and light waves [2]. The idea of such metamaterials was first conceptualised by

R. Xu (✉)
Science Research Programme (SRP), Raffles Institution (JC), Singapore, Singapore
e-mail: ruochenn@gmail.com

Z. Lu
Aeroscience Division, Temasek Laboratories, National University of Singapore, Singapore, Singapore
e-mail: tslluz@nus.edu.sg

© Springer Nature Singapore Pte Ltd. 2019
H. Guo et al. (eds.), *IRC-SET 2018*,
https://doi.org/10.1007/978-981-32-9828-6_23

275

Veselago in 1967, where he proposed a theory of probable materials which were able to possess negative magnetic permeability (μ) and electric permittivity (ε), which gives rise to a negative refractive index [3]. This was further supported by Pendry, expanding on the theoretical development of electromagnetic metamaterials and allowing negative magnetic permeability and negative electric permittivity to be applied to many phenomena [4], such as superlensing, cloaking and negative refractive index [5].

Acoustic metamaterials, a less well-developed counterpart of metamaterials, have negative bulk modulus and/or negative mass density [6]. In order to bend sound waves and create a cloaking effect, an acoustic metamaterial would have to possess both negative bulk modulus and negative mass density [7]. Acoustic metamaterials also have the potential to absorb sound at lower frequencies, and are significantly more effective than typical systems of sound insulation with similar weight and dimensions [8]. This is applied to fields like the aerospace industry, where lightweight materials are preferred when manufacturing bodies like aircrafts to retain the aerodynamic nature and to reduce its overall mass [9].

Materials with negative mass density to effectively absorb and insulate sound include those of phononic crystals that comprise of solid spherical materials with high density and a soft elastic coating [10]. The membrane-type acoustic metamaterial (MAM) postulated by Yang et al. is able to provide for a thinner elastic membrane of up to 15 mm thickness and a density of less than 3 kg/m^2 held in place by an rigid frame. To adjust the resonances, a small weight is positioned in the middle of the elastic membrane [11], and the MAM panel is able to screen around 20 dB of sound at a frequency of 200 Hz. To aim for a broader range of sound absorption and isolation, different designs of central masses and ring masses attached to the elastic membrane were investigated to increase the effectiveness of the MAMs at screening sound. In order to study the optimised eccentric mass designs of the MAMs, MAMs with eccentric masses were designed to investigate and compare their acoustic performances in terms of sound insulation, and seek the optimised mass distribution which can effectively increase the frequency band of sound insulation in the low-frequency range.

2   Hypothesis

It is hypothesised that an eccentric mass distribution of two circular masses would be more effective than one due to new anti-resonance peaks induced with increased mass distribution. Separations between circular or split ring eccentric masses attached to the elastic membrane may be able to induce new anti-resonance transmission loss peaks, thus allowing parts of the masses as well as the elastic membrane to vibrate independently [12, 13].

3 Materials and Methods

The MAMs with eccentric masses were each designed using SolidWorks, a computer programme designed to facilitate solid modelling as well as 3-D engineering [14]. Each of the designs were drawn within a 100 mm by 100 mm membrane plate, and extruded with a thickness of 0.4 mm. Figures 1 and 2 include the various designs of the mass configurations on the MAMs drawn using SolidWorks.

The various MAMs with eccentric masses were then run through a COMSOL Multiphysics simulation to predict and determine their respective acoustic performances [15]. COMSOL Multiphysics is a computer software that will allow the following setup to be simulated, generating relevant data for the analysis of effectiveness of sound insulation of each eccentric mass design. In order to determine the effectiveness in terms of sound insulation of the respective MAMs, a rectangular duct with a loudspeaker is installed at one end and the MAM installed at the centre is simulated in the COMSOL Multiphysics simulation [16]. Hyper elastic material dielectric elastomer (DE) film of thickness 1.0 mm (3M VHB 4910) is used as the

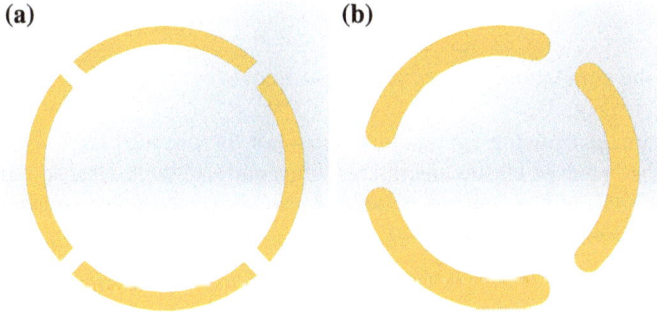

Fig. 1 Circular split ring mass with different distances between each section, **a** circle1 and **b** circle2

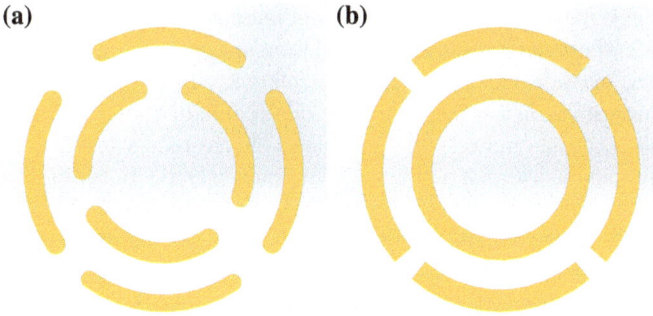

Fig. 2 One circular split ring mass and one circular ring mass, **a** 2-circles1 and **b** 2-circles2

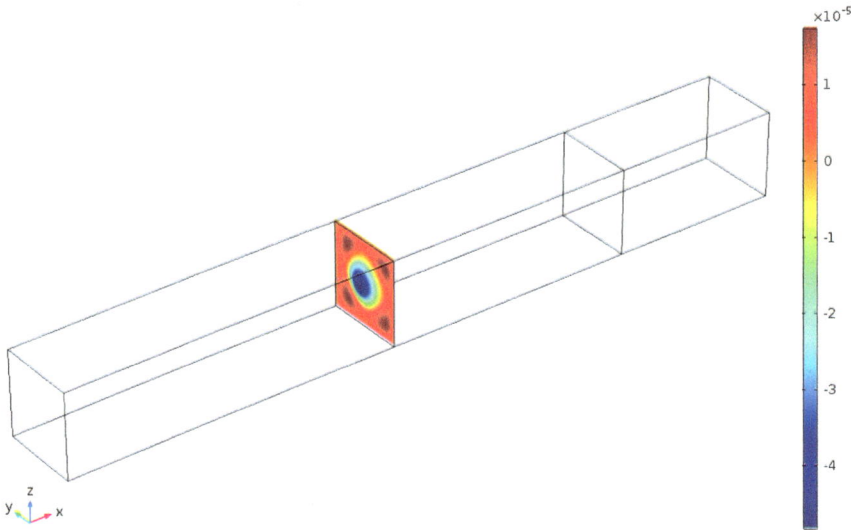

Fig. 3 Diagram of setup simulated

MAM for the simulation. The measurement frequency range of the simulated duct is 2–1000 Hz (Fig. 3).

An advantage of using simulations instead of the physical membrane and duct would be that a tensed elastic membrane with a uniform inner-stress distribution for the membrane would be guaranteed without using pre-stretch mechanisms while still retaining accuracy of results.

4   Results and Discussion

To investigate the acoustic performance of MAMs in both the low- and mid- frequency ranges, the aforementioned mass configurations were used, as illustrated in Figs. 1 and 2. These masses were embossed from a copper plate of thickness 0.4 mm. The eccentric mass designs on the elastic membrane will be able to induce new anti-resonance transmission loss peaks, and the different radii of the circular ring masses (i.e. 2-circles1 and 2-circles2) may be able to alter certain vibration modes of the elastic membrane to alter the anti-resonance peaks induced. In addition, separations between circular or split ring masses may allow part of the mass and the elastic membrane to vibrate independently and may induce new anti-resonance peaks. Therefore, each proposed eccentric mass design would have different acoustic performances and experience different surface displacements.

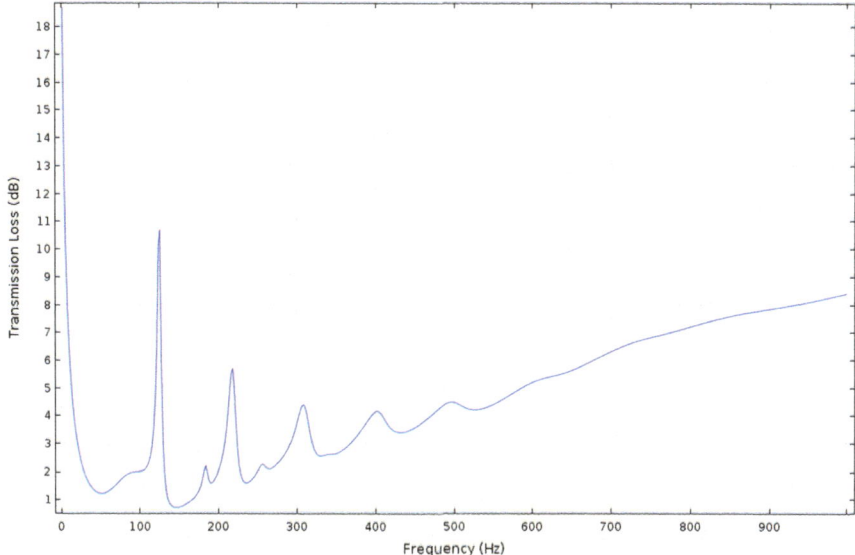

Fig. 4 Graph of transmission loss for membrane frame only

4.1 Membrane and Frame Only

Firstly, only the membrane and frame were run through the simulation and the results were tabulated and graphed in Fig. 4.

The transmission loss for the membrane plate peaked at a frequency of 126 Hz at a transmission loss of 10.7 dB. The surface displacement of the membrane plate was also recorded at the frequency of 126 Hz in order to get a visual representation of the membrane plate when sound waves are passed through it, as shown in Fig. 5.

Thereafter, the various MAM designs as shown in Figs. 1 and 2 were run through the simulation as well. The insights gained from the effectiveness of these eccentric mass configurations at sound insulation can contribute to drawing conclusions about the optimised eccentric mass design.

4.2 Circular Split Ring Masses Circle1 and Circle2

With the mass configurations as shown in Fig. 1 attached to the DE membrane, the surface displacement at the peak of transmission loss is derived using the simulation, as can be seen in Fig. 6.

As shown in Fig. 6a, circle1 seems to act similarly to a circular ring mass, as its surface displacement diagram does not show much separation between each of the four sections. In Fig. 6b, since the distance between each of the three sections is

freq(63)=126 Hz Surface: Displacement field, X component (m)

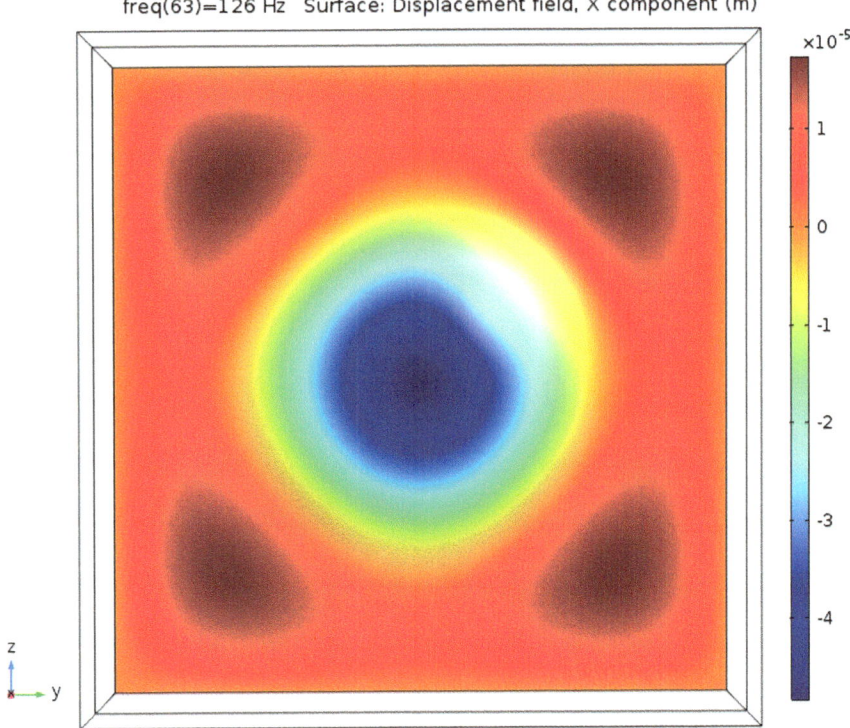

Fig. 5 Surface displacement diagram of membrane only

much more pronounced, there seems to be a more distinct separation between each of the sections in the surface displacement diagram. However, it can be noted that the four sections in circle1 and the three sections in circle2 have the same or similar surface displacement, as they are equidistant from the centre of the membrane.

The transmission loss for both mass distributions is tabulated and graphed in Fig. 7. It can be observed from that the transmission loss (TL) graphs that new peaks were induced. circle1 has two new peaks (at 78 Hz and 23.1 dB TL; 168 Hz and 23.2 dB TL), while of circle2 has three (at 72 Hz and 30.9 dB TL; at 92 Hz and 24.6 dB TL; at 156 Hz and 22.4 dB TL). 5 dB transmission loss was chosen as a benchmark for sound insulation due to most noises being able to be filtered out when that point is reached.

It can be observed that at lower frequencies, the first peak as well as the other smaller spike in transmission loss only make up about a very narrow band of 8 Hz range sound insulation (the band ranges from 122 to 128 Hz, and 216 to 218 Hz). circle1 increases the 5 dB-TL band significantly to a range of around 70 Hz, a 775% increase from the TL of the original membrane plate (the band ranges from 66 to 86 Hz, 122 to 124 Hz, and 140 to 188 Hz). circle2 increases the 5 dB-TL band up to around 74 Hz, a 825% increment from the TL of the original membrane plate

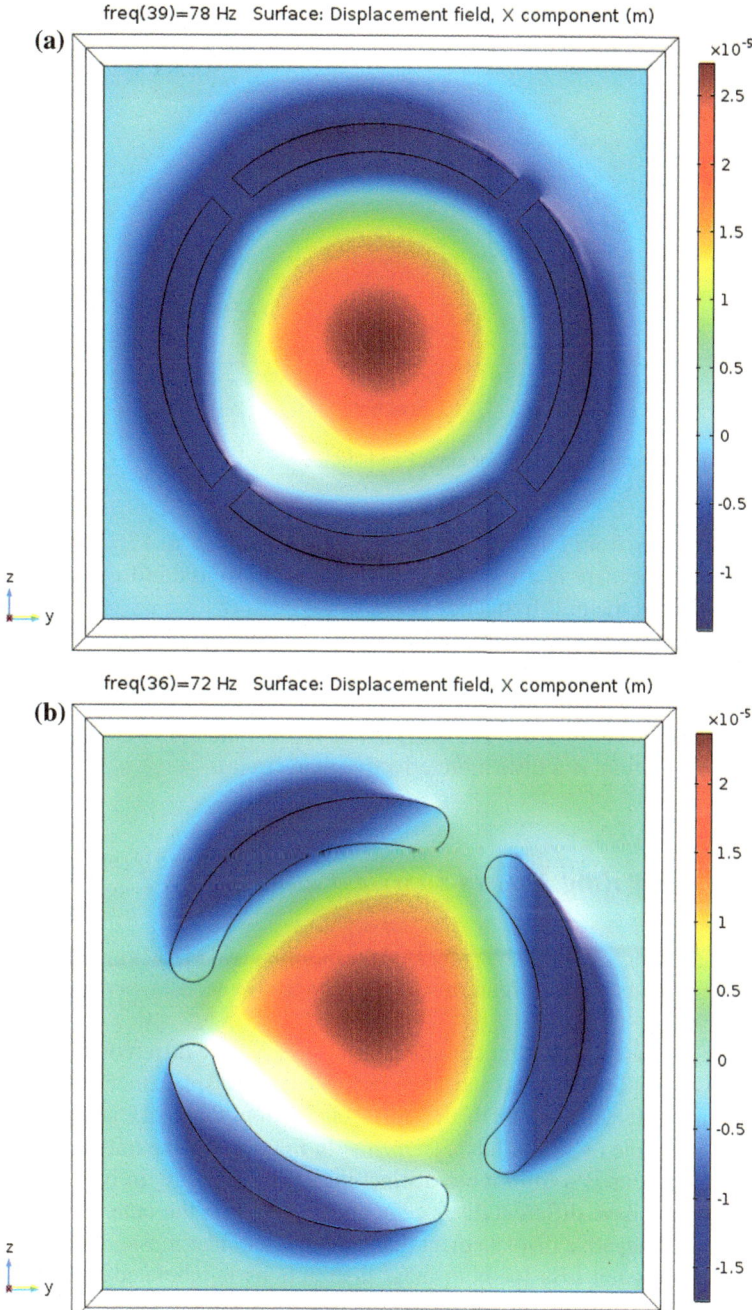

Fig. 6 **a** Surface displacement diagram of circle1; **b** surface displacement diagram of circle2

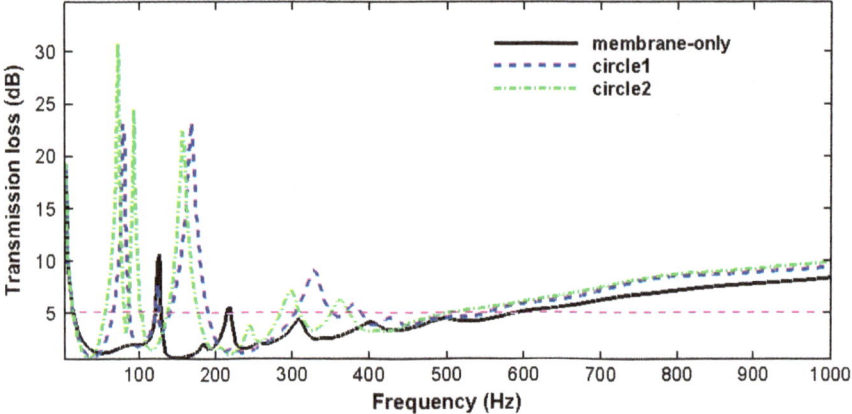

Fig. 7 Graph of transmission loss for circle1 and circle2

(the band ranges from 56 to 78 Hz, 86 to 100 Hz, and 138 to 176 Hz). It can also be observed that there is a very slight increase in transmission loss for both mass configurations at the mid-frequency range.

Therefore, the additional masses added to the membrane are able to alter the anti-resonance transmission loss peaks of the membrane, and even add new peaks, especially at lower frequencies. The two mass configurations significantly increased the attenuation band of 5 dB-TL with similar results, though circle2, with a larger split ring separation, is a little more effective than circle1.

4.3   Circular Ring Masses 2-circles1 and 2-circles2

The simulation process was then repeated for the other two MAMs in Fig. 2 to further study the optimal mass configuration to further improve acoustic performance of the MAMs.

In Fig. 8a, the two different circular masses clearly have different displacements, with the inner split ring mass having a more negative displacement than the outer split ring mass. It can be noted that the different sections that make up each split ring still have the same or similar displacements. In Fig. 8b, the separation between the four outer sections is a little more distinct than that of circle1 in Fig. 6a. The two circular masses have different displacements as well, with the outer split ring mass having a more negative displacement than the inner circular mass.

The transmission loss for both mass distributions is tabulated and graphed in Fig. 9. It can be observed from that the transmission loss (TL) graphs that more new peaks were induced when there were two split ring or circular masses present than when there was only one, as can be seen in comparison to Fig. 7. 2-circles1 has four new peaks (at 68 Hz and 16.7 dB TL; 100 Hz and 34.7 dB TL; at 132 Hz and 18.0 dB

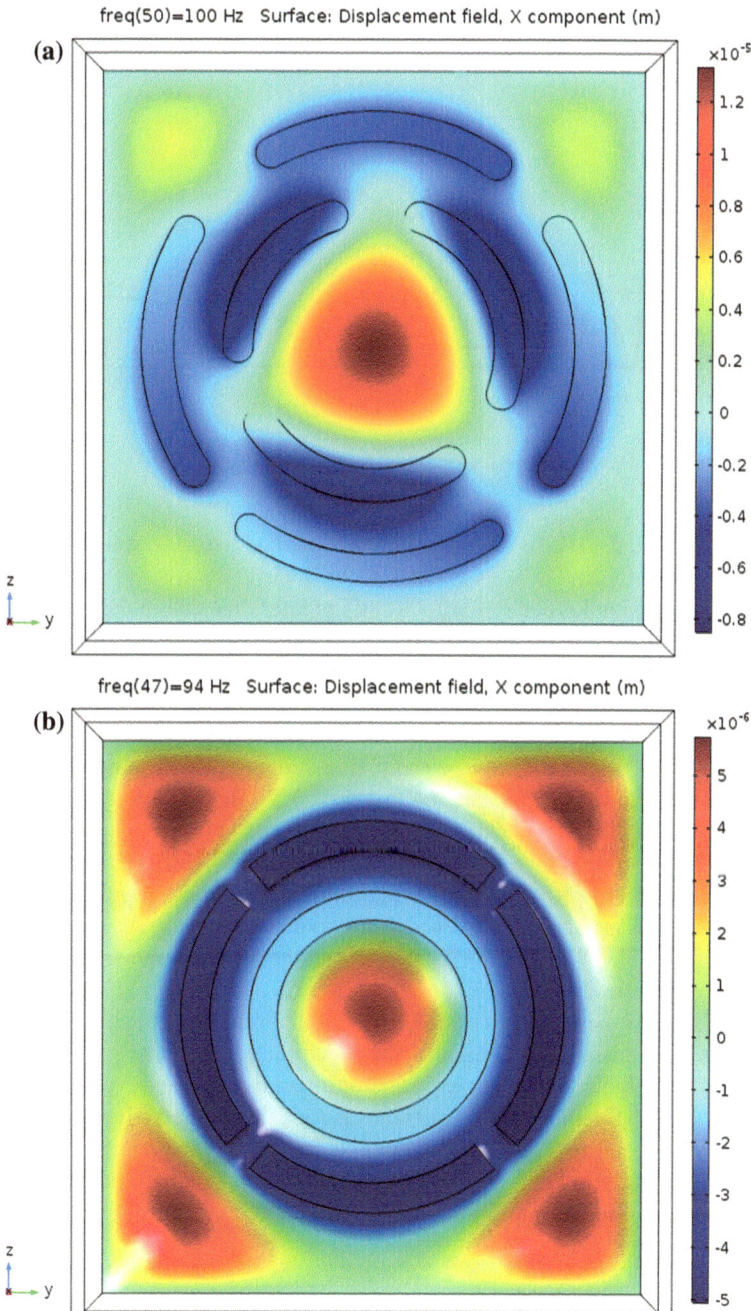

freq(50)=100 Hz   Surface: Displacement field, X component (m)

freq(47)=94 Hz   Surface: Displacement field, X component (m)

Fig. 8 **a** Surface displacement diagram of 2-circles1; **b** surface displacement diagram of 2-circles2

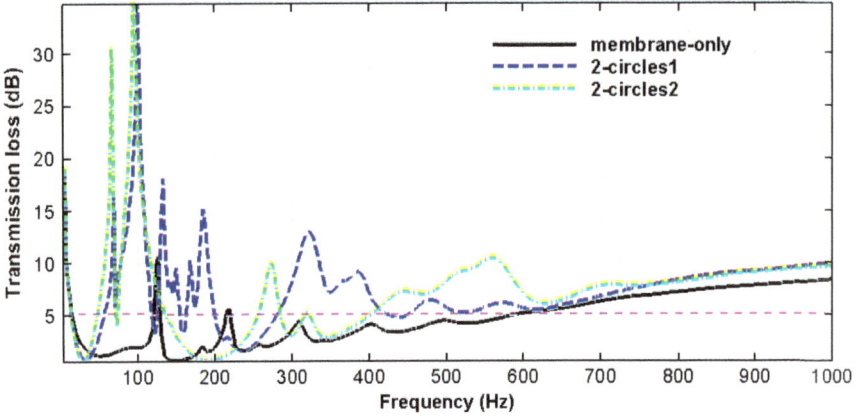

Fig. 9 Graph of transmission loss for 2-circles1 and 2-circles2

TL; at 184 Hz and 15.1 dB TL), while 2-circles2 has two new, much higher peaks (at 66 Hz and 30.5 dB TL; at 94 Hz and 37.1 dB TL).

The transmission loss here is significantly greater than that in the previous simulations with circle1 and circle2. As previously mentioned, the transmission loss for the membrane plate at lower frequencies make up only a 8 Hz narrow range of sound insulation. This is tremendously improved with the presence of the newer mass distributions, with 2-circles1 increasing the 5 dB band to a range of around 120 Hz, a 1400% increase from the TL of the original membrane plate (the band ranges from 56 to 116 Hz, 126 to 152 Hz, and 164 to 198 Hz). 2-circles2 increases the 5 dB-TL band up to around 82 Hz, a 925% increment from the TL of the original membrane plate (the band ranges from 48 to 70 Hz, and 74 to 134 Hz). It can also be noted that apart from the improvement in sound insulation for lower-range frequencies, the transmission loss mid-range frequencies have increased as well, such as the 5 dB-TL range from 280 Hz to 420 Hz for 2-circles1 and the 5-dB-TL range from 256 Hz to 282 Hz for 2-circles2.

Therefore, the new masses added to the membrane are able to alter the anti-resonance transmission loss peaks of the membrane even more significantly than before, adding higher and more new peaks at lower frequencies, even increasing the transmission loss for mid-range frequencies as well. This therefore provides evidence that the presence of two split ring or circular mass distributions, as can be seen in 2-circles1 and 2-circles2 has a more effective acoustic performance than having only one split ring or circular mass distribution, with 2-circles1 having the more effective eccentric mass distribution out of the two.

5 Conclusion

The MAMs with eccentric masses designed have proven to provide much more effective sound insulation as can be seen in their respective acoustic performances. The results agree with previous experimentation [10] that MAMs with optimised eccentric mass distribution are able to drastically increase the effectiveness of sound absorption at a wider range of frequencies. Findings conclude that having two split ring or circular masses instead of one can further tremendously increase the transmission loss of the MAMs, and even have potential to improve acoustic performance for mid-range frequencies as well. The optimised mass configuration for the present paper is 2-circles1, with two circular split ring masses of different radii.

For future studies, more research work can be proposed to test out these MAMs with eccentric masses experimentally to confirm the findings from the simulation. In addition, further experimentation can be done to investigate the specific variables tested in this paper (i.e. radii and number of split ring/circular masses), while keeping all other variables constant such as surface area and mass of eccentric mass. This would corroborate the findings made in this paper, investigating the effects of the different eccentric mass designs across a wider range. Furthermore, other different mass designs of different natures (e.g. polygons, spirals, ellipses, etc.) can also be further looked into to seek more optimised mass configuration for the MAMs. Further investigation can also be made into the statistical differences of each design.

Acknowledgements I would like to thank Dr. Lu Zhenbo from the Aero-science Division of Temasek Laboratories at the National University of Singapore (NUS) for his support and invaluable guidance as my research mentor.

Appendix

See Figs. 10, 11, 12 and 13.

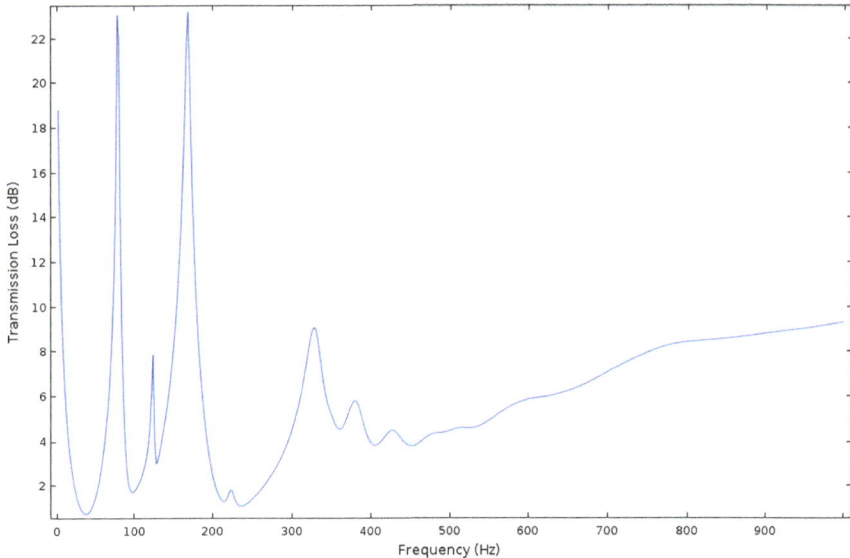

Fig. 10 Individual graph of transmission loss for circle1

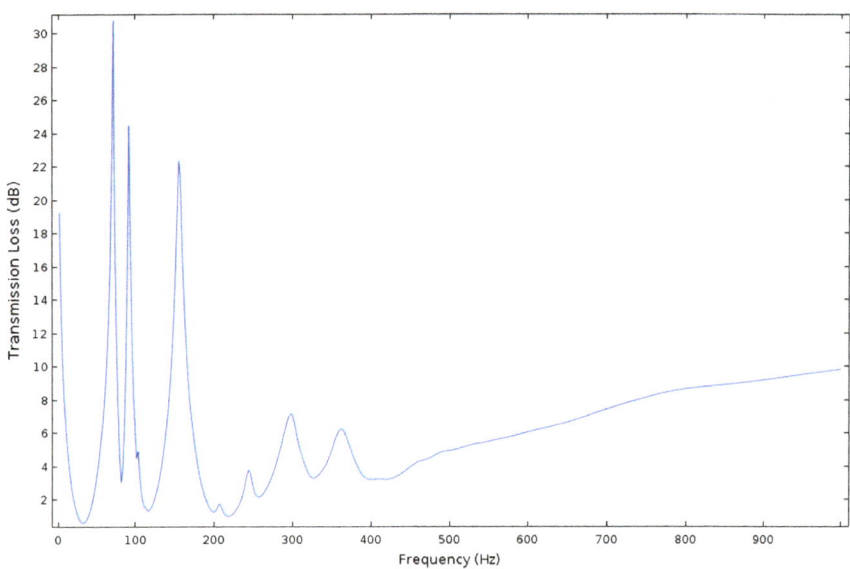

Fig. 11 Individual graph of transmission loss for circle2

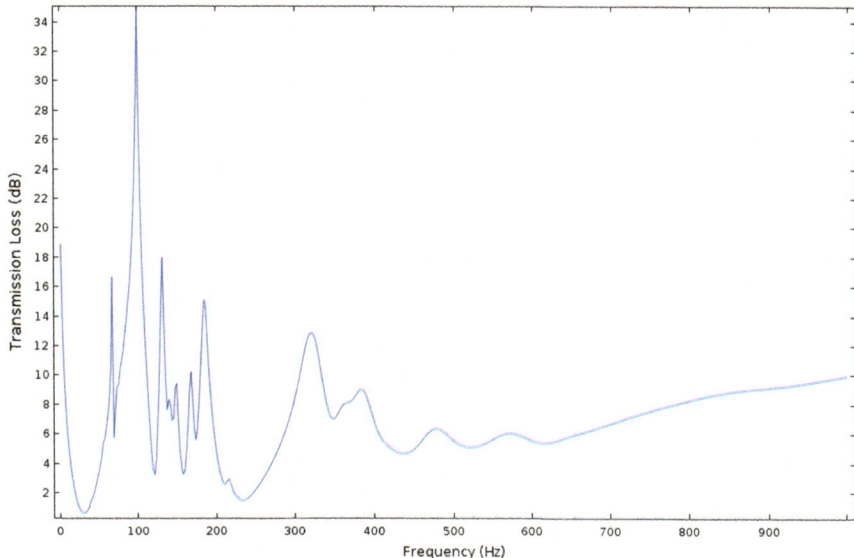

Fig. 12 Individual graph of transmission loss for 2-circles1

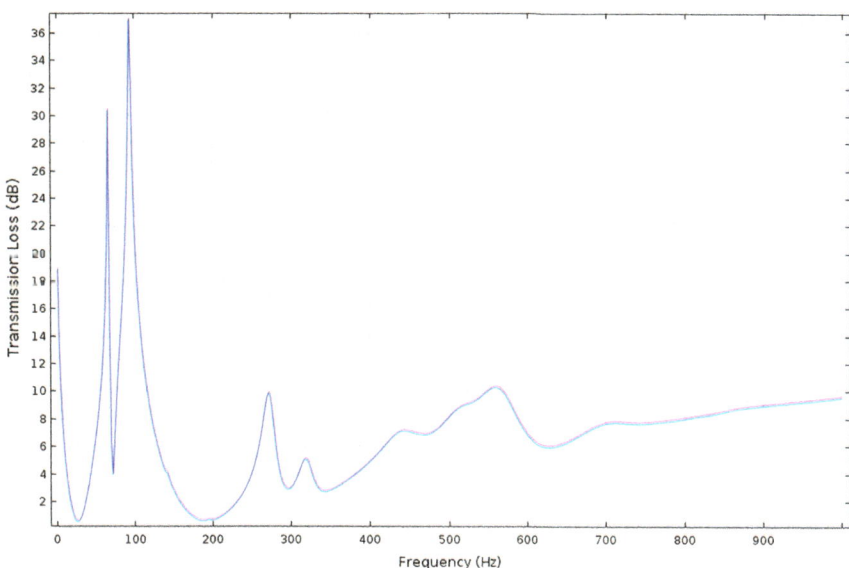

Fig. 13 Individual graph of transmission loss for 2-circles2

References

1. Cummer, A., Christensen, J., & Alù, A. (2016). Controlling sound with acoustic metamaterials. *Nature Reviews Materials, 1*(16001).
2. Shu, Z. (2010). *Acoustic metamaterial design and applications*, pp. 51–77.
3. Veselago, V. G. (1968). The electrodynamics of substances with simultaneously negative values of ε and μ. *Soviet Physics Uspekhi, 10*(4), 509–514.
4. Pendry, J. B. (2000). Negative refraction makes a perfect lens. *Physical Review Journals, 85*(18), 3966–3969.
5. Wong, Z. J., Wang, Y., O'Brien, K., Rho, J., Yin, X., Zhang, S., et al. (2017). Optical and acoustic metamaterials: Superlens, negative refractive index and invisibility cloak. *Journal of Optics, 19*(8).
6. Ma, G., & Sheng, P. (2016). Acoustic metamaterials: From local resonances to broad horizons. *Science Advances, 2*(2), 1–16.
7. Li, J., & Chan, C. T. (2004). Double-negative acoustic metamaterial. *Physical Review, 70*(055602), 1–4.
8. Yang, M., & Sheng, P. (2017). Sound absorption structures: from porous media to acoustic metamaterials. *Annual Review of Materials Research, 47,* 83–114.
9. Naify, C. J., Chang, C., McKnight, G., & Nutt, S. (2011). Transmission loss of membrane-type acoustic metamaterials with coaxial ring masses. *Journal of Applied Physics, 110*(124903).
10. Lau, S., Lu, Z., Liu, Y., Yu, X., Endot, M. Z. B., & Khoo, B. C. (2017). Experimental investigations of membrane-type acoustic metamaterials with eccentric masses for broadband sound isolation. Paper presented at the 46th International Congress and Exposition on Noise Control Engineering, Inter-Noise 2017, Hong Kong.
11. Dassault Systèmes. (2017). *Discover SolidWorks*. Retrieved December 28, 2017 from Dassault Systèmes Website: https://www.3ds.com/products-services/solidworks/.
12. COMSOL Inc. (2017). *Understand, predict, and optimize engineering designs with the COMSOL Multiphysics® Software*. Retrieved December 28, 2017 from COMSOL Inc. Website: https://www.comsol.com/comsol-multiphysics.
13. Lau, S., Lu, Z., Lim, K. H., Liu, Y., Khoo, B. C., & Yu, X. (2017). Numerical investigations of membrane-type acoustic metamaterials with eccentric masses for broadband sound isolation. Paper presented at the 24th International Congress on Sound and Vibration, ICSV 24, London.
14. Ang, L. Y. L., Koh, Y. K., & Lee, H. P. (2016). Acoustic metamaterials: A potential for cabin noise control in automobiles and armored vehicles. *International Journal of Applied Mechanics, 8*(5), 1–23.
15. Lu, M., Feng, L., & Chen, Y. (2009). Phononic crystals and acoustic metamaterials. *Materials Today, 12*(12), 34–41.
16. Yang, Z., Dai, H. M., Chan, N. H., Ma, G. C., & Sheng, P. (2010). Acoustic metamaterial panels for sound attenuation in the 50–1000 Hz regime. *Applied Physics Letters, 96*(4).

Investigating the Aerodynamic Performance of Biomimetic Gliders for Use in Future Transportation

Jiwei Wang and Yiyang Wang

Abstract This study focuses on the concept of biomimicry, looking to nature's best flyers and gliders and comparing the flight performances of the structures of eight chosen species. The species were 3D modelled then analysed by the Xfoil method. Generated results were verified by the Computational Fluid Dynamics method (CFD) in ANSYS Fluent. The top-two models in terms of aerodynamic performance were 3D printed and tested in an open return wind tunnel. Javan Cucumber performs the best among the chosen species. This model also outperforms a reputable current commercial model, XT912, in terms of lift. In addition, the successful test of such wing design manufactured through Computer Numerical Control (CNC) on a Micro Air Vehicle (MAV) platform also highlights its flying capability.

Keywords Biomimicry · Aerodynamics · Glider · Xfoil · CFD · ANSYS Fluent · Javan Cucumber · CNC · MAV

1 Background and Purpose of Research

1.1 Importance of Research

Personal flying vehicles have long been a staple of science fiction and imagination. Unlike land transport, aircrafts use air space. Thus, flying vehicles can allow for better use of space instead of relying on land area. Personal flying vehicles would also help in reducing congestion as commuters can fly at different heights and even use routes that would otherwise be inaccessible by road.

J. Wang (✉) · Y. Wang
Temasek Junior College, Singapore, Singapore
e-mail: wangjiweisteven@gmail.com

Y. Wang
e-mail: kalvinwangyiyang@gmail.com

© Springer Nature Singapore Pte Ltd. 2019
H. Guo et al. (eds.), *IRC-SET 2018*,
https://doi.org/10.1007/978-981-32-9828-6_24

289

1.2   Literature Review

Companies such as Uber are already working on flying cars which are set to be complete by 2020 [1]. However, apart from flying cars, there are also other flying vehicles that are currently already in use but can be further improved [2]. One such vehicle is the powered glider, which requires less distance to take off and land, can take 1–2 people and use gliding as its primary power source [3]. Currently, it is widely used as a sport activity, especially in the United States. Although it has been used for many years, its overall structure has remained mostly unchanged. Thus, it can be improved so that it can also be used in future transportation. Biomimicry has been long applied in aircraft design [4]; however, these aircrafts usually have a tiny load capacity [5], or there has not been a broad comparison across species to discover which offers the best structure for flight [6, 7].

1.3   Rationale for Research

We thus embarked on a research that seeks to find a better design for powered gliders using biomimicry [criteria refer to Chapter "Study of Bird Feathers to Improve Design of Absorbent Pads for Greater Efficiency of Oil Spill Removal (A)"]. This research is anticipated to greatly contribute to the research of personal flying vehicles for future transport.

1.4   Flying and Gliding Species Studied

This research studies the features of some of nature's most notable flying and gliding creatures and examines their structures. The eight species chosen for study are Atlantic Flying fish (*Cheilopogon melanurus*), Spotted Eagle Ray (*Aetobatus narinari*), Snowy Owl (*Bubo scandiacus*), Albatross (*Diomedeidae*), Malaya Colugo (*Galeopterus variegatus*), Golden-capped Fruit Bat (*Acerodon jubatus*), Flying Dragon (*Draco Volans*) and Javan Cucumber (*Alsomitra macrocarpa*) (refer to Fig. 2).

2   Hypothesis

Among all the species to be studied, we hypothesised that the Javan Cucumber would perform the best, i.e. be the most effective at having a high lift, a high coefficient of lift (CL) to coefficient of drag (CD) ratio (CL/CD) and be able to sustain them across a wide range of Angles of Attacks (AoA). The Javan Cucumber features elliptical wings, which was seen in the advanced Spitfire plane during the Second World War. Furthermore, according to *Model Aircraft Aerodynamics* by Martin Simons, the elliptical wing produces a constant downwash at any speed with the minimum induced drag for a given lift [8].

3   Method and Materials

This research was done in seven stages. Firstly, we shortlisted the species to study. Secondly, we scaled all the models to have a wingspan of 15 cm which would be convenient for our investigation. Thirdly, we utilised Autodesk CAD 2016 to sketch out their planforms and imported them into XFLR5. We then did a computational simulation to analyse the aerodynamic characteristics of these models. Fourthly, we constructed the 3D models using CATIA V5 R20, then meshed the models in ICEM-CFD 18.0 and conducted the verification test using ANSYS 18.0 Fluent. Fifthly, the best two models are selected, 3D-printed, and tested in an open return wind tunnel to validate the results obtained from the simulations. Sixthly, we chose a reputable commercial model, XT912 Tundra-Merlin, to compare and evaluate the performance of the best model. Lastly, the prototype of the best model was made using CNC and applied on a MAV to test its flying capability.

3.1   Choosing of Species

The species were chosen such that they allowed us to broadly cover most types of flying species in nature, allowing us identify which type of flying species has the best structure that we can apply to powered gliders. The process of choosing the species to study can be divided into three phases. In the first phase, we shortlisted all flying or gliding species. In the second phase, we then looked at whether the structures of these species could be applied to human use. It was at this phase that we removed the flying snake from the list of species to be modelled. The flying snake has to continuously move side to side throughout the flight and it can only glide at high speeds [9], making it unsuitable to be adapted to personal flying vehicles. Lastly, in the third stage, we grouped similar species and selected one to represent each group. This step was to ensure that there were no repeats in structures and that we could get a broad range of structures to study. It was at this phase that the flying squirrel was removed from the shortlist as it shared a very similar structure with the Malaya Colugo (Figs. 1 and 2).

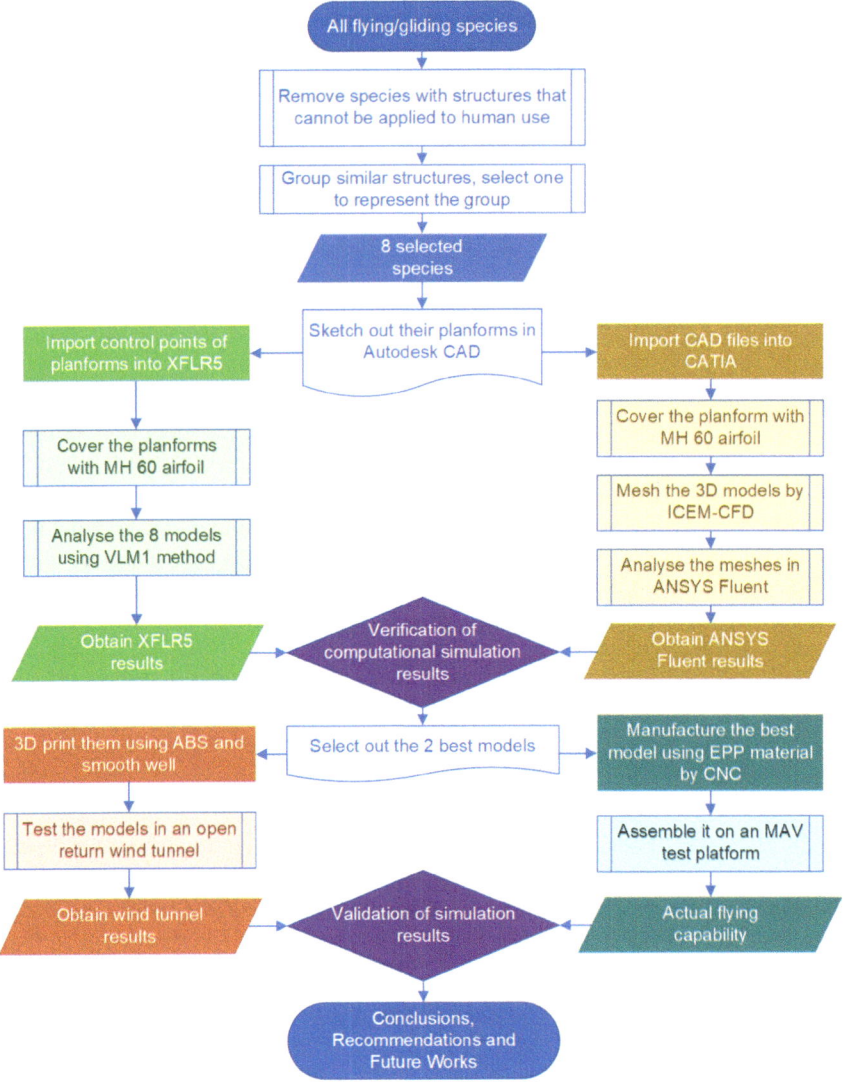

Fig. 1 Overall methodology

3.2   Scale Modelling and Realisation of Distorted Models

Due to the feasibility of realisation in XFLR5, we had to scale our models so as to comply with the software limitations. The official guidelines for XFLR5 [10] suggest that this software cannot handle a full-scale test, hence necessitating scaling. In order to keep the consistency with physical wind tunnel testing, we fixed the wingspan at 15 cm and let their wing areas vary. This is because we consider

Fig. 2 Eight different flying species selected

their wing areas, which are determined by the species' body shapes, as a feature of those species selected. Additionally, according to Similitude requirements, the dimensionless quantities, e.g., Reynold numbers should remain the same before and after scaling [11]. However, this will lead to a huge wind speed for scale models when we set the application velocity as 20 ms^{-1}, approximately 1333 ms^{-1}, or 3.922 Mach equivalent (taking the Javan Cucumber as the example). According to Scaling laws:

$$\text{Re} = \left(\frac{\rho VL}{\mu}\right) \rightarrow V_{\text{Model}} = V_{\text{application}} \times \left(\frac{\rho_a}{\rho_m}\right) \times \left(\frac{\text{MAC}_a}{\text{MAC}_m}\right) \times \left(\frac{\mu_m}{\mu_a}\right) \quad (1)$$

$$V_{\text{Model}} = 20 \times \frac{3.721}{0.05581} = 1333 \text{ ms}^{-1} = 3.922 \text{ Mach}$$

Thus, as we cannot have the models satisfy first-order similarities, we chose to realise distorted models rather than adequate models under the working conditions of full-scale ones. We would then conduct a horizontal analysis across scale models, selecting the best one to conduct the full-scale test in a more advanced and sophisticated software—ANSYS Fluent. The full-scale simulation will suggest the real performance, while scale models are only used in horizontal analysis, i.e. select the ones with better performance among species.

3.3 3D Modelling and Analysis Using XFLR5

(1) Pre-validation of XFLR5 by Testing NACA 0015 Airfoil Section

The first wind tunnel simulation software used was XFLR5. Prior to the simulations, to ensure that the data generated by the software was valid, we first simulated the airfoil NACA 0015 and compared the results obtained from the simulations with existing literature and experimental data of the airfoil [12]. The results matched in all aerodynamics characteristics, proving the validity of XFLR5 algorithms.

(2) *3D-Modeling Using Autodesk CAD and XFLR5*

We searched suitable pictures of the eight species online. Using these pictures, we then plotted the outline of all these models in Autodesk CAD. The outline was separated into several parts. The coordinates of the turning points and corresponding chord lengths were obtained and then imported into XFLR5. All models used MH60 airfoil, exported from Profili 2.21, since this airfoil has a high *Cl/Cd* over a wide range of AoA. This airfoil is also self-stablising as it has an "S" shape thickness distribution. This characteristic is helpful in levelling a blended body-wing design aircraft by significantly reducing its pitching moment (Fig. 3).

Thus, this airfoil is widely used in fly wing designs. Considering the Reynold Number (Re) of physical powered gliders, it will perform better than most airfoils when the wingspan is relatively larger [13]. (All models refer to Fig. 4).

(3) *Analysis of All Models Using XFLR5*

Firstly, we ran the Xfoil Direct Analysis and collected polar data of MH60 under different Re numbers. The polar curves were obtained through Multi-threaded Batch Analysis from Re 0 to 300,000 under Mach 0.059 (20 ms^{-1}). The increment for Re was 10, and the testing AoA was from $-20°$ to $20°$ with increment of 0.5°. The speed of 20 ms^{-1} is comparable with the average speed of a Mass Rapid Transit (MRT) train [14] and a current personal flying vehicle, EHANG 184. As its speed is higher, less time is taken to transport commuters, thus fewer vehicles would be using the roads, resulting in less congestions.

We then tested the models with a fixed speed of 20 ms^{-1} by using the Horseshoe Vortex (VLM1) analysis method. This method was chosen as the models generally had a low aspect ratio (AR), which was a hindrance of applying Lifting-Line Theories (LLT) method. The analysis sequence started at $-10.0°$ and ended at 10.0° with an increment of 0.5°. The air data was extracted from the U.S. Standard Atmosphere 1976 [15]. The set of data we selected, at a temperature of 15 °C and altitude of 25 m, is the one closest to Singapore's environment. All polars are exported and then analysed in Microsoft Excel. The values of their *CL* divided by their *CD* gave their *CL/CD* values (refer to Eqs. 3 and 4). The higher the ratio, the greater the capability of the models to generate lift and reduce drag. This feature, together with relatively high lift force, indicates the better aerodynamic performance of such a design. The various sets of *CL/CD* data were then plotted on a graph against the respective AoA.

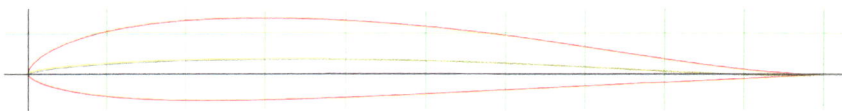

Fig. 3 MH 60 airfoil exported from Profili V2.21

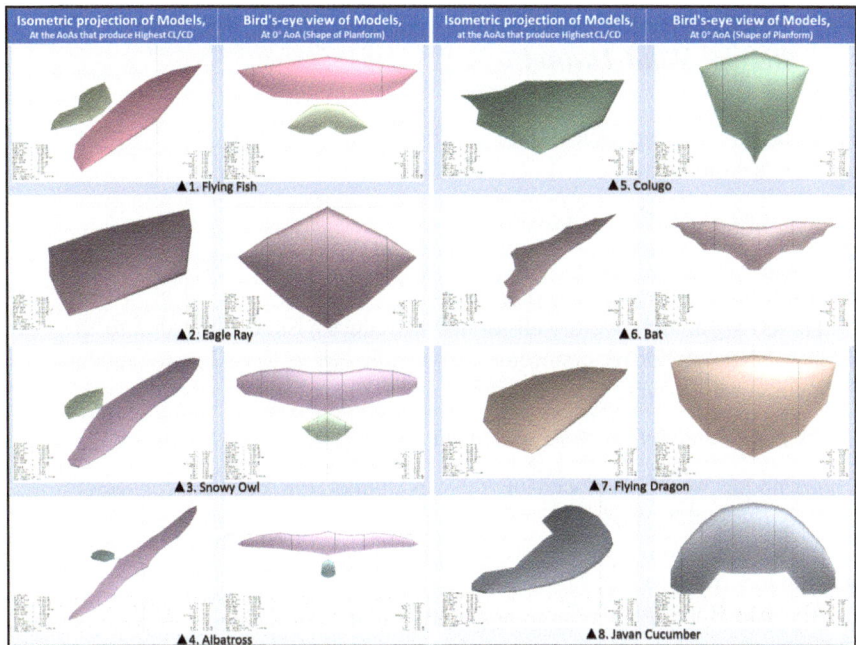

Fig. 4 Isometric projection and Bird's-eye view of eight models

3.4 Verification of Results Using ANSYS Fluent

After modelling and analysing all models in XFLR5, the results obtained were proven to be accurate [16]. However, a verification process is needed because there might be potential software flaws. Apart from using wind tunnel data to validate the results obtained from XFLR5, we repeated the simulations in another software, ANSYS Fluent. We re-modelled all species using CATIA with the same control points and MH 60 airfoil used in XFLR5. The 3D files in .STP format were imported in ICEM-CFD to create meshes using a structured Cartesian–Hexahedron (C–H) grid topology. The y+ value for each mesh was controlled below 1 to satisfy the requirement of Spalart–Allmaras (S–A) turbulence model which resolves viscous sublayers. Later, the meshes were imported into ANSYS Fluent to test their performance under the same condition in XFLR5 simulations. The AoA was given as $0°$, $\pm 2°$, $\pm 4°$, $\pm 5°$, $\pm 8°$ and $\pm 10°$. The results of lift, drag and *CL/CD* were recorded and compared together with the data generated from XFLR5.

3.5 Validation of Top 2 Models' Results Using the Open Return Wind Tunnel

From the first two experiments, we shortlisted the two best performing models to be tested in a wind tunnel.

(1) *3D Printing of Top 2 Models*

The models were 3D printed using Acrylonitrile Butadiene Styrene (ABS) by Fused Deposition Modelling (FDM) (Fig. 5).

These two models were well-sanded to reduce viscous drag due to the rough surface. Their supportive connectors, which allowed us to change the models' AoA in the test section, were also 3D-printed and well-sanded.

(2) *Wind Tunnel Set-up and Experiment*

Those models were then placed into the wind tunnel (Fig. 6) by the connector (Fig. 5) at the bottom of the models (Fig. 7).

A digital micromanometer was used to measure the change in air pressure in the wind tunnel (refer to Fig. 7).

Applying Bernoulli's Velocity equation (Eq. 2), we then calculated the wind speed in the tunnel.

$$V = \sqrt{\frac{2 \times \Delta P}{\rho}} \tag{2}$$

where V is the wind speed, ΔP is the change of air pressure, and ρ is density of the air. In the experiment, the rotational speed of the suction turbine motor in the wind tunnel was adjusted till the micromanometer hit the reading of 245 Pa, which by Bernoulli's Velocity equation would mean the wind speed is 20 ms^{-1}. We then waited for another 5 s for the reading to stabilise. If the micromanometer reading deviated from 245 Pa, we made corrections by adjusting the rotational frequency of

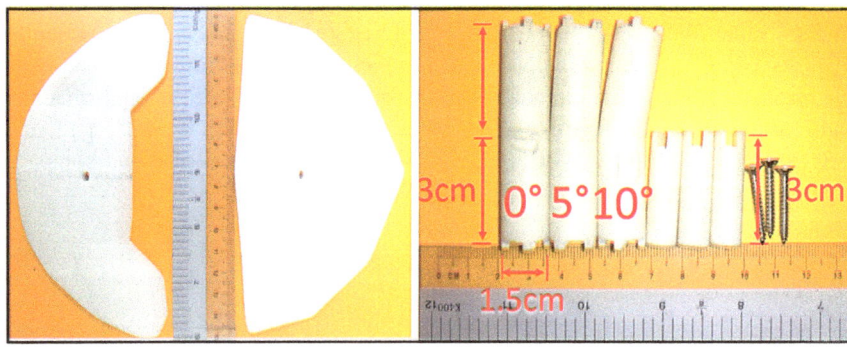

Fig. 5 3D printed models and connectors

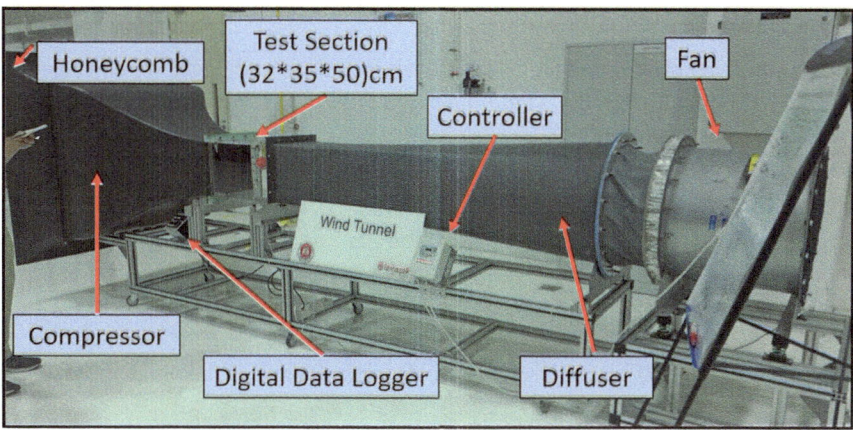

Fig. 6 Open return wind tunnel set up

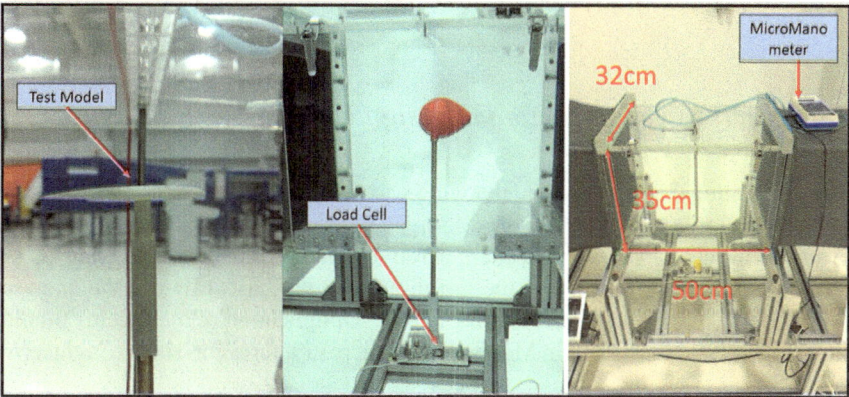

Fig. 7 Fixed model, close ups of the test section and digital micromanometer

the motor. If the micromanometer reading remained at 245 Pa after 5 s, the lift and drag was then recorded for the next 10 s. After each test we changed the connector to change the AoA of models. The AoA was given as 0°, ±5°, and ±10°.

(3) *Collecting Data from Wind-Tunnel Sensors*

The lift and drag readings were displayed on two data loggers, and then recorded by videos. After which we took down the data from videos every 0.345 s (29 sets in total). Then we calculated the mean of the data from each of the two sets of wind tunnel tests and took the mean value of both. Subtracting the reading without model i.e. the forces of supporting structure, would give the lift and drag of models, which will be compared with the simulation results.

(4) *Calculation of CL/CD*

We first derived the *CL* and the *CD* of tested models respectively. We can calculate the *CL* with the lift from the wind tunnel and the Eq. (3). We can also calculate the *CD* with the drag and Eq. (4).

$$CL = \frac{L}{\rho V^2 \frac{A}{2}} \tag{3}$$

$$CD = \frac{D}{\rho V^2 \frac{A}{2}} \tag{4}$$

where L is the lift, D is the drag, ρ is the density of air, V is the wind speed, and A is the reference area of models.

From here, the *CL/CD* of each model can be found for the respective AoA. The different sets of *CL/CD* for each of the tested models were then plotted against the AoA in respective graphs. The graphs were then compared to the results obtained from the computational simulations.

3.6 Analysis of Conventional Design Using ANSYS Fluent

The conventional design of the delta wing was also analysed in ANSYS Fluent. We took the example of XT912 Tundra-Merlin, a popular powered glider that is readily available in the market with excellent aerodynamic performance. The wingspan of the 3D model was 10.0 m and the airfoil applied was also MH 60. We analysed this model using the same method in Chapter "Optimization of the Electrospinning Process to create Pure Gelatin Methacrylate Microstructures for Tissue Engineering Applications (D)".

3.7 Full-Scale Model Simulation Using ANSYS Fluent

Due to the limitation in software algorithm mentioned in Chapter "Optimization of the Electrospinning Process to create Pure Gelatin Methacrylate Microstructures for Tissue Engineering Applications (B)", XFLR5 is not suitable to simulate full-scale models. Thus, we decided to use ANSYS Fluent for the full-scale simulation. Taking current commercial powered gliders as references, we designed the full-scale model of the best model with a wingspan of 10.0 m. The 3D model of the best one was further refined by adding new control points in its planform in CATIA. It was also meshed in ICEM-CFD using C-H grid topology (Fig. 8 left) and later simulated in ANSYS Fluent under the same condition experienced by its scaled counterpart and the conventional design. The post-processing was done in Tecplot 360 (Fig. 8 right).

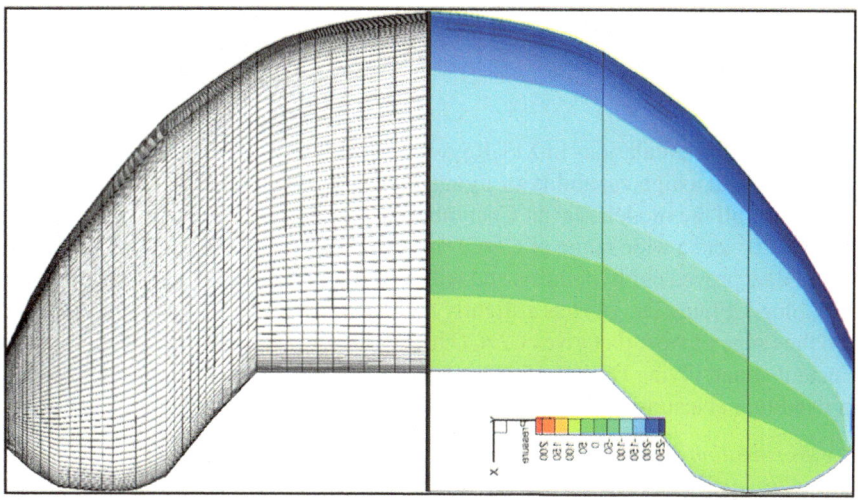

Fig. 8 Meshing in ICEM-CFD and pressure distribution in Tecplot

3.8 Prototyping and Testing of the Best Model

After previous testing and experiments, we manufactured a prototype of the best model with a wingspan of 50 cm. We used Computer Numerical Control (CNC) to mill the model out from Expanded Polypropylene (EPP). This prototype was installed on an MAV (refer to Fig. 15) as the test platform to examine the flying capability of such a design.

4 Results and Discussion

4.1 Criteria for Comparing and Selecting Models with Better Performance

Considering the purpose and future potential of our research, we think that the new design of a powered glider for transportation use should have a relatively high CL/CD, be capable to sustain high CL/CD under its working conditions and be able to generate a relatively large lift. Taking the gliding theories into consideration, the wind speed was fixed at 20 ms^{-1} and their performance was examined by plotting their CL/CD against the AoA. A better structure would be the one with higher CL/CD and maintain a high ratio over a broad range of AoA with a relatively large lift force.

4.2   Characteristics and Comparison of Results

(1)   XFLR5

From our initial simulations in XFLR5, we obtained the polar of all models. Overall, all models had a high capability of flying due to their high *CL/CD* (refer to Fig. 9).

Among all the models, Javan Cucumber has the highest *CL/CD* of 10.823 when AoA is 7.0° and a wide range of *CL/CD* above 10.0, from 4.2° to beyond 10°. These sets of data proved our hypothesis and will be further discussed later. Compared with the Colugo, Flying Dragon has a slightly higher *CL/CD* of 10.423 when AoA is 6.5°, and the range of AoA that gives *CL/CD* higher than 10.0 is also wider. Hence, Javan Cucumber and Flying Dragon were the top two models that would be further studied in wind tunnel testing.

(2)   *Verification Using ANSYS Fluent*

The lift data produced by ANSYS Fluent accurately matched that produced by XFLR5 (refer to Fig. 10).

However, the drag data has a small deviation between XFLR5 results and ANSYS Fluent results. Despite this, those two sets of data, after plotting against AoA, show consistent trends. The deviation, which may be due to the inviscid assumption of Horseshoe Vortex method of XFLR5 in calculating drag, was also in an acceptable range (average difference is less than 10%). The inaccuracy with reference to the

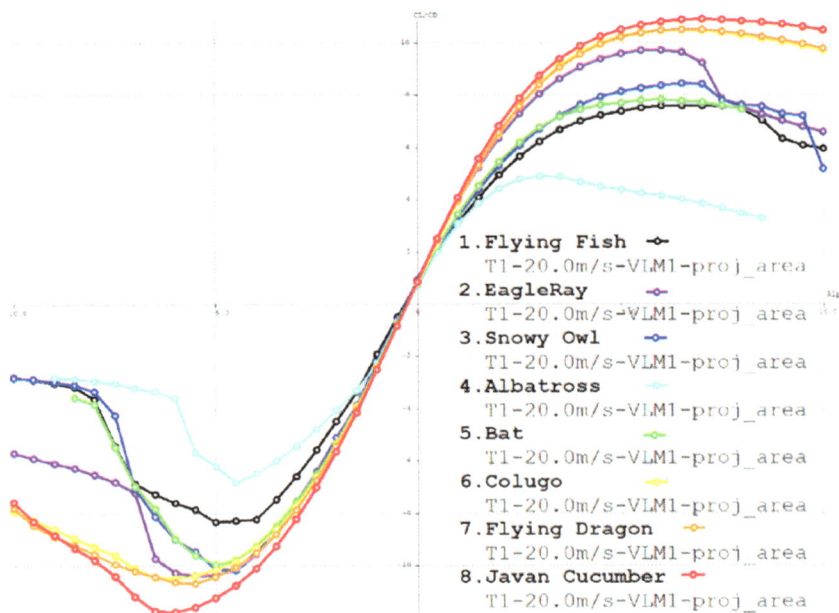

Fig. 9 Simulated results of all eight species in XFLR5

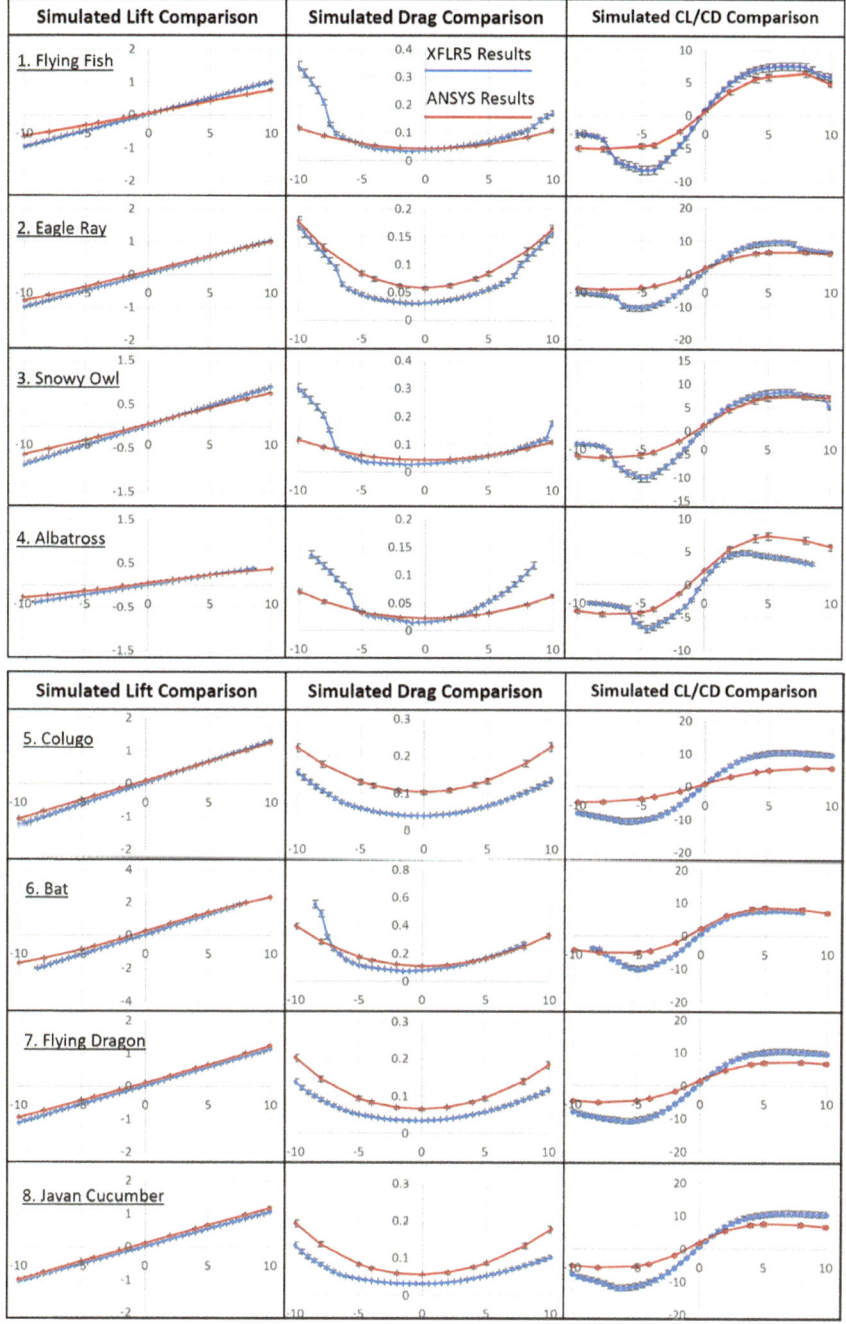

Fig. 10 Simulated data from XFLR5 and ANSYS Fluent for comparison

deviation can be considered negligible as all the results were affected to the same extent due to the limitations of the software used. Overall, the XFLR5 data was proven to be accurate and convincing [16], which made our comparison reasonable and reliable.

(3) *Results from Open Return Wind Tunnel Testing*

Due to the limitation of the wind tunnel, we did not manage to get valid lift data. However, we still obtained the set of drag data which were comparable (refer to Figs. 11 and 12). Although these data were not accurate enough quantitatively compared with the simulated drag force, they both showed the same trend of the change of drag against the AoA. This gave a certain credit to our simulations, and we may infer that the lift may also follow the trend of the simulated results as the lift is correlated to the drag.

Fig. 11 Javan Cucumber drag comparison

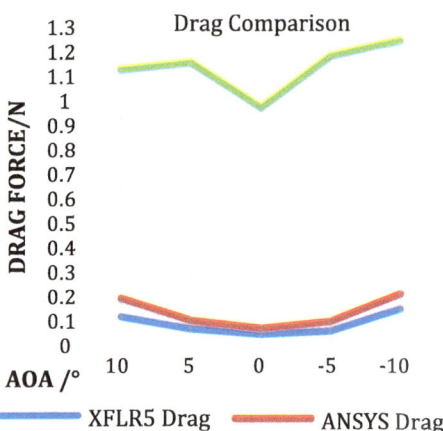

Fig. 12 Flying Dragon drag comparison

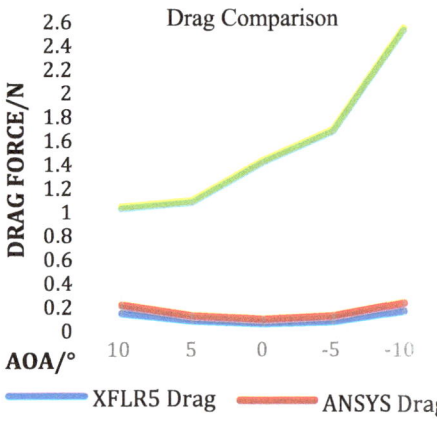

4.3 Comparison between Javan Cucumber and Conventional Design

(1) *Scale Models in XFLR5*

Javan Cucumber had a much higher polar of *CL/CD* against AoA compared with the conventional delta wing design with the same wingspan of 15 cm. (Refer to Fig. 4) The maximum *CL/CD* of the delta wing is only 7.9 when AoA is 5.5°. Meanwhile, the polar of it is not as smooth as that of Javan Cucumber, which suggested that the Delta Wing design might not have a stable flying capability under various AoA compared to the Javan Cucumber. These illustrate that Javan Cucumber had the potential to be further studied to improve on the current flying models for the powered glider.

(2) *Full-Scale Testing of the Best Model and the XT912 Tundra-Merlin Using ANSYS Fluent*

Compared with the current commercially available powered glider, such as XT912 Tundra-Merlin [17] with a 10.0-m wingspan, both data and graph showed that Javan Cucumber has a slightly low *CL/CD* polar curve and greater drag (refer to Figs. 13 and 14).

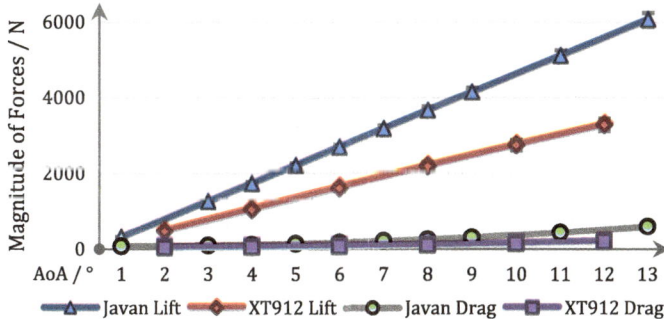

Fig. 13 Javan Cucumber and XT912 comparison

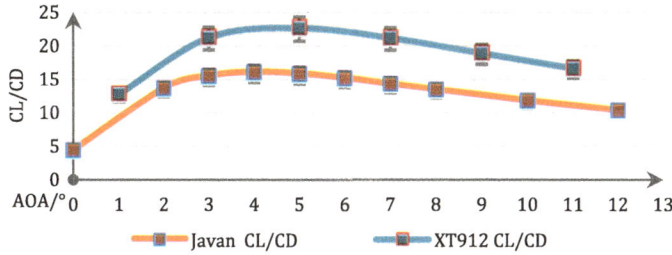

Fig. 14 Javan Cucumber versus XT912 CL/CD

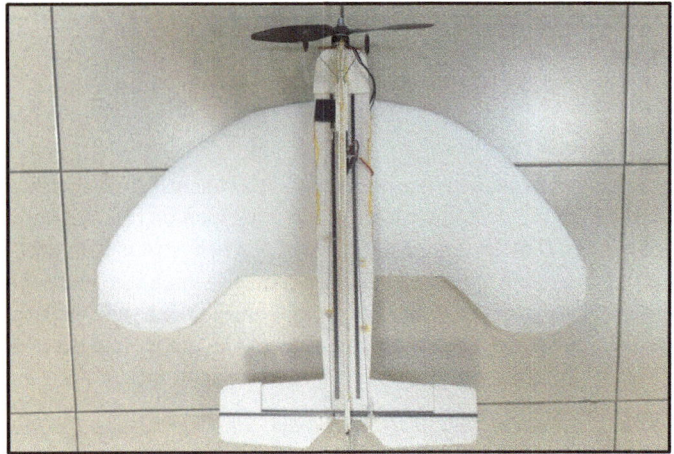

Fig. 15 MAV with Javan Cucumber wing

However, the lift that the Javan Cucumber can generate at the speed of 20 ms^{-1} was about 66% greater than the lift that the XT912 can generate at the same AoA (refer to Fig. 13). Hence, the greater drag is acceptable given this model's advantage in lift force. Meanwhile, the Javan Cucumber maintains a high *CL/CD* over a broad range of AoA, above 10.0° from 1.3° to 12.1°, with a relatively large lift force. This full-scale testing showed that the Javan Cucumber potentially performed better than the currently used wing model for powered gliders. However, physical and detailed experiments were needed to achieve such an engineering goal since the actual flying performance greatly depends on the manufacturing technology.

4.4 Testing Prototype on a MAV

The testing of the prototype on a MAV test platform was successful. The plane took off at a slower speed and required less throttle to maintain its altitude compared to that with conventional wings. The success of this trial showed the great potential of Javan Cucumber to be further studied and commercialised (Fig. 15).

4.5 Limitations

(1) *Software–XFLR5*

We used the Horseshoe Vortex method (VLM1) to analyse all the models. This method was derived from non-viscous assumptions of the fluids. Hence, results from this method ignored the viscous drag and were independent of speed. However, due to the small dimension and low speed of our model aircraft, the viscous drag cannot be ignored. Due to the lack of theory for low Re numbers 3D analysis, the XFLR5 software had to extrapolate it from the 2D analysis.

In XFLR5 testing, due to the limitation of the algorithm of Xfoil, we did not manage to fix the problem that some points could not be interpolated in the Xfoil direct analysis. For example, the Albatross model with Span pos = 0.07 m, Re = 3333, $CL = -1.18$ could not be interpolated. That error was because we had reached the limits of the 2D approximation for the viscous drag. This limitation would lead to approximation and affected the accuracy of a simulated result. These suggested that the models needed more accurate wind tunnel tests to validate the analysis.

(2) *Limitations of the Open Return Wind Tunnel Testing*

While using the wind tunnel, we encountered multiple challenges and limitations, one of which was that the wind tunnel's lift sensors were not sensitive enough. Another was that the structure of the supporting bar of the model was connected to a pivot instead of a wind tunnel balance (refer to Fig. 7). The drag that the model and the bar experienced is much greater than the lift the model experienced. Hence, the bar cannot be lifted, thus the lift sensor was unable to produce reliable data.

Besides, yet another limitation was the accuracy of the wind tunnel. For the Institute that allowed us to use this wind tunnel, faculties usually only use it for qualitative demonstration but seldom used it for quantitative analysis. Hence, the calibration done for this wind tunnel might not have been sufficient to produce accurate data.

5 Conclusion and Recommendations

In conclusion, the hypothesis was proven to be true: Javan Cucumber performs the best among all the selected species. It could sustain high CL/CD across a wide range of AoA. This excellent performance was because the Javan Cucumber features elliptical wings and could experience constant lift under different AoA. The models of Flying Dragon and Colugo also performed well in the simulations and testing. The similar structure of Colugo was applied in the wing-suit currently and those two designs should also have the potential to be developed in the motor glider area.

In comparison with the existing wings for motor gliders, our best model Javan Cucumber has comparable or even better aerodynamic performance. The Javan Cucumber model can carry theoretically 373 kg of mass under a low speed of 20 ms^{-1} when its AoA is 7° while the XT912 Tundra-Merlin can only carry 223 kg of mass at the same speed. Although this is mainly because of the bigger wing area of Javan Cucumber, such design features can still possibly be applied in future power glider designs.

6 Future Work

In this research, we have identified the structure of the Javan Cucumber to be one of the most suitable structures to mimic for flight or gliding. However, as this study was done on a range of very different structures from nature, further studies could be done to investigate plant structures similar to that of the Javan Cucumber to determine an ideal structure. Further studies could also be done to investigate the stability and manoeuvrability of such structures as such attributes would also be important to flight, especially in the context of future traffic.

The geometry of these designs can be further explored in other personal air vehicle development aided by the future improvement of the technologies, such as thrust vector control engine, new material like aerogel and full-scale 3D printing. All these revolutionary technologies can be utilised in our designing process to create more efficient powered gliders or other flying vehicles.

Acknowledgements We would like to thank Mr. Khoh Rong Lun of Temasek Junior College for his unwavering support and invaluable guidance through this year-long research. We would also like to thank Ms. Koh Poh Lee of Temasek Polytechnic for the use of the wind Tunnel and the lab technicians for their kind support.

References

1. Uber Elevate. Uber Elevate | The future of urban air transport. https://www.uber.com/info/elevate/.
2. EHANG184. EHANG | Official Site-EHANG 184 autonomous aerial vehicle. http://www.ehang.com/ehang184.
3. Powered hang glider. Powered hang glider—Wikipedia. https://en.wikipedia.org/wiki/Powered_hang_glider.
4. McMasters, J. *A legacy of sustaining innovations in biomimetic aircraft design and engineering education*. PPT.
5. Mueller, T. J. (2001). *Fixed and flapping wing aerodynamics for micro air vehicle applications*. Reston, VA: American Institute of Aeronautics and Astronautics.
6. Barrett, R. M., & Barrett, C. M. (2014). Biomimetic FAA-certifiable, artificial muscle structures for commercial aircraft wings. *Smart Materials and Structures, 23*(7), 074011. https://doi.org/10.1088/0964-1726/23/7/074011.
7. Marks, C. R., James J. J., & Gregory, W. R. (2013). A reconfigurable wing for biomimetic aircraft. In *54th AIAA/ASME/ASCE/AHS/ASC Structures, Structural Dynamics, and Materials Conference*. https://doi.org/10.2514/6.2013-1511.
8. Simons, M. (2000). *Model aircraft aerodynamics*. Chris Lloyd Sales & Marketing.
9. Socha, J. J., & Michael, L. (2005). Effects of size and behavior on aerial performance of two species of flying snakes (Chrysopelea). *Journal of Experimental Biology, 208*(10), 1835–1847. https://doi.org/10.1242/jeb.01580.
10. Drela, M., & Youngren, H. (2013). *XFLR5 v6.02 guidelines: Analysis of foils and wings operating at low Reynolds numbers*. PDF. February 28, 2013.
11. Similitude requirements and scaling relationships as applied to model testing. NASA. https://ntrs.nasa.gov/search.jsp?R=19790022005.

12. Bertagnolio, F. *NACA0015 measurements in LM wind tunnel and turbulence generated noise.* Report. Aeroelastic Design, Wind Energy Division, Risø National Laboratory for Sustainable Energy, Technical University of Denmark. Denmark.
13. Kaufmann, W., & Wohlfahrt, M. *Airfoils for fly wings. Airfoils for flying wings and tailless airplanes.* https://www.mh-aerotools.de/airfoils/foil_flyingwings.htm.
14. UrbanRail.net. UrbanRail.Net > Asia > Singapore > Singapore MRT (Metro). http://www.urbanrail.net/as/sing/singapore.htm.
15. U.S. Standard Atmosphere. (1976). PDF. *National Oceanic and Atmospheric Administration, National Aeronautics and Space Administration*, United States Air Force, October 1976.
16. Morgado, J., Vizinho, R., Silvestre, M. A. R., & Páscoa, J. C. (2016). XFOIL vs CFD performance predictions for high lift low Reynolds number airfoils. *Aerospace Science and Technology, 52*, 207–214. https://doi.org/10.1016/j.ast.2016.02.031.
17. MICROLIGHT WING GUIDE. PDF. Airborne Australia Pty Ltd.

Building Nanostructured Porous Silica Materials Directed by Surfactants

Wayne W. Z. Yeo, Su Hui Lim, Connie K. Liu and Gen Yong

Abstract Silica nanomaterials have found prevailing use in biomedical applications due to their biocompatibility and non-toxicity. The use of a structure-directing agent in silica sol-gel synthesis enables us to direct the formation of silica nanostructures into forms that are otherwise difficult to obtain, allowing the exertion of a fine degree of control over the morphology, dimensions and architecture of the nanostructures. Single-tailed surfactants have been used extensively as soft templates to produce mesoporous silica materials. This study investigates the use of a double-tailed surfactant, a didodecyldimethylammonium phosphate surfactant ($DDAH_2PO_4$) as a structure-directing agent in the sol-gel synthesis of silica at ambient conditions in aqueous solution. The effects of varying reaction parameters such as surfactant concentration and solution temperature on resulting silica morphology are presented. Morphological transitions from nanobeads to hexagonal plates and toroidal concave particles are observed with increasing surfactant concentrations, as well as a gradual loss in templating ability at elevated solution temperatures (up to 25 °C). This allows us to access different morphologies and dimensions of nanostructures within the same synthesis scheme templated with $DDAH_2PO_4$.

Electronic supplementary material The online version of this chapter (https://doi.org/10.1007/978-981-32-9828-6_25) contains supplementary material, which is available to authorized users

W. W. Z. Yeo
Victoria Junior College, Singapore, Singapore
e-mail: wayne.yeo.wei.zhong.2017@vjc.sg

S. H. Lim · C. K. Liu (✉)
Institute of Materials Research and Engineering, Agency for Science, Technology and Research, Singapore, Singapore
e-mail: connie-liu@imre.a-star.edu.sg

S. H. Lim
e-mail: limsuh@imre.a-star.edu.sg

G. Yong (✉)
Cancer Early Detection Advanced Research Center, Knight Cancer Institute, Oregon Health & Science University, Portland, OR, USA
e-mail: yong@ohsu.edu

© Springer Nature Singapore Pte Ltd. 2019
H. Guo et al. (eds.), *IRC-SET 2018*,
https://doi.org/10.1007/978-981-32-9828-6_25

309

1 Introduction

Silica is of central importance in nanomedicine. The suitability of silica nanoparticles in various *in vivo* biomedical applications, including drug delivery, encapsulation [1] and fluorescence imaging [2] has been established in previous work [3]. This is largely by virtue of its biodegradability, biocompatibility and low toxicity [4]. The thermal and mechanical stability of silica, with low density and high specific surface area, lends itself to numerous applications [5]. We foresee that silica of different morphology and sizes have novel biomedical uses that have yet to be established, such as artificial platelets.

The synthesis of colloidal silica, consisting of monodisperse nano- to micrometer sized silica nanoparticles, was first described in 1968 [6] using a sol-gel process where a silicon alkoxide precursor undergoes hydrolysis with water in an alcohol solution containing ammonia as a catalyst. Recent developments to silica sol-gel chemistry has enabled the synthesis of silica using aqueous solvents at neutral pH for convenient and environmentally sound processing [7]. In addition, modifying the sol-gel approach through composition doping can incorporate various ions into the silica nanostructure [8], modifying material properties such as biodegradability.

Different strategies for making silica particles with diverse morphologies have been previously reported. Templated synthesis has been a notably effective technique for the controlled synthesis of nanostructured materials [9]. Here, a pre-existing template, or structure-directing agent (SDA), with desired nanoscale features, directs the formation of silica into unique structural forms that are otherwise difficult to obtain without the directing effect of the template. As a result, templated synthesis is capable of fabricating nanomaterials with an abundance of consistent and well-defined structures, morphologies and properties. Inanc et al. demonstrated the templated synthesis of a complex and functional silica nanostructure in the development of dual-porosity hollow silica nanospheres as a vehicle to deliver nonhuman enzyme therapies [11]. Such morphological precision was attained through the use of colloidal polystyrene nanoparticles as templates and nanomasks [11, 12]. This demonstrates the utility of the synthesis described by Yang et al. [12] with appropriate hard colloidal physical templates.

In addition to hard templating strategies such as the polystyrene nanospheres described in the previous paragraph, a myriad of soft template methods ranging from emulsions, micelles or vesicles, to polymers and biological molecular assemblies have been developed. The principal advantage of soft template synthesis is the capacity to synthesize different materials with various non-traditional morphologies [13], as demonstrated with MCM-41 [10, 20]. The fabrication of amorphous hexagonal silica platelets using polypeptides such as poly-L-lysine (PLL) [14, 15] has been reported. However, the synthesis described by Tomczak et al. is dependent on the unstable secondary structure transition of PLL coils to alpha helices. Failing to undergo a secondary structure transition would result in the loss of any templating ability. A robust and flexible fabrication approach that allows for the synthesis of a wide variety of nanostructures, over the variation of a broad range of precise reaction conditions has yet to be developed and remains a key objective that this work is centered upon.

Surfactants are promising candidate SDAs as they self-assemble to form different aggregate structures resulting in a diverse range of lyotropic liquid crystal phases. Microstructures in surfactant systems are governed by the interplay of geometrical constraints [16] and for cationic surfactants, electrostatic interactions due to different counterions [17]. In addition, microstructure formation can also be influenced by the addition of salts and other electrolytes [18]. Indeed, the first incidence of mesoporous silica structures was successfully produced in 1992 by templating on liquid crystalline surfactant phases [10, 20]. MCM-41, is prepared by a hydrothermal sol-gel reaction with cetyltrimethylammonium bromide (CTAB) surfactant as the SDA, yielding a three-dimensional ordered mesoporous silica material with non-intersecting hexagonal channels.

In this work, we explore the soft template synthesis of silica at ambient conditions using double-chained cationic surfactant, didodecyldimethylammonium phosphate ($DDAH_2PO_4$) (Fig. 1). The preparation, self-assembly and phase behavior of double-tailed didodecyldimethylammonium (DDA^+) surfactants have been extensively reported in the literature [21]. The microstructure formation of DDA^+ surfactants have been found to be sensitive to the surfactant concentration and hydrolysis state of its counterions [22]. In this research, we chose DDA^+ phosphate systems as a candidate structure-directing agent, primarily because double-tailed surfactants have largely been unexplored as soft templates for the synthesis of silica nanostructures. In addition, DDA^+ phosphate form prolate micelles that are stabilised by hydrolysable phosphate counterions [22]. This project aims to study the effects on the morphology of silica nanostructures templated by DDA^+ phosphate surfactant when varying reaction conditions such as surfactant concentration and reaction temperature. We posit that accessing a plethora of varied silica morphologies will result in unveiling a wealth of previously unattainable biomedical applications, including targeted delivery vehicles and artificial platelets through biofunctionalisation.

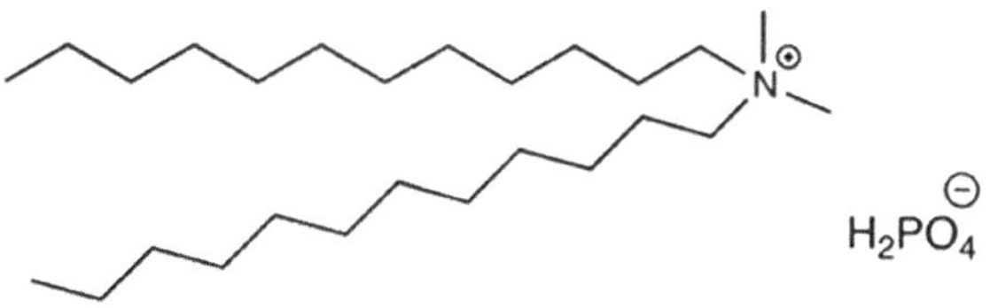

Fig. 1 Molecular structure of $DDAH_2PO_4$, the major species of DDA^+ phosphate present in synthesis

2 Methods

2.1 Materials

Milli-Q water prepared by a SGwater UltraClear Basic TWF system with a resistivity of 18.2 MΩ cm was used for all reactions and dilutions in this study. Amberlite IRA-400(OH) resin and orthophosphoric acid (85 wt%) were obtained from Sigma-Aldrich. pH measurements were made with a Metrohm 827 pH lab meter. Tetramethoxysilane (TMOS) was used as received from Merck. Methanol was obtained from J. T. Baker. Didodecyldimethylammonium bromide (DDAB, 98%) was used as purchased from TCI.

2.2 Preparation of Starting Materials

Dilute phosphoric acid was prepared by diluting concentrated orthophosphoric acid with water. DDAB was dissolved to form a colorless solution in a methanol-water solvent. Didodecyldimethylammonium hydroxide (DDAOH) was prepared by substituting bromide ions with hydroxide ions from DDAB using the Amberlite ion exchange resin for 24 h. The successful conversion of the counterion from bromide to hydroxide was verified using pH measurements to confirm that the DDAOH is highly alkaline (pH 12.8). Methanol from the DDAOH was removed through rotary evaporation, which yielded a clear, viscous solution of aqueous DDAOH. The final $DDAH_2PO_4$ was then prepared by titrating dilute phosphoric acid into DDAOH solution to an appropriate equivalence to pH 7.02, which indicates that the hydroxide ions are successfully neutralized to a buffer system of phosphate species ($H_2PO_4^-$/HPO_4^{2-}). The clear, viscous solution of aqueous $DDAH_2PO_4$ was then freeze-dried and isolated as a fine white powder. In this study, we denote the phosphate surfactant used in synthesis as $DDAH_2PO_4$, because the predominant counteranion was later determined to be $H_2PO_4^-$, though it coexists with a minor fraction of HPO_4^{2-}.

2.3 Synthesis of Templated Silica Nanostructures

5 wt% stock $DDAH_2PO_4$ solution was prepared by dissolving the previously prepared anhydrous powdered $DDAH_2PO_4$ in water, which was then equilibrated under mixing for 48 h. Different required concentrations were drawn from sub-dilutions of different batches of the stock solution. 1 mL of a desired concentration of surfactant solution was equilibrated in a water bath at the desired temperature for 30 min. A silica alkoxide source, in our case, tetramethoxysilane (TMOS) was added into the above surfactant solution in one portion. The reaction was conducted on a vortex agitator on a high speed, which provides rapid mixing of reactants. The reaction mixture was

then returned into the water bath and retrieved after 24 h. The silica nanostructures were separated from the supernatant by centrifugation as a pellet and resuspended in deionised water for further characterization.

2.4 Design of the Experimental Matrix

The two reaction variables are the surfactant concentration and solution temperature, as the phase behaviour of similar DDA^+ surfactant systems are known to be responsive to these variables [17]. As such, a range of concentrations and temperatures have been generated from what is feasible, afforded by the viscosity of the solution to ensure homogeneous mixing. The sol-gel synthesis experiments were conducted at temperatures ranging from 0, 4, 8, 12, 16 and 20 °C and concentrations from 0.5, 1.0, 2.0, 3.0 and 5.0 wt%.

2.5 Scanning Electron Microscopy

High resolution SEM images were taken using a JEOL JSM-6700F Field Emission Scanning Electron Microscope. SEM samples were prepared by drop casting an aqueous suspension of nanosilica onto a silicon substrate, and dried before being sputtered with gold for microscopy.

3 Results and Discussion

3.1 Preparation of the $DDAH_2PO_4$ Structure-Directing Agent

$DDAH_2PO_4$ was prepared from commercially available DDAB as in Scheme 1. DDAOH was titrated against phosphoric acid, with the resulting pH tracked with a pH meter.

A typical triphasic titration curve was obtained from the titration, shown in Fig. 2. For the purpose of the following synthesis steps, an endpoint of pH 7 was chosen for the titration. During the titration, numerous observable changes to the titre were seen. At pH 13, the titre turned cloudy, with bubbles forming at the surface of the titre. At pH 11.6, the titre adopted the appearance of a milky emulsion. From pH 10.9 onwards, the titre remains a clear, viscous solution until the experimental endpoint of pH 7.02.

$DDAH_2PO_4$ was determined to be the major surfactant species present, making up 60.6% of all surfactant species in solution. A minor species of DDA_2HPO_4 made

Scheme 1 Preparation of DDA$^+$ phosphate from DDAB through an ion exchange and titration

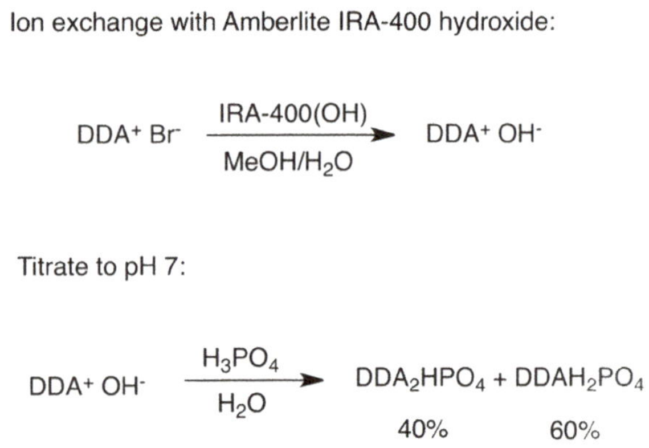

Ion exchange with Amberlite IRA-400 hydroxide:

$$DDA^+ Br^- \xrightarrow[\text{MeOH/H}_2\text{O}]{\text{IRA-400(OH)}} DDA^+ OH^-$$

Titrate to pH 7:

$$DDA^+ OH^- \xrightarrow[\text{H}_2\text{O}]{\text{H}_3\text{PO}_4} DDA_2HPO_4 + DDAH_2PO_4$$
$$\qquad\qquad\qquad\qquad\qquad\quad 40\% \qquad\qquad 60\%$$

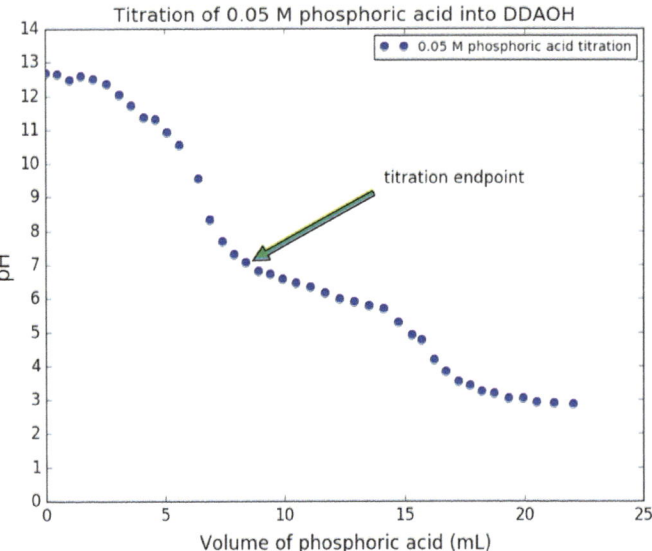

Fig. 2 Triphasic titration curve of phosphoric acid into DDAOH, with the titration endpoint of 7.02 indicated by the arrow

up the remaining 39.4%. This is because hydrolysis leads to a distribution of different phosphate counterion species in solution ($H_2PO_4^-$, HPO_4^{2-}, PO_4^{3-}) which can be expressed as a function of solution pH. This is done by firstly considering the K_a expressions for the three deprotonations of phosphoric acid and expressing the initial concentration of phosphoric acid as the sum of the concentration of all species.

This can be solved as a system of four equations, with the species fraction being expressed as a function of the hydronium ion concentration (see Derivation 1 in the Supplementary Information section). The graph showing the fractional speciation of

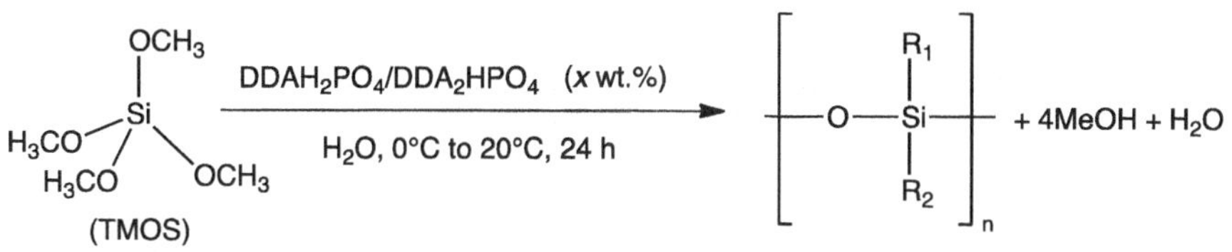

Scheme 2 Templated sol-gel synthesis of silica with DDA$^+$ phosphate surfactant from TMOS, in an aqueous solution at ambient conditions

the phosphoric acid system as a function of pH is displayed as Fig. S1. The fractions of $H_2PO_4^-$ and HPO_4^{2-} are indicated in the plot.

3.2 Synthesis of Surfactant-Directed Silica Nanostructures

Silica made with DDAH$_2$PO$_4$ was synthesised according to Scheme 2, and the method described in Sect. 2.3. In the sol-gel synthesis, TMOS is hydrolysed in aqueous solution to give silicic acid, which acts as a precursor for the polycondensation reaction at the nucleation point. The initial chemistry of the process is detailed below [7].

The hydrolysis of TMOS is an acid-catalysed reaction, whereas the formation of the silicate species and further polymerization are base-catalysed. Previous studies indicate that the precipitation reaction can potentially be controlled through preferential precipitation at the high positive charge density at the surface of self-assembled structures of cationic surfactant

This leads to a high concentration of silicate species at the positively charged surfactant headgroup interface, resulting in favourable silicate polymerisation at the surfactant-silicate interface [23]. The mechanism of the templating employed in this synthesis is still a topic of much debate by the community, but there is a general consensus that the occurrence of direct geometrical templating is improbable [19]. Instead of geometrical templating, the electrostatic interactions between cationic headgroups and counterions of the DDA$_2$HPO$_4$ surfactant are postulated to play a significant role in the control of silica formation.

3.3 Silica Morphologies as a Function of Concentration

At a reaction temperature of 0.5 °C, the nanostructures are discrete particles and do not form a continuous network. Increasing surfactant concentration from 0.5 to 2.0 wt% results in the silica morphology changing from nanobeads to hexagonal plates (Fig. 3q, m, i). Thus, we can conclude that by varying the concentration of surfactant in solution morphologies of silica can be changed.

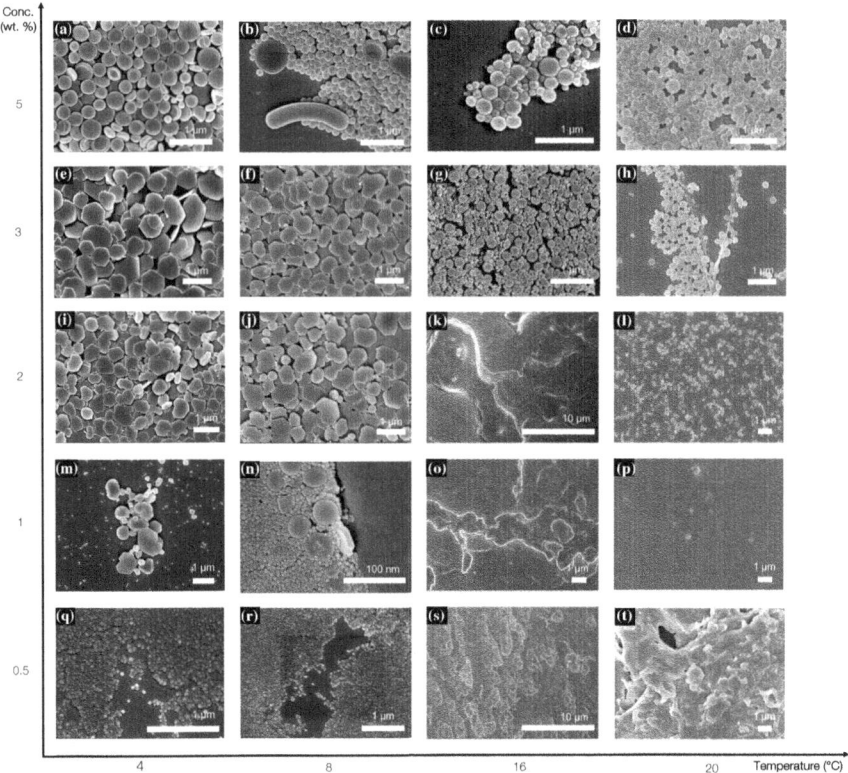

Fig. 3 a–t A matrix of representative SEM images of silica structures made with the surfactant template, varying concentration and temperature

When the surfactant concentration is increased further, at low temperatures, a second transition in morphology from hexagonal plates (Fig. 3i, e) to concave toroidal structures (Fig. 3a) was observed. These concave structures are approximately 0.5 times smaller than the hexagonal plates that were produced at lower concentrations. It has been previously reported that silica condensation causes the charge density across the silicate network to decrease, causing an increase in the effective head group area of the surfactant. As a result, it is more favourable to form surfactant microstructures with high surface curvatures [16, 25]. This increase in surfactant concentration will provide a greater headgroup charge density which drives the transformation from plates to concave structures in a similar fashion.

3.4 Silica Morphologies as a Function of Temperature

At elevated reaction temperatures, above 16 °C, we observed a loss in structure-directing ability (Fig. 3l, t) by the surfactant. This effect was especially apparent at a lower range of surfactant concentrations. The pellet of silica that could be retrieved from the supernatant and re-suspended in water became vanishingly small, and the wavy streaks (Fig. 3k, o, s) in the micrograph resemble colloidal silica or traces of surfactant residue that could not be removed by centrifugation.

Increasing temperature at high surfactant concentrations produced a range of intermediate morphologies ranging from spheroids, to large toroidal and larger nanorod structures. The wide range of structures could be attributed to changes in the aggregate structure of the surfactant that the silica can interact with. At high concentrations and temperatures, we see gradual shifts toward more consistent morphologies and size distributions. In this region, we see two main morphologies, nanobeads and larger spheroids, of which some can be identified as concave (Fig. 3c, d, g, h). A number of trends can be identified from the data collected and are summarised below.

At low temperatures, an increase in surfactant concentration will result in clearly defined morphological transitions from nanobeads, to hexagonal plates and concave toroidal particles. Across the range of concentrations that were investigated in this study, an increase in temperature results in a gradual loss in structure-directing ability by the surfactant, yielding colloidal silica at low surfactant concentrations and spheroidal nanoparticles at high surfactant concentrations.

4 Conclusion and Future Work

In summary, we have shown that the sol-gel method and the use of double-tailed cationic surfactant, $DDAH_2PO_4$ as a structure-directing agent can be developed into a promising platform for producing a diverse range of functional silica nanostructures, with varying morphological, architectural and dimensional properties. Different morphologies and dimensions of silica nanostructure, such as hexagonal plates and concave toroidal particles can be made through controlling the concentration of $DDAH_2PO_4$ in the reaction system. Furthermore, an increase in the solution temperature results in a decrease in the structure-directing ability of the surfactant, yielding interesting intermediate morphologies. This work has demonstrated that varying easily accessible reaction parameters, such as solution temperature and surfactant concentration, can allow researchers to modulate the structure of silica produced from the same surfactant-templated synthesis, displaying consistent trends in morphology and can be applied to various biomedical applications.

As it was previously established that the phase behaviour of DDA^+ surfactants are affected by the hydrolysis state of the counterion [22], pH is another dimension of control over the structure-directing behaviour of $DDAH_2PO_4$ that can be investigated in the future. Another parameter for exploration is doping quantities of different cations into the surfactant solution as it is well-known that the addition of salts and electrolyte change the self-assembly behaviour of cationic surfactants.

Acknowledgements We acknowledge the Institute of Materials Research and Engineering, A*STAR for providing the resources to conduct this study and Ms. Zhang Nan for collecting the titration data used in Fig. 2. W.W.Z. Yeo is grateful to Victoria Junior College and Mr. Wong Shiongwei for advice, support and guidance.

References

1. Slowing, I. I., Vivero-Escoto, J. L., Wu, C.-W., & Lin, V. S.-Y. (2008). Mesoporous silica nanoparticles as controlled release drug delivery and gene transfection carriers. *Advanced Drug Delivery Reviews, 60*, 1278–1288.
2. Ow, H., Larson, D. R., Srivastava, M., Baird, B. A., Webb, W. W., & Wiesner, U. (2005). Bright and stable core-shell fluorescent silica nanoparticles. *Nano Letters, 5*, 113–117.
3. Liberman, A., Mendez, N., Trogler, W. C., & Kummel, A. C. (2014). Synthesis and surface functionalization of silica nanoparticles for nanomedicine. *Surface Science Reports, 69*, 132–158.
4. Popplewell, J., King, S., Day, J., Ackrill, P., Fifield, L., Cresswell, R., Di Tada, M., & Liu, K. (1998). Kinetics of uptake and elimination of silicic acid by a human subject: A novel application of ^{32}Si and accelerator mass spectrometry. *Journal of Inorganic Biochemistry, 69*, 177–180.
5. Xu, Z. P., Zeng, Q. H., Lu, G. Q., & Yu, A. B. (2006). Inorganic nanoparticles as carriers for efficient cellular delivery. *Chemical Engineering Science, 61*, 1027–1040.
6. Stöber, W., Fink, A., & Bohn, E. (1968). Controlled growth of monodisperse silica spheres in the micron size range. *Journal of Colloid and Interface Science, 26*, 62–69.
7. Yang, J., Lind, J. U., & Trogler, W. C. (2008). Synthesis of hollow silica and titania nanospheres. *Chemistry of Materials, 20*, 2875–2877.
8. Pohaku Mitchell, K. K., Liberman, A., Kummel, A. C., & Trogler, W. C. (2012). Iron(III)-doped, silica nanoshells: A biodegradeable form of silica *Journal of the American Chemical Society, 134*, 13997–14003.
9. Liu, Y., Goebl, J., & Yin, Y. (2013). Templated synthesis of nanostructured materials. *Chemical Society Reviews, 42*, 2610–2653.
10. Beck, J. S., Vartuli, J. C., Roth, W. J., Leonowicz, M. E., Kresge, C. T., Schmitt, K. D., ... & Higgins, J. (1992). A new family of mesoporous molecular sieves prepared with liquid crystal templates. *Journal of the American Chemical Society, 114*(27), 10834–10843.
11. Ortac, I., Simberg, D., Yeh, Y.-S., Yang, J., Messmer, B., Trogler, W. C., Tsein, R. Y., & Esener, S. (2014). Dual-porosity hollow nanoparticles for the immunoprotection and delivery of nonhuman enzymes. *Nano Letters, 14*, 3023–3032.
12. Trogler, W. C., Esener, S. C., Messmer, D., Lind, J. U., Mitchell, K. K., & Yang, J. (2013, April). Hollow silica nanospheres and methods of making same. US Patent 20130230570A1.
13. Wan, Y., & Zhao, D. (2007). On the controllable soft-templating approach to mesoporous silicates. *Chemical Reviews, 107*, 2821–2860.
14. Tomczak, M. M., Glawe, D. D., Drummy, L. F., Lawrence, C. G., Stone, M. O., Perry, C. C., Pochan, D. J., Deming, T. J., & Naik, R. R. (2005). Polypeptide-templated synthesis of hexagonal silica platelets. *Journal of the American Chemical Society, 127*, 12577–12582.
15. Bellomo, E. G., & Deming, T. J. (2006). Monoliths of aligned silica-polypeptide hexagonal platelets. *Journal of the American Chemical Society, 128*, 2276–2279.
16. Israelachvili, J. N., Mitchell, D. J., & Ninham, B. W. (1976). Theory of self-assembly of hydrocarbon amphiphiles into micelles and bilayers. *Journal of the Chemical Society, Faraday Transactions 2: Molecular and Chemical Physics, 72*, 1525–1568.

17. Kang, C., & Khan, A. (1993). Self-assembly of systems of didodecyldimethylammonium surfactants: Binary and ternary phase equilibria and phase structures with sulphate, hydroxide, acetate and chloride counterions. *Journal of Colloid and Interface* Science, *156*, 218–228.

18. Thalberg, K., Lindman, B., & Karlstroem, G. (1991). Phase behavior of a system of cationic surfactant and anionic polyelectrolyte: The effect of salt. *The Journal of Physical Chemistry*, *95*, 6004–6011.

19. Cölfen, H., Page, M. G., Dubois, M., & Zemb, T. (2007). Mineralization in complex fluids. *Colloids and Surfaces A: Physicochemical and Engineering Aspects, 303*, 46–54.

20. Kresge, C., Leonowicz, M., Roth, W. J., Vartuli, J., & Beck, J. (1992). Ordered mesoporous molecular sieves synthesized by a liquid-crystal template mechanism. *Nature*, *359*, 710.

21. Warr, G. G., Sen, R., Evans, D. F., & Trend, J. E. (1988). Microemulsion formation and phase behavior of dialkyldimethylammonium bromide surfactants. *The Journal of Physical Chemistry*, *92*, 774–783.

22. Liu, C. K., & Warr, G. G. (2014). Self-assembly of didodecyldimethylammonium surfactants modulated by multivalent, hydrolyzable counterions. *Langmuir*, *31*, 2936–2945.

23. Monnier, A., Schuth, F., Huo, Q., Kumar, D., Margolese, D., Maxwell, R., Stucky, G., Krishnamurty, M., Petroff, P., Firouzi, A., Janicke, M., & Chmelka, B. F. (1993). Cooperative formation of inorganic-organic interfaces in the synthesis of silicate mesostructures. *Science, 261*, 1299.

24. Yong, G., Xu, W., & Liu, C. (2017, July). Hexagonal silica platelets and methods of synthesis thereof., WO Patent App. PCT/SG2017/050,025.

25. Che, S., Li, H., Lim, S., Sakamoto, Y., Terasaki, O., & Tatsumi, T. (2005). Synthesis mechanism of cationic surfactant templaitng mesoporous silica under an acidic synthesis process. *Chemistry of Materials, 17*, 4103–4113.

Buccal Delivery of Curcumin to Address Its Poor Gastrointestinal Stability

Bing Lim and Mak Wai Theng

Abstract Curcumin has numerous health benefits but has low bioavailability in the digestive system due to its low aqueous solubility, high metabolism rate caused by bile (from the liver), and degradation of curcumin under the basic conditions of the small intestine. The alternative way of delivering curcumin into the systemic circulation is through the buccal mucosa, but curcumin has to be delivered quickly before it undergoes degradation under the neutral conditions of the mouth. Oral disintegrating films (ODFs) are investigated as a pathway to release nanoparticles across the buccal mucosa to improve bioavailability. Chitosan, a mucoadhesive polymer, is used to coat nanoparticles. Pregelatinized starch (PGS), granular hydroxypropyl starch (GHS), hydroxypropylmethylcellulose (HPMC) and polyvinyl alcohol (PVA) are mucoadhesive polymers used in films, and propylene glycol (PG) as plasticizer. Mass, thickness, percentage recovery, surface pH, disintegration time and dissolution are assessed for each film. All films have a surface pH of around 6.7, making them non-irritant. Films with formulation of GHS and HPMC has highest maximum percentage dissolution, and has high percentage recovery, giving high drug load. It has released more curcumin than other films in the same time frame, allowing for curcumin to be released quickly. The average disintegration time for it is 53 minutes, which is long for ODFs. More drug can be released across the buccal mucosa over a longer timeframe. Films with PVA may be the thinnest and lightest, but break upon handling, making them hard to apply onto the buccal mucosa and unsuitable as an oral film.

Keywords Curcumin · Oral disintegrating films · Nanoparticles · Bioavailability

B. Lim · M. W. Theng (✉)
National Junior College, Singapore, Singapore
e-mail: waitheng8@gmail.com

B. Lim
e-mail: limbing09@gmail.com

© Springer Nature Singapore Pte Ltd. 2019
H. Guo et al. (eds.), *IRC-SET 2018*,
https://doi.org/10.1007/978-981-32-9828-6_26

1   Introduction

Curcumin is a polyphenol derived from turmeric. It is known for its health benefits, including anti-tumour, anti-inflammatory and antioxidant properties [1]. It is also found that curcumin has a cytotoxic effect on different types of cancerous cells [2–11]. However, the aqueous solubility of curcumin is low (approx. 11 ng/ml in pH 5 buffer solution) [12] and curcumin experiences high metabolism rate due to bile from the liver. [13]. It also degrades in basic conditions of the small intestine. This results in poor bioavailability of curcumin and a higher dosage of curcumin is required for sufficient amount of curcumin to enter systemic circulation. An alternative location of systemic absorption of curcumin is in the mouth. However, curcumin is also found to degrade quickly in the neutral condition of the mouth [13]. As such, for curcumin to possess high bioavailability in the mouth, it is required for the curcumin to be delivered quickly to the systemic circulation before it undergoes degradation in the mouth.

In the recent years, nanomedicine has been recognised as a means to improve dissolution of poorly soluble drugs, hence improving bioavailability. The size of the nanoparticles makes it suitable for administration through the mucosa [14]. Curcumin is found to have cytotoxic effects on cells, hence it is important to have targeted delivery of curcumin to avoid destruction of healthy cells [15]. To achieve the desired response and reduce cytotoxic effects, the drug then has to be encapsulated within nanoparticles [16]. Studies has shown that the solubility of curcumin nanoparticles in water was approximately 640-fold that of crude curcumin, proving nanoparticles of curcumin as a suitable form of drug delivery [13]. The choice of polymer for polymeric nanoparticles is then important such that it improves the solubility, mucoadhesiveness and hence bioavailability of curcumin.

In traditional methods of administration, the process of producing tablets and capsules lead to agglomeration, poor-redispersion and poor recovery of nanoparticles. The conversion of nanoparticles in patient dosage form is then time consuming [12]. This leads to using Oral Disintegrating Films (ODFs) as an alternative administration method to deliver curcumin. Specific types of ODFs can possess high mucoadhesive ability [17–23] and permeability, releasing great amount of nanoparticles through the mucosa [13, 21]. This allows for quick systemic absorption of curcumin through the mouth, and brings us to the possibility of administration of curcumin through the oral mucosa in a film. Pregelatinized starch (PGS), Granular hydroxypropyl starch (GHS), hydroxypropylmethylcellulose (HPMC), polyvinyl alcohol (PVA) and chitosan are all mucoadhesive polymers, given their high hydrogen bonding capacity, which is important for their mucoadhesive ability. As such, these polymers are used in the films. Our aim is to create a film that releases a large amount of curcumin that passes through the buccal mucosa within the shortest time, into the bloodstream. This way, an effective film with high bioavailability is created to prevent the need for higher drug intake and drug wastage.

2 Hypothesis

The formulation with PVA will be the most thin, and most lightweight. Furthermore, both starch and PVA films will have high drug release rates, as PVA and starch have improved drug release rates for films in other studies [24, 25]. However, with the inclusion of HPMC to increase mucoadhesive ability of films, starch and PVA films containing HPMC can have lower drug release rates, as HPMC controls drug release [26].

3 Methodology

3.1 Nanoplex

500 mg of curcumin was dissolved in 100 ml of 0.1 M KOH, giving a mixture with a concentration of 5 mg/ml. 100 ml of chitosan, which was 5.835 mg/ml dissolved in 1.2% (volume per volume) acetic acid, was added to the mixture and mixed well. The mixture was sonicated for 6 min and poured equally into six falcon tubes. Each tube contains 33.3 ml of mixture. All falcon tubes were centrifuged at 13,000 rpm for 90 min at 4 °C. The supernatant in all tubes were decanted. Resuspension was conducted with 15 ml of deionised (DI) water added into each falcon tube. They were centrifuged again at 13,000 rpm for 60 min at 4 °C, followed by decantation. Resuspension was conducted again with 5 ml of DI water. The contents in the falcon tube which underwent the second resuspension were transferred to another falcon tube. This process was repeated for the remainder of the falcon tubes. The final falcon tube would contain all the nanoplex. 8000 times serial dilution with 80% ethanol was conducted onto the nanoplex. The absorbance value was measured with a Ultraviolet spectrophotometer (UV-VIS) blanked with 80% ethanol at 423 nm wavelength. Concentration of curcumin in nanoplex was calculated using the equation

$$y = 0.0079\,x - 0.0002, \tag{1}$$

where x = absorbance value × dilution factor and y = concentration of nanoplex in mg/ml.

3.2 Film Matrix and Plating

All formulations contain 5% (weight per volume) propylene glycol as plasticiser. Plasticisers can overcome the brittleness and soften the rigidity of the film structure by reducing the intermolecular forces [27]. For starch films, they contain 15% (weight per volume) starch. For PVA films, they contain 0.5% (weight per volume) PVA. The theoretical concentration of curcumin for each film is 1 mg/cm^2. Films that contain HPMC as sole polymer will act as a control.

3.2.1   Starch

4050 mg of starch (PGS or GHS) was added into a falcon tube, followed by 23.4 ml of DI water, then vortexed, giving a 15% starch formulation. The test tubes were heated in preheated water bath at 80 °C until starch mixture was translucent. 1.37 ml of propylene glycol was added to each tube. All tubes were vortexed. To find the volume of nanoplex to be added into each tube, take

$$\frac{\text{mass of curcumin needed (mg)}}{\text{concentration of curcumin in nanoplex (mg/ml)}}. \tag{2}$$

The tubes were vortexed and degassed (to remove bubbles for plating of films). For films containing HPMC, was added until total volume had reached 27 ml.

3.2.2   PVA

135 mg of PVA was added into a falcon tube, followed by 12.77 ml of DI water. The resultant 0.5% PVA mixture was vortexed. 1.37 ml of propylene glycol was added, followed by 1.61 ml of vortexed nanoplex. For PVA HPMC, HPMC was added until total volume had reached 27 ml. 22.7 ml of matrix was pipetted into each petri dish. All plates were then left in the drying oven at 60 °C for approximately 16 h.

3.3   Characterisation of Film

3.3.1   Mass and Thickness

The mass (mg) of each film was measured with an electronic balance (to 0.1 mg). The thickness of each film (mm) was measured at three different sides with an electronic vernier caliper (to 0.01 mm). The thickness of film was obtained by taking the average of three sides measured.

3.3.2   Percentage Recovery

One curcumin film was dissolved in 5 ml of DI water in each falcon tube. After 20 min, 20 ml of 100% ethanol was added to make the mixture to 80% ethanol. All tubes were centrifuged at 10,000 rpm for 10 min. 10 times serial dilution was conducted on supernatant with 80% ethanol. With the UV-VIS blanked with 80% ethanol at

423 nm wavelength, absorbance value was measured. Using Eq. (1), concentration of curcumin in film was calculated, then used to derive percent recovery of curcumin via the following equation

$$\frac{\text{concentration of curcumin (mg/ml)} \times \text{Total volume of solvent used to dissolve the film (ml)}}{\text{concentration of curcumin by area of film (mg/cm}^2)} \times 100\% \qquad (3)$$

3.3.3 Disintegration Time

Simulated salivary solution (SSS) at pH 6.75 was prepared by adding 4.494 g of Na_2HPO_4:7H2O, 0.19 g of KH_2PO_4, 8 g of NaCl to 1000 ml DI water, and adjusting with 540 μl of concentrated H_3PO_4. Petri dishes were filled with 10 ml of SSS and left to warm in an incubator set at 37 °C. One 1 cm by 1 cm film was added to each dish. The time required for each film to disintegrate was recorded. Films were not adhered to glass slides in order to eliminate the possibility that sticking films on glass slides will affect the duration the film takes to disintegrate. This serves as an estimation to residence time of films in patients' mouth.

3.3.4 Surface pH

The film-SSS mixture was stirred slightly after the film had disintegrated. With a calibrated pH meter, pH values were recorded. pH values of each type of film is determined by taking the average of the first set of film triplicates and the respective second set of film triplicates from duplicate film.

3.3.5 Dissolution

A variable amount of film was weighed, then adhered onto a glass side with double sided tape. Each glass slide is placed into a one Pyrex bottle. The mass of the film was then used to calculate the volume of SSS to be added to the Pyrex bottle with the following equations

$$\text{drug final concentration} = 1.5 \ \mu\text{g/ml} \qquad (4)$$

$$\text{mass of film (mg)} \times \text{mass of curcumin for 1 cm}^2 \text{ of film} = x \ \mu\text{g of curcumin} \qquad (5)$$

$$\frac{x}{1.5} = \text{volume of simulated salivary solution} \qquad (6)$$

Up to a total of four hours, 2 ml of solution was transferred from Pyrex bottle to cuvette. With the UV-VIS blanked with SSS, the concentration of curcumin and

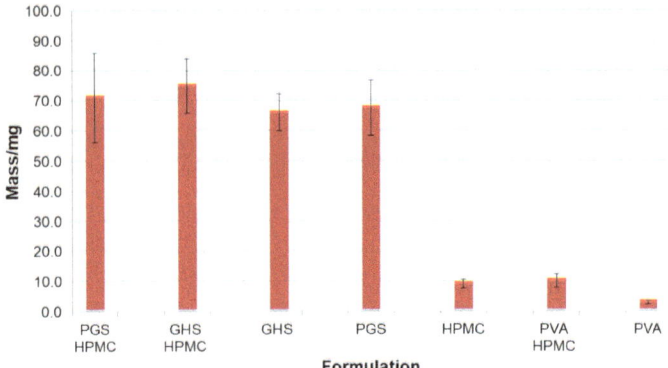

Fig. 1 Mass/mg of films

starch in starch films, or PVA in PVA films was measured at 423 nm wavelength, and turbidity, exhibited by starch and PVA, [28] at 600 nm wavelength. The difference between the two values is taken to find concentration of curcumin, to derive percentage dissolution.

4 Results and Discussion

4.1 Characterisation of Films

4.1.1 Mass

From Fig. 1, formulations containing PGS or GHS Starch have similar mass, and are heavy as compared to films that do not contain starch. Films containing HPMC also weigh more than their counterparts which do not contain HPMC. It can also be seen that PVA is a lightweight polymer as compared to starch, as PVA films are lightest. With HPMC, PVA HPMC and PVA films being significantly lighter, they are more desirable films.

4.1.2 Thickness

From Fig. 2, films that contain starch are thicker compared to other films without starch. HPMC increases the thickness of films. However, with PGS HPMC being of similar thickness to GHS, we can tell that GHS results in thicker films, as GHS films are similar in thickness to PGS HPMC films even though they do not contain HPMC. Thinner films, which are those without starch, or the thinnest film, which contains solely PVA, are better films.

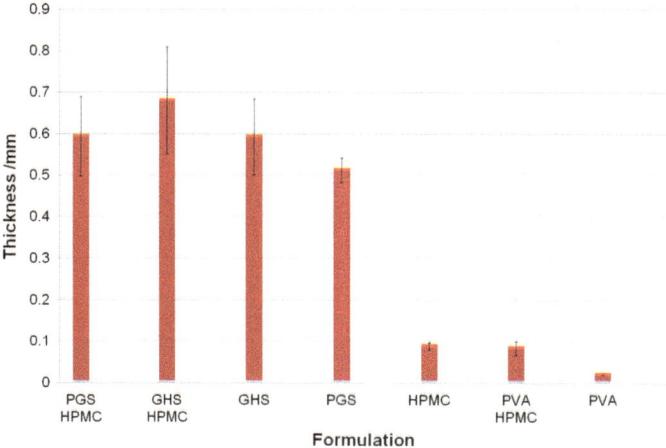

Fig. 2 Thickness/mm of films

4.1.3 Percentage Recovery

From Fig. 3, all films have high percentage recovery, with a minimum of about 80%. Films containing HPMC combined with another polymer (i.e. GHS or PGS or PVA), have greater percentage recovery than other films. It can be deduced that without either of the polymers, the percentage recovery will be reduced, as seen from formulations containing only one type of polymer. The exception is PVA films, which still have percentage recovery as high as PGS HPMC, a formulation with two types of polymers. However, when compared with PVA HPMC, we can tell that having two types of polymers and using PVA results in higher percentage recovery. HPMC, PVA HPMC and PVA films have a very high standard deviation of above 20%, which can be caused by low drug uniformity in the films as they are formed. As for the other films, the standard deviation is significantly lower. This can be

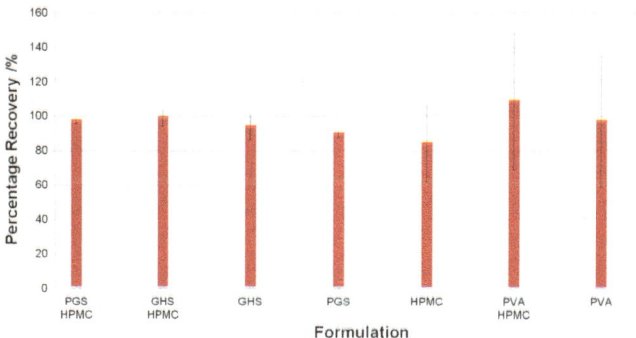

Fig. 3 Percentage recovery/% of films

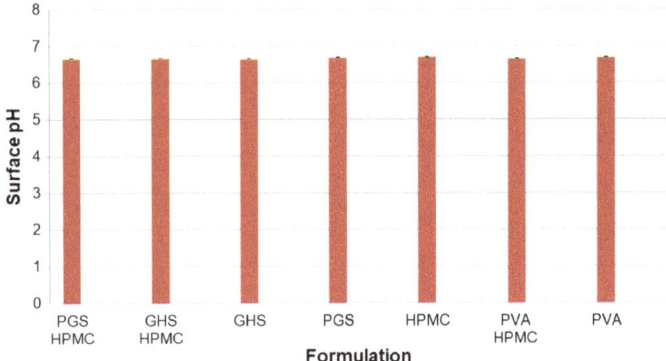

Fig. 4 Surface pH of films

attributed to greater difference in particle sizes between curcumin nanoparticles and PVA particles, leading to lower drug uniformity, as compared to that of curcumin nanoparticles and starch particles [29].

4.1.4   Surface pH

From Fig. 4, films of all formulations have very similar pH level of around 6.7, a suitable pH to be adhered to patient's buccal mucosa so no discomfort will be felt. The result has also taken into account the pH of SSS, since the films are dissolved in SSS to simulate the pH of films when placed in the oral cavity.

4.1.5   Disintegration Time

HPMC and PVA HPMC films did not disintegrate within 240 min, so it is given the maximum time span of the test. From Fig. 5, HPMC plays a large part in lengthening disintegration time for the film, since HPMC films have the highest disintegration time, and films containing HPMC have longer disintegration time than films that do not. PGS and GHS starch promotes disintegration, as both films have the shortest disintegration time. While PGS and GHS films, with or without HPMC are hard, these films become soft shortly after being submerged in SSS. On the patient, these films can become flexible to follow curvature of mouth, increasing comfort. While PVA films have the third shortest disintegration time, they are very fragile and easily break apart during handling, making it difficult for patients to apply the film. With addition of HPMC, PVA films do not break apart as easily. While keeping its flexibility, this also makes it easy for patients to apply it. PVA HPMC films also have a long disintegration time of over 240 min, allowing it to release a large amount of curcumin before it degrades in the mouth. The time when curcumin starts to degrade due to neutral conditions [13], leading to lower concentration is also the time percentage

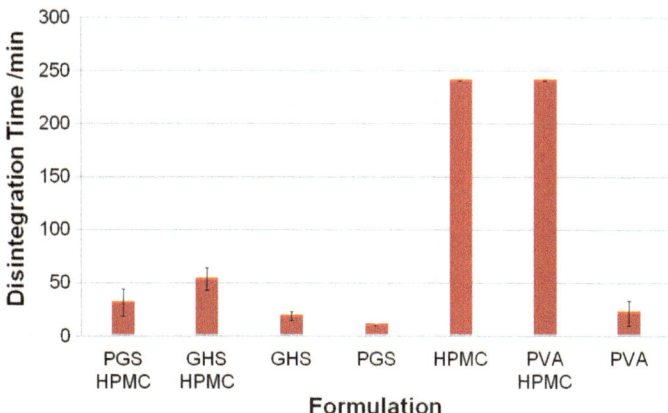

Fig. 5 Disintegration times/min of films

dissolution drops, shown by the decreasing slope of most line graphs in Figs. 6 and 7.

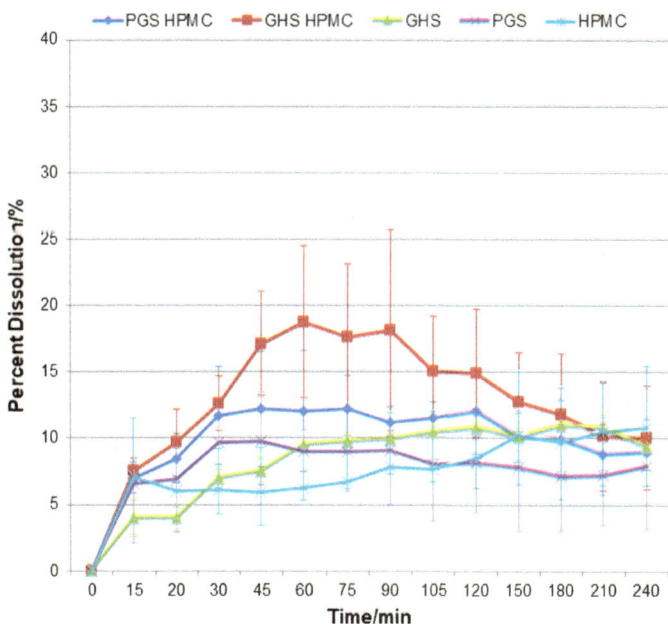

Fig. 6 Percentage dissolution/% of starch and HPMC films

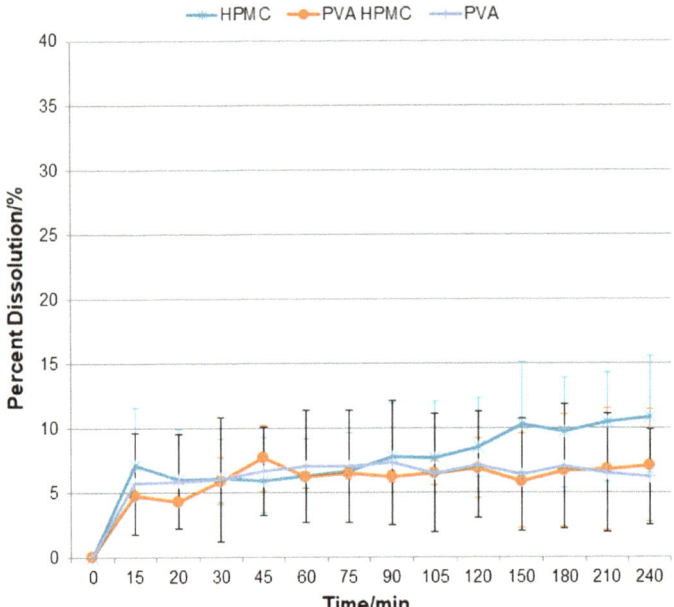

Fig. 7 Percentage dissolution/% of PVA and HPMC films

4.1.6   Dissolution

From Figs. 6 and 7, the percentage dissolution of all films increase to a maximum, then remain constant or decrease. This is attributed to disintegration time, of which the film stops releasing curcumin after it has completely disintegrated. There is also a standard deviation of 0–5% for percentage dissolution of a sample at a certain time point, as the UV-VIS used to measure absorbance is not sensitive, giving such errors. However, the percentage dissolution for PGS, GHS, HPMC, HPMC PVA and PVA films still increase beyond their respective average disintegration time. This can be due to the films being adhered to a glass slide during dissolution and hence lengthening disintegration time, enabling the film to release curcumin for a longer period of time, as compared to disintegration tests where the film is submerged in saliva without a glass slide. As such, it is recommended that the films have a backing layer, so the films are not exposed to saliva, and hence lengthening disintegration time for continued drug release [30–32]. From Fig. 6, it can be seen that both GHS and PGS help improve drug release when added to HPMC formulation. However from Fig. 7, PVA restricts drug release when added to HPMC formulation. With the assumption that all curcumin released passes through the buccal mucosa to enter bloodstream, GHS HPMC films release the most curcumin, and at fastest rate, and given the second highest percentage recovery of 98.8, this makes it a very desirable film for drug load and drug release. Even though PVA HPMC films have a higher percentage recovery, their drug release is low as compared to GHS HPMC film, making them an undesirable film for drug release.

5 Conclusion

GHS HPMC films not only have the highest maximum percentage dissolution (18.7% at 60 min), but also have the second highest percentage recovery after PVA films (98.8%), with low standard deviation ($\pm 4.28\%$). This shows that drug uniformity in GHS HPMC films is good as well. Despite being the heaviest and thickest of all films, during disintegration tests, it was found that GHS HPMC films soften upon contact with saliva. This allows patients to feel comfortable while the film is adhered to their buccal mucosa. In contrast to the hypothesis, starch films containing HPMC have greater drug release than those that do not as seen from Fig. 6, and HPMC does not greatly affect drug release rates of PVA films as seen from Fig. 7. As HPMC controls and delays drug release [26], there may be concerns that GHS HPMC films have not released all of its drug across buccal mucosa before disintegrating at 53 min. It is hence recommended to attach a backing layer to the drug film to prevent curcumin from undergoing degradation in the basic condition of the mouth when the film is exposed to saliva. This ensures that residence time of film in the patients' mouth is sufficient for it to release as much curcumin as possible across buccal mucosa. This backing layer should also be thin enough to not greatly increase film's thickness, reducing patients' comfort.

Acknowledgements We thank Associate Professor Kunn Hadinoto Ong, Mr. William Phua, and Lim Li Ming for helping us throughout our project as our supervisor, teacher mentor and student mentor respectively.

References

1. Adahoun, M. A., Al-Akhras, M. H., Jaafar, M. S., & Bououdina, M. (2016). Enhanced anti-cancer and antimicrobial activities of curcumin nanoparticles. *Artificial Cells, Nanomedicine, and Biotechnology, 45*(1), 98–107.
2. Maheshwari, R. K., Singh, A. K., Gaddipati, J., & Srimal, R. C. (2006). Multiple biological activities of curcumin: A short review. *Life Sciences, 78*(18), 2081–2087.
3. Duvoix, A., Blasius, R., Delhalle, S., Schnekenburger, M., Morceau, F., Henry, E., et al. (2005). Chemopreventive and therapeutic effects of curcumin. *Cancer Letters, 223*(2), 181–190.
4. Aggarwal, B. B., Kumar, A., & Bharti, A. C. (2003). Anticancer potential of curcumin: pre-clinical and clinical studies. *Anticancer Research, 23*, 363–398.
5. Wang, Z., Zhang, Y., Banerjee, S., Li, Y., & Sarkar, F. H. (2006). Notch-1 down-regulation by curcumin is associated with the inhibition of cell growth and the induction of apoptosis in pancreatic cancer cells. *Cancer, 106*, 2503–2513.
6. Lev-Ari, S., Zinger, H., Kazanov, D., Yona, D., Ben-Yosef, R., Starr, A., et al. (2005). Curcumin synergistically potentiates the growth inhibitory and pro-apoptotic effects of celecoxib in pancreatic adenocarcinoma cells. *Biomedicine and Pharmacotherapy, 59*(Suppl 2), S276–280.

7. Aggarwal, B. B., Shishodia, S., Takada, Y., Banerjee, S., Newman, R. A., Bueso- Ramos, C. E., et al. (2005). Curcumin suppresses the Paclitaxel induced nuclear factor-kappaB pathway in breast cancer cells and inhibits lung metastasis of human breast cancer in nude mice. *Clinical Cancer Research, 11,* 7490–7498.
8. Khor, T. O., Keum, Y. S., Lin, W., Kim, J. H., Hu, R., Shen, G., et al. (2006). Combined inhibitory effects of curcumin and phenethyl isothiocyanate on the growth of human PC-3 prostate xenografts in immunodeficient mice. *Cancer Research, 66,* 613–621.
9. Bisht, S., Feldmann, G., Soni, S., Ravi, R., Karikar, C., Maitra, A., et al. (2007). Polymeric nanoparticle-encapsulated curcumin ('nanocurcumin'): A novel strategy for human cancer therapy. *Journal of Nanobiotechnology, 5,* 3.
10. Deeb, D., Jiang, H., Gao, X., Hafner, M. S., Wong, H., Divine, G., et al. (2004). Curcumin sensitizes prostate cancer cells to tumor necrosis factor-related apoptosis inducing ligand/Apo2L by inhibiting nuclear factor-kappaB through suppression of IkappaBalpha phosphorylation. *Molecular Cancer Therapeutics, 3,* 803–812.
11. Lev-Ari, S., Strier, L., Kazanov, D., Madar-Shapiro, L., Dvory-Sobol, H., Pinchuk, I., et al. (2005). Celecoxib and curcumin synergistically inhibit the growth of colorectal cancer cells. *Clinical Cancer Research, 11,* 6738–6744.
12. Mahmood, K., Zia, K. M., Zuber, M., Nazli, Z., Rehman, S., & Zia, F. (2016). Enhancement of bioactivity and bioavailability of curcumin with chitosan based materials. *Korean Journal of Chemical Engineering, 33*(12), 3316–3329.
13. Liu, W., Zhai, Y., Heng, X., Che, F. Y., Chen, W., Sun, D., et al. (2016). Oral bioavailability of curcumin: Problems and advancements. *Journal of Drug Targeting, 4*(8), 694–702.
14. Rajalakshmi, R., Indira, Y., Aruna, U., Vinesha, V., Rupangada, V., & Krishna, S. B. (2014). Chitosan nanoparticles—An emerging trend in nanotechnology. *International Journal of Drug Delivery, 6*(3), 204–229.
15. Rajan, S. S., Pandian, A., & Palaniappan, T. (2016). Curcumin loaded in bovine serum albumin–chitosan derived nanoparticles for targeted drug delivery. *Bulletin of Materials Science, 39*(3), 811–817.
16. Aggarwal, B. B., Kumar, A., & Bharti, A. C. (2003). Anticancer potential of curcumin: Preclinical and clinical studies. *Anticancer Research, 23*(1A), 363–398.
17. Hanif, M., & Zaman, M. (2017). Thiolation of arabinoxylan and its application in the fabrication of controlled release mucoadhesive oral films. *DARU Journal of Pharmaceutical Sciences, 25*(1), 6.
18. Patel, R., & Shah, D. (2015). Nanoparticles loaded sublingual film as an effective treatment of chemotherapy induced nausea and vomiting. *International Journal of PharmTech Research, 8*(10), 77–87.
19. Mazzarino, L., Borsali, R., & Lemos-Senna, E. (2014). Mucoadhesive films containing chitosan-coated nanoparticles: A new strategy for buccal curcumin release. *Journal of Pharmaceutical Sciences, 103*(11), 3764–3771. https://doi.org/10.1002/jps.24142.
20. Mašek, J., Lubasová, D., Lukáč, R., Turánek-Knotigová, P., Kulich, P., Plocková, J., et al. (2017). Multi-layered nanofibrous mucoadhesive films for buccal and sublingual administration of drug-delivery and vaccination nanoparticles - important step towards effective mucosal vaccines. *Journal of Controlled Release, 249,* 183–195.
21. Laffleur, F., Schmelzle, F., Ganner, A., & Vanicek, S. (2016). In vitro and ex vivo evaluation of novel curcumin-loaded excipient for Buccal delivery. *AAPS PharmSciTech.*
22. Mazzarino, L., Coche-Guérente, L., Lemos-Senna, P. L., & Borsali, R. (2014). On the mucoadhesive properties of chitosan-coated polycaprolactone nanoparticles loaded with curcumin using quartz crystal microbalance with dissipation monitoring. *Journal of Biomedical Nanotechnology, 10*(5), 787–794.
23. Mazzarino, L., Travelet, C., Ortega-Murillo, S., Otsuka, I., Pignot-Paintrand, I., Lemos-Senna, E., et al. (2012). Elaboration of chitosan-coated nanoparticles loaded with curcumin for mucoadhesive applications. *Journal of Colloid and Interface Science, 370*(1), 58–66.
24. Ramdayal, G., Suresh, P., & Kr, S. U. (2011). Comparative study on novel pregelatinized garadu starch PGGS and starch 1500 as a direct compression/wet-granulation tableting excipient. *Journal of Pharmacy Research, 4,* 2406–2410.

25. Muppalaneni, S., & Omidian, H. (2013). Polyvinyl alcohol in medicine and pharmacy: A perspective. *Journal of Developing Drugs, 02*(03). https://doi.org/10.4172/2329-6631.1000112.
26. Karki, S., Kim, H., Na, S., Shin, D., Jo, K., & Lee, J. (2016). Thin films as an emerging platform for drug delivery. *Asian Journal of Pharmaceutical Sciences, 11*(5), 559–574. Retrieved from https://doi.org/10.1016/j.ajps.2016.05.004.
27. Muralisrinivasan, N. S. (2017). *Polymer blends and composites: Chemistry and technology.* Hoboken, NJ: John Wiley & Sons Inc.
28. Okaya, T., Kohno, H., Terada, K., Sato, T., Maruyama, H., & Yamauchi, J. (1992). Specific interaction of starch and polyvinyl alcohols having long alkyl groups. *Journal of Applied Polymer Science, 45*(7), 1127–1134. https://doi.org/10.1002/app.1992.070450701.
29. Rane, S. S., Hamed, E., & Rieschl, S. (2012). An exact model for predicting tablet and blend content uniformity based on the theory of fluctuations in mixtures. *Journal of Pharmaceutical Sciences, 101*(12), 4501–4515. https://doi.org/10.1002/jps.23313.
30. Choi, D., & Hong, J. (2013). Layer-by-layer assembly of multilayer films for controlled drug release. *Archives of Pharmacal Research, 37*(1), 79–87. https://doi.org/10.1007/s12272-013-0289-x.
31. Lindert, S., & Breitkreutz, J. (2017). Oromucosal multilayer films for tailor-made, controlled drug delivery. *Expert Opinion on Drug Delivery*, 1–15. https://doi.org/10.1080/17425247.2017.1276899.
32. Choi, M., Kim, K., Heo, J., Jeong, H., Kim, S. Y., & Hong, J. (2015). Multilayered graphene nano-film for controlled protein delivery by desired electro-stimuli. *Scientific Reports, 5*(1). https://doi.org/10.1038/srep17631.

Developing Modified Peptide Nucleic Acids to Regulate Dysregulated Splicing

Samuel Foo Enze, Tristan Lim Yi Xuan and Jayden Kim Jun-Sheng

Abstract Peptide Nucleic Acid (PNA) can stabilise the Tau Exon 10-Intron 10 self-regulatory RNA hairpin, known to regulate the alternative splicing of exon 10 in Microtubule-Associated Protein Tau (MAPT) transcript [1], and thus rescue the aberrant ratio of protein isoforms (4R/3R), preventing tauopathy. We used antisense PNA (ASPNA) and triplex-forming PNA (TFPNA) methods, and compared the effectiveness of both. We synthesised PNA oligomers using solid phase peptide synthesis (SPPS) and tested binding affinity to the RNA hairpin using polyacrylamide gel electrophoresis (PAGE). TFPNA is less costly and more efficient to synthesise, but has low binding affinity, whereas ASPNA costs more to synthesise but has a higher binding affinity.

Hypothesis—Binding a PNA strand to the Tau Exon 10-Intron 10 self-regulatory hairpin can rescue the aberrant ratio of 3R:4R isoforms and thus cure tauopathy.

Keywords Antisense strands · Alternative splicing · Hairpins · Hoogsteen base pair · Missense mutation · Peptide nucleic acid (PNA) · Steric hindrance · Isoform · U1snRNP · ASPNA · TFPNA

1 Introduction

PNA is an artificially synthesised nucleic acid composed of a peptide backbone which replaces the negatively-charged phosphate backbone in RNA and has nitrogenous bases (Fig. 1). PNA consists of a neutral N-(2-aminoethyl)-glycine (AEG) backbone, a methylene carbonyl linker and nucleobases. While many labs in the world are using traditional antisense strands and oligonucleotides to achieve similar aims of stabilising RNA, PNA has the following advantages: [2–4]

Mentors: Mr. Alan Ong, Asst. Prof. Chen Gang
School of Physical and Mathematical Sciences, Division of Chemical and Biological Sciences,
Nanyang Technological University
Singapore, Singapore

S. F. Enze (✉) · T. L. Y. Xuan · J. K. Jun-Sheng
Raffles Institution, Singapore, Singapore
e-mail: samuelfooenze@yahoo.com

© Springer Nature Singapore Pte Ltd. 2019
H. Guo et al. (eds.), *IRC-SET 2018*,
https://doi.org/10.1007/978-981-32-9828-6_27

335

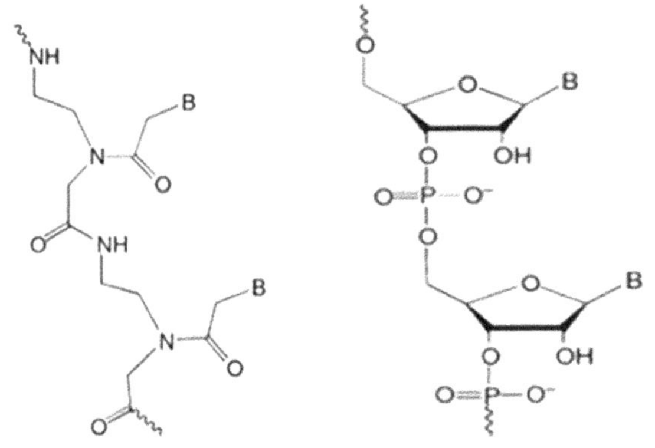

Fig. 1 Structures of PNA (left) and RNA (right) [3, 4]

- They are stable thermally and chemically, and are resistant to enzymes.
- Electrically, the peptide backbone has no charge, thus there is no electrostatic repulsion between PNA and RNA, making the bonds stronger [2].
- PNA can be hybridised at lower salt concentrations than DNA or RNA which leads to higher stringency.
- PNA has high selectivity yet a broader binding range, as PNA can be artificially altered to bind to many different specific targets.

This project deals with 2 types of PNA, ASPNA and TFPNA.

ASPNA consists of unmodified, natural monomers in conventional nucleotides (A, C. G and T). In binding, it disrupts the preformed RNA duplex and substitutes the existing strand, forming a PNA-RNA duplex instead of a PNA-RNA$_2$ triplex (known as strand-invasion), producing a stray nucleotide strand in the process. The process of synthesis is often costly because most labs are not equipped with machines to synthesise the natural monomers needed for antisense binding, and they have to buy commercially. The sequence used is also often very long, which increases the cost of synthesis and makes it harder for the strand to penetrate into the cell nucleus. Another disadvantage is that strand invasion results in a stray nucleotide strand being produced. This strand may bind with other complementary single-stranded RNA in the cell or may cause further strand invasions in other RNA duplexes. This poses the danger of side mutations that may have adverse effects. However, antisense binding is a tried and tested method used in many labs, and it has shown to have a high binding affinity due to it being complementary to one of the RNA strands. Therefore, our use of ASPNA is not only to evaluate its effectiveness up against TFPNA, but also to test our hypothesis (mentioned in the abstract) that binding PNA to the target hairpin can regulate the 3R:4R ratio.

TFPNA is made up of artificial monomers (T, L, S, Q) and binds to duplexes to form a PNA-RNA$_2$ triplex. A RNA triplex is a complex structure stabilised by multiple base triples formed between an RNA duplex and a third strand of nucleic acid (PNA in this case). A triplex structure usually forms through tertiary interactions in the major or minor groove of a Watson Crick base-paired hairpin. When consecutive major-groove base triples such as U·A-U and C+·G-C are formed, a major-groove

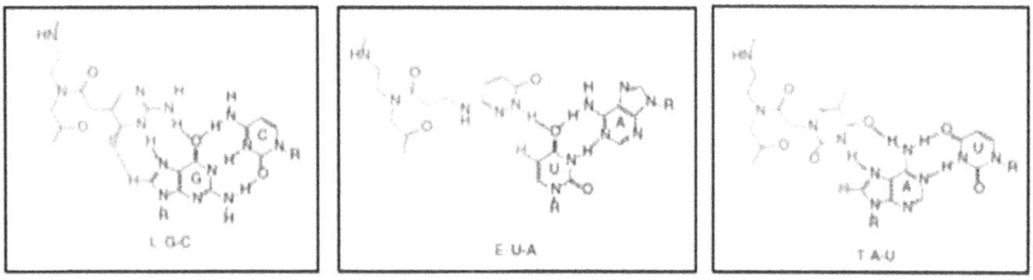

Fig. 2 Chemical structure of several base triples like TAU, EUA, QCG

RNA triplex becomes stable in isolation. These are formed without disrupting the preformed duplex structure. Using the four letter code of artificial TFPNA monomers, we can recognise the RNA duplexes by forming T·AU, L·GC, S·UA, and Q·CG base triples. The four modified nucleobases are subsequently attached to the Peptide Nucleic Acid (PNA) [2].

The use of artificial monomers gives TFPNA several advantages. Firstly, it has a wider binding range compared to strands with natural monomers. For example, only the PNA monomer E can recognise the U-A Watson-Crick base pair (Fig. 2), while no natural nucleotide is found to have such an affinity to the U-A pair. Secondly, as these artificial monomers facilitate the targeting of both the sequence and secondary structure of RNA (as opposed to only sequence in the case of natural monomers), this results in high selectivity. For example the L PNA monomer only recognises G in duplex but not G in single strand (Fig. 2). Thirdly, as a triplex is formed, there is no stray nucleotide strand being produced, the related disadvantages of antisense thus do not apply here. Fourthly, artificial monomers can be easily synthesised using liquid phase and solid phase organic synthesis with raw materials. Combined with the fact that TFPNA strands are often only 7 or 8 monomers long, the cost of TFPNA is much lower and synthesis is faster.

To the best of our knowledge, our lab is currently the first group utilizing TFPNA for the specific application of stabilising the tau hairpin. TFPNA is our main focus as it is the novelty of our project and has the above-stated advantages over the conventional method of ASPNA.

2 Understanding Tauopathy

Dysregulated RNA splicing can result in a multitude of diseases, ranging from musculoskeletal ones such as Duchenne Muscular Dystrophy or X-linked dilated cardiomyopathy, to neurodegenerative ones such as Alzheimer's Disease or dementia

In this project, we are seeking to stabilise the tau exon 10-intron 10 self-regulatory pre-mRNA hairpin to regulate dysregulated splicing of the MAPT transcript, by using PNA to target the corresponding RNA hairpin in order to restore the correct 4R:3R isoform ratio.

RNA alternative splicing is exhibited in regulating the formation of the 3R and 4R tau protein isoforms. Tau is a microtubule-associated protein found in the axons of the brain, important for morphogenesis and helps to stabilise axonal microtubules, assemble the tubulins and regulate axonal transport via phosphorylation by kinases in neurons, thus playing a pivotal role. However, point mutations in the regulatory hairpin often reduces its thermal stability by inducing mismatched base pairs (Fig. 3), resulting in the overproduction of 4R. When there is aberrant isoform ratio, the general structure of the proteins is disrupted, leading to the tau proteins being hyperphosphorylated (as kinases are no longer able to remove the phosphoryl groups, the phosphorylation sites are fully saturated). Hyperphosphorylated tau disassembles microtubules and sequesters normal tau into tangles of paired helical filaments (PHFs). The tau proteins are thus no longer able to bind to the microtubules. The previously highly soluble tau protein now aggregates due to insolubility, resulting in the formation of neurofibrillary tangles and insoluble PHFs (Fig. 4) [5]. These are deposited in the cytosol of neurons and glial cells and are cytotoxic, eventually killing neuron cells, resulting in supranuclear palsy (gradual deterioration and death of specific volumes of the brain). Furthermore, as normal tau proteins also help to stabilise the microtubules, the lack of normal tau proteins bound to microtubules will destabilise them, causing disintegration (Fig. 4) [6–8]. 4R aggregation is thus the predominant cause for the class of neurodegenerative diseases known as tauopathy.

Research shows that a 4R:3R ratio should ideally be 1, whereas an aberrant ratio is a 4R:3R ratio that greatly exceeds 1, beyond a currently uncertain threshold value [5, 6].

Fig. 3 Tau pre-mRNA regulatory RNA hairpin. The natural occurring mutations in FTDP-17 are underlined. The 4R and 3R beside each mutation refers to the isoform preferred

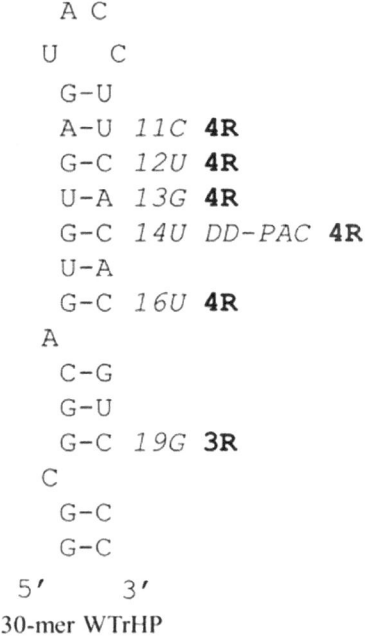

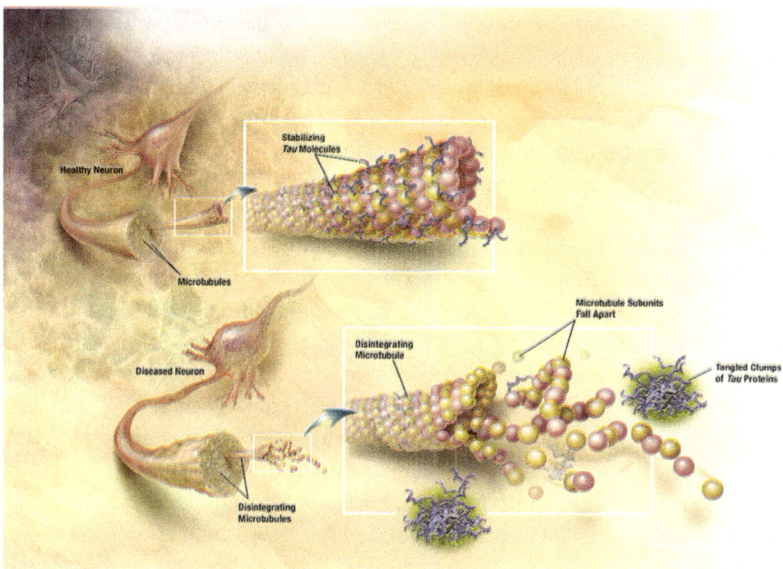

Fig. 4 A demonstration of how dysregulated splicing of the tau protein results in tau aggregation and the formation of toxic neurofibrillary

Tau protein is encoded by the MAPT gene on chromosome 17q21 which consists of 16 exons. Alternative splicing of the MAPT gene results in 6 tau isoforms being created, translated by exons 2, 3 and 10. Exon 10 inclusion leads to a 4R isoform, its exclusion a 3R isoform (Fig. 5).

There is potential to synthesise a PNA oligomer that targets and stabilises the tau RNA self-regulatory hairpin, found in the tau pre-mRNA exon 10-intron 10 junction site. The biological function of this RNA hairpin is to regulate the splicing of exon 10. In order for the splicing to occur, the splicing machinery U1 snRNP has

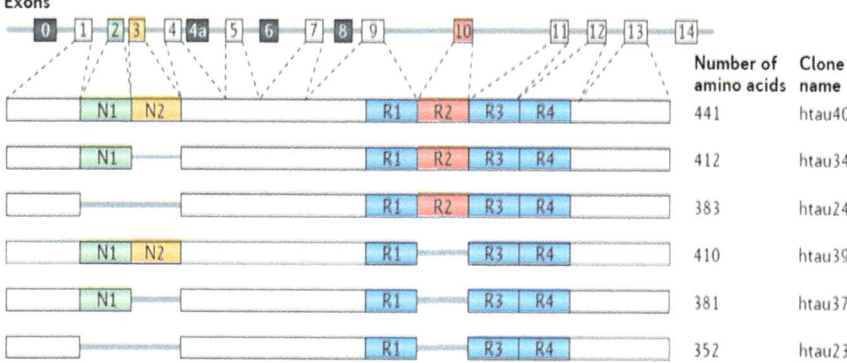

Fig. 5 The six tau isoforms produced by alternative splicing of exon 2, 3 and 10 [1]

to recognise the 5′ splice site (5′ss) by base pairing. When this particular hairpin is formed, the 5′ss is masked, thus the U1 may not be able to recognise the splice site. Point mutations found in the hairpin reduce the thermal stability of this hairpin resulting in the equilibrium shifting to the side where the hairpin is not formed (unfolded state). The hairpin conformation is in a dynamic equilibrium regulated by polypyrimidine tract protein associated splicing factor (PSF) and the trans-acting DDX5 helicase, both modulating a stem loop structure at the 5′ss. When the hairpin is not formed, the 5′ss is exposed and then U1 can easily recognise the 5′ss and splicing will occur, including exon 10 in splicing. In essence, when the hairpin is folded, U1 is blocked from reading the 5′ss and exon 10 is excluded, resulting in a 3R isoform (3 microtubule binding domains). Whereas when the hairpin is unfolded, U1 reads the 5′ss and exon 10 is included, resulting in a 4R isoform (4 microtubule binding domains, one of which is exon 10) [8–10]. If there is an excess amount of exon 10 inclusion in the splicing, there will be an aberrant 4R:3R ratio.

Thus, binding a PNA strand to the hairpin, will stabilise it and allow it to stay in the folded state longer without interference by the point mutations. The 5′ss will thus be masked and exon 10 will not be incorporated into the production of the protein isoform. This results in a 3R isoform being formed instead, reducing the 4R:3R ratio. Targeting and stabilising the tau regulatory exon 10 hairpin helps to induce steric hindrance against the binding in the region of exon 10 and intron 10 of the trans-acting proteins to the cis-acting sites, preventing U1 from excessively removing exon 10 and thus limits the overproduction of 4R isoform (Figs. 6 and 7) [9–11].

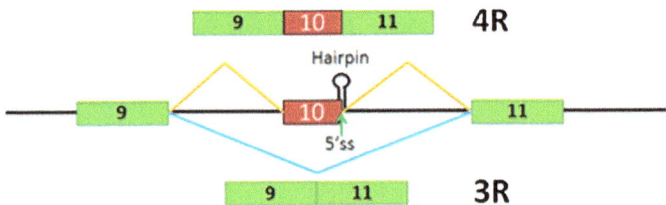

Fig. 6 The location of 5′ss at the intron 10 and exon 10 junction and the result of inclusion and exclusion of exon 10

Fig. 7 The left shows the recognition of the 5′ss by the U1 snRNP hairpin and the right shows the hairpin inducing steric blockage at the exon 10-intron 10 junction

3 Methods and Materials

The synthesis of TFPNA and ASPNA follow the same steps, except that most of the monomers of TFPNA were synthesized using solid phase organic synthesis (SPOS), but the monomers for ASPNA were bought commercially.

We synthesised the PNA oligomers through solid-phase peptide synthesis (SPPS), carried out in a nitrogen-based condition (a necessary condition for the formation of peptide bonds). Thereafter, to maximise the yield and to ensure that our desired compound was obtained, we would carry out aqueous workup and various methods of chromatography like thin-layer chromatography (TLC) and reversed phase-high performance liquid chromatography (HPLC).

3.1 Synthesis of PNA Monomers by Solution-Phase Organic Synthesis

PNA T monomer is purchased commercially. PNA L, E and Q monomers are synthesised based on the procedure below.

1. Set up the reaction by adding starting material, reagents and solvent based on the reported procedures.
2. Monitor the reaction progress by Thin Layer Chromatography (TLC).
3. After the reaction is complete, perform aqueous workup (purification, removal of excess reagents, etc.) if necessary.
4. Purify the product using Flash Column Chromatography, Recrystallization, etc.
5. Characterise the product using Mass Spectrometry and Nuclear Magnetic Resonance.

3.2 Synthesis of PNA Oligomers by Solid-Phase Peptide Synthesis

1. Synthesise the PNA using 0.3 mmol/g of mBHA resin as a solid support to synthesise the PNA oligomers.
2. Apply tert-butyloxycarbonyl protecting group chemistry (Boc) and use PyBop as the activating agent.
3. Monitor the coupling of each step by the Kaiser test.
4. Couple the PNA monomers on the solid support and cleave the PNA from the solid support by using the trifluoroacetic acid (TFA)/trifluoromethanesulfonic acid (TFMSA) system in the ratio 10:1.
5. Carry out cold ether precipitation before dissolving the oligomers in water.

6. Purify the desired PNA by Reversed-Phase High Performance Liquid Chromatography (RP-HPLC).
7. Characterise and rapidly identify that the desired PNA has been obtained by Matrix Assisted Laser Desorption/Ionization (MALDI-TOF).

3.3 Testing the Affinity of the Designed PNA to the Target RNA Hairpin (PAGE)

1. 12 wt% (weight percent, 6 g in 50 ml) Polyacrylamide Gel was used and all samples contain 1 μM of RNA hairpin.
2. PNAs with increasing concentrations were titrated into 20 μL of RNA hairpin each
3. Snap cool RNA hairpin at 95 °C.
4. Annealing of PNA was carried out at 65 °C to room temperature.
5. Incubation at 4 °C overnight with the incubation buffer of 200 mM NaCl, 0.5 mM EDTA, 20 mM HEPES (pH 7.5).
6. Mix the samples with 35% glycerol and load them into the wells.
7. Gel was run at 4 °C with a constant voltage of 250 V for 5 h with the running buffer: 1X Tris-Borate EDTA (TBE) buffer at pH 8.3.
8. Ethidium bromide staining was carried out for 30 min.
9. Gels were imaged using the Typhoon Trio Variable Mode Imager and observed.

4 Results and Discussions

We evaluated the binding affinity of both ASPNA and TFPNA with the tau RNA hairpin through PAGE. Shown above are the sequences with the best binding affinity for each approach. From Fig. 8, the duplex RNA hairpin band concentration (shown by the black band at the bottom) decreases as the TFPNA concentration increases. We are unable to stain the triplex band (which will appear above the RNA hairpin band) as we used ethidium bromide (EtBr), which could only intercalate and show a staining with Watson-Crick base pairs. In our case, the Watson-Crick base pairs left after binding was insufficient to show observable staining as our RNA oligomer has quite a short sequence. However, we can deduce the formation of the PNA•RNA$_2$ triplex as the bands start to disappear as the TFPNA concentration increases, this implying that there is less intercalation of base pairs occurring due to the decrease in duplex RNA concentration.

By observation, there is a significant drop of duplex RNA concentration from sample 8 (10 μM of duplex concentration) onwards, until it almost fully disappears in sample 12 (50 μM). This means that the PNA's binding affinity is strong enough to

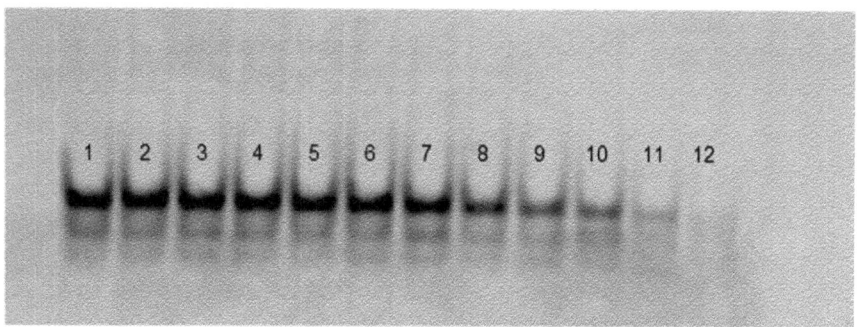

Fig. 8 Effect of TFPNA concentration on intensity of RNA hairpin. Sample 1 to 12 contains TFPNA with increasing concentrations of 0, 0.2, 0.4, 1, 1.6, 2, 4, 10, 16, 20, 28 and 50 μM respectively. This is the PAGE result of 7-mer Tau TFPNA versus 30-mer Tau WT RNA Hairpin model

bind to the RNA duplex at a concentration of 10 μM and above. While the triplex is successfully formed, the binding affinity of this particular TFPNA is unsatisfactory. For cell culture studies, TFPNA with stronger binding affinity is required, taking into account the unpredictability of the PNA's specificity and affinity towards the desired target sequence within the cell. This will be further discussed in our future extensions.

Looking at Fig. 9, binding is also seen between ASPNA and the target RNA hairpin. The risen band shows the PNA-RNA duplex after strand invasion. EtBr is able to intercalate as the product is a duplex. Observable strand invasion first occurs at 0.05 μM, whereas complete strand invasion occurs at 0.1 μM. Thus, the sequence

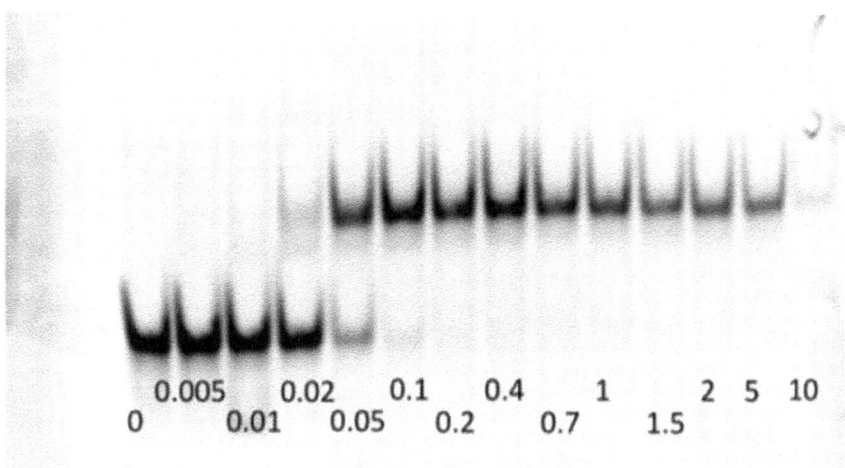

Fig. 9 Effect of ASPNA concentration on intensity of target hairpin and duplex after strand invasion. ASPNA concentrations is shown in μM. This is the PAGE result of 10-mer Tau ASPNA versus 32-mer Tau WT RNA Hairpin

Table 1 Cell culture studies with ASPNA

3R:4R ratio before and after injection[a]		
Phase	Percentage of 3R/%	Percentage of 4R/%
Before injection	30	70
After injection	50	50

[a]Carried out with immortalized cells. Raw data not shown due to lab approval issues. Figures as reported by affiliated department within NTU

can move on to cell culture testing (since PNA attachment is observable through PAGE at ≤1 μM).

Table 1 shows cell culture testing results. Our hypothesis that binding a PNA with the target hairpin can regulate the 3R:4R ratio. This is critical for our future extensions as it gives us confidence in further developing both ASPNA and TFPNA methods.

From Fig. 10, shows the theoretical interaction between the PNA strands and the target hairpin. For TFPNA, monomer L interacts with Watson Crick base pair G-C and G-U, Q interacts with C-G, E interacts with U-A and T interacts with A-U and A bulge by Hoogsteen base pairing.

Fig. 10 On the left shows strand invasion by ASPNA, with sequence CGCCGTCACA. The right shows TFPNA binding with sequence N-terminal Lys-LLQTLEL

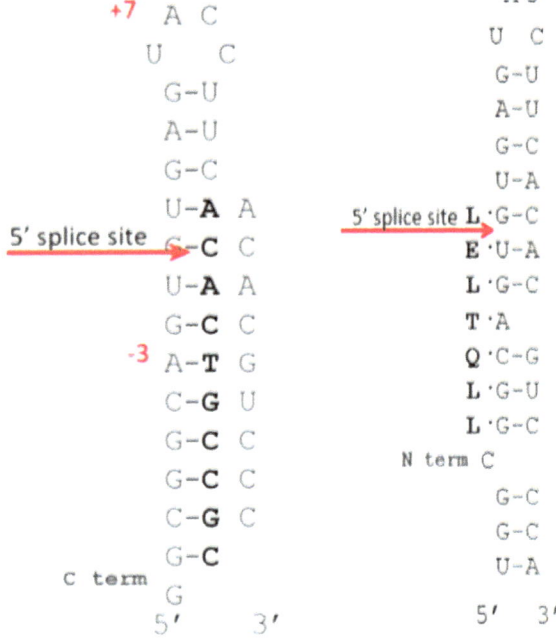

5 Conclusion and Future Work

PNA has tremendous potential and we should further continue our studies until we find a sequence that shows strong binding. However, few labs around the world use PNA for antisense applications, even fewer use TFPNA for stabilising hairpins. PNA could still be further explored as since it is artificially made, many modifications could be made to change its properties, especially to increase binding affinity, and thus surpass effectiveness of natural oligonucleotides.

We intend to take the following steps in future.

- Explore other staining methods such as fluorophores to improve the visibility of our PAGE bands, especially for TFPNA
- Programme a computer simulation to visualise the interaction between PNA and the target
- Try out other sequences in future to get better binding.
- Try out a possible approach by injecting both TFPNA and antisense PNA
- Establish a cost and time efficient method of synthesising PNA with high binding affinity
- Move on to advanced cell culture studies, animal testing and human subjects
- Due to time constraints, we could only test out the PNA sequences in Fig. 10 once, so we can repeat our experiments several more times to get more reliable, accurate and conclusive results.
- For TFPNA, our mentor suggested that ideal selectivity is most often achieved by a TFPNA of length 8 (we currently only have 7). We could thus increase the length of the TFPNA sequence. In extending the TFPNA sequence, it would not be easy extending the PNA upwards of L as the TFPNA can only bind to the RNA duplex downwards in sequence from the 5'ss. Extending above L would require resequencing the entire PNA sequence. Thus, the better alternative is to find out which PNA monomer binds best to the C monomer in RNA, for which there is currently no publication about.
- Eventually deliver PNA into cells through carriers such as aminoglycosides and polypeptides, across the plasma and nuclear membranes of the cells, and then observe the 4R to 3R isoform ratio within living cell.

Acknowledgements This project would not have been possible without the guidance and help of individuals who have extended their invaluable assistance in the completion of this project:

• Our external mentor, Mr. Alan Ong, for his immense help and advice throughout the course of our research

• Our internal teacher-mentor, Mrs. Elizabeth Foo, for guiding us through our internal school deadlines and keeping us up to date

• Assistant Professor Chen Gang, for regularly sending us informative emails related to our research topic as well as allowing us to use his lab for research

• NTU, RI and MOE-GEB for giving us this unique opportunity to participate in the programme

• Our friends and family who have morally supported us throughout our project.

References

1. Svoboda, S., & Di Cara, A. (2006, April). *Hairpin RNA: A Secondary Structure of Primary Importance*. Retrieved February 25, 2017 from https://www.ncbi.nlm.nih.gov/pubmed/16568238.
2. Koh, W. (n.d.). *Peptide Nucleic Acid (PNA) and Its Applications*. Yuseong-gu, Daejeon: Panagene Inc.
3. Nielsen, P. E., Egholm, M., Berg, R. H., & Buchardt, O. (1991). Sequence-selective recognition of DNA by strand displacement with a thymine-substituted polyamide. *Science, 254,* 1497–1500.
4. Wang, Y., & Mandelkow, E. (2016). Tau in physiology and pathology. *Nature Reviews Neuroscience, 17,* 5–21.
5. Roca, X., Krainer, A. R., & Eperon, I. C. (2013). Pick one, but be quick: 5′ splice sites and the problems of too many choices. *Genes and Development, 27,* 129–144.
6. Kar, A., Fushimi, K., Zhou, X., Ray, P., Shi, C., Chen, X., et al. (2011). RNA helicase p68 (DDX5) regulates tau exon 10 splicing by modulating a stem-loop structure at the 5' splice site. *Molecular and Cellular Biology, 31,* 1812–1821.
7. Ray, P., Kar, A., Fushimi, K., Havlioglu, N., Chen, X., & Wu, J. Y. (2011). PSF suppresses tau exon 10 inclusion by interacting with a stem-loop structure downstream of exon 10. *Journal of Molecular Neuroscience, 45,* 453–466.
8. Chen, G. (n.d.). *RNA Folding and Therapeutics*. Retrieved February 25, 2017 from http://www.ntu.edu.sg/home/rnachen/research.htm.
9. Devi, G., Zhou, Y., Zhong, Z., Toh, D. K., & Chen, G. (2014). RNA triplexes: From structural principles to biological and biotech applications. *John Wiley & Sons, Ltd.*. https://doi.org/10.1002/wrna.1261.
10. Fredericks, A. M., Cygan, K. J., Brown, B. A., & Fairbrother, W. G. (2015, June). *RNA-Binding Proteins: Splicing Factors and Disease*. Retrieved February 25, 2017 from https://www.ncbi.nlm.nih.gov/pmc/articles/PMC4496701/.
11. Nielsen, P. E., & Egholm, M. (1999). An introduction to peptide nucleic acid. *Current Issues in Molecular Biology, 1,* 89–104.
12. Cleveland, D. W., Hwo, S. Y., & Kirschner, M. W. (1977). Purification of tau, a microtubule-associated protein that induces assembly of microtubules from purified tubulin. *Journal of Molecular Biology, 116*(2), 207–225.
13. Goedert, M. (2005). Tau gene mutations and their effects. *Movement disorders: Official journal of the Movement Disorder Society, 20*(Suppl 12), S45–52.
14. Gaillard, F. (n.d.). *Neurofibrillary tangles | Radiology Reference Article*. Retrieved August 06, 2017 from https://radiopaedia.org/articles/neurofibrillary-tangles.
15. Hoogsteen Base Pairing (Molecular Biology). (n.d.). Retrieved February 25, 2017 from http://what-when-how.com/molecular-biology/hoogsteen-base-pairing-molecular-biology/.
16. Leontis, N. B., & Westhof, E. (2001). Geometric nomenclature and classification of RNA base pairs. *RNA, 7*(4), 499–512. https://doi.org/10.1017/s1355838201002515 (n.d.). Retrieved February 16, 2017 from http://rna.bgsu.edu/triples.
17. PNA: Peptide Nucleic Acid as a more stable alternative to DNA and RNA for many applications. (n.d.). Retrieved February 25, 2017 from http://www.crbdiscovery.com/newsletter/august2013.html.
18. Poulos, M. G., Batra, R., Charizanis, K., & Swanson, M. S. (2011, January). *Developments in RNA Splicing and Disease*. Retrieved February 25, 2017 from https://www.ncbi.nlm.nih.gov/pmc/articles/PMC3003463/.
19. Sanders, R. (2015, April 6). New Target for Anticancer Drugs: RNA. *Berkeley News*. Retrieved February 25, 2017 from http://news.berkeley.edu/2015/04/06/new-target-for-anticancer-drugs-rna/.
20. Wang, G., & Xu, X. S. (n.d.). *Peptide Nucleic Acid (PNA) Binding-Mediated Gene Regulation*. https://doi.org/10.1038/sj.cr.729020.

21. Warf, M. B., & Berglund, J. A. (2010, March). *The Role of RNA Structure in Regulating pre-mRNA Splicing*. Retrieved February 25, 2017 from https://www.ncbi.nlm.nih.gov/pmc/articles/PMC2834840/.

22. Jubilut, G. N., Cilli, E. M., Tominaga, M., Miranda, A., Okada, Y., & Nakaie, C. R. (2001, September). *Evaluation of the Trifluoromethanosulfonic Acid/Trifluoroacetic Acid/Thioanisole Cleavage Procedure for Application in Solid-Phase Peptide Synthesis*. Retrieved June 10, 2017 from https://www.ncbi.nlm.nih.gov/pubmed/11558592.

Making 2D Nanolayers Visible by Optical Imaging

Xiaohe Zhang, Yang Jing, Hiroyo Kawai and Kuan Eng Johnson Goh

Abstract Recent developments have proved optical imaging to be a promising method to identify and locate 2D materials efficiently and non-invasively. By putting a 2D material on a substrate, the nanolayer will add to an optical path and create a contrast with the case when the nanolayer is absent, which can be used to identify the 2D material and its number of layers. To make the optical imaging process in the laboratories more convenient, this report uses Fresnel Law as a model to simulate the optical imaging results of various 2D materials (graphene, MoS_2, $MoSe_2$) on top of different thickness of SiO_2 and Si wafer. The results provide details of the optimal conditions (the optimal light wavelength and optimal thickness of SiO_2) to identify and locate the 2D nanolayer, which can be used directly in laboratories. The model used in this report was benchmarked by simulating the system of graphene on top of SiO_2 and Si to ensure its accuracy and comparing with existing literature. The model was then used to simulate the optical contrasts of 1–5 layers of MoS_2 and $MoSe_2$, the latter of which has not been reported in previous literature. In particular, we highlight the sensitivity of the used model on the accuracy of the refractive indices used. In conclusion, we show through computational modelling that optical contrast can in principle allow effective determination of layer numbers in few layered 2D materials.

Keywords 2D materials · Optical imaging · Substrate · Fresnel law

1 Background and Purpose of Research

Since the research paper published in 2004 by Novoselov et al. [11] reported the successful isolation of graphene, great interests has been attracted to the field of two-dimensional (2D) materials, which are crystalline materials consisting of a single layer of atoms. They are expected to have a significant impact on a large variety of applications, ranging from electronics, gas storage or separation, catalysis, high

X. Zhang (✉)
Victoria Junior College, Singapore, Singapore
e-mail: zhang.xiaohe.2017@vjc.sg

Y. Jing · H. Kawai · K. E. J. Goh
Institute of Materials Research and Engineering, A*STAR, Singapore, Singapore

© Springer Nature Singapore Pte Ltd. 2019
H. Guo et al. (eds.), *IRC-SET 2018*,
https://doi.org/10.1007/978-981-32-9828-6_28

performance sensors, support membranes to inert coating [8]. Graphene, for example, exhibits high crystal quality [7] and can be used to accurately mimic massless Dirac fermions [16]. Transition metal dichalcogenides, with the common formula MX2, where M stands for transition metals and X for chalcogens, are a new group of promising 2D materials which exhibit a large variety of electronic behaviors such as semiconductivity, superconductivity or charge density wave. For example, semiconducting dichalcogenides such as MoS_2 are promising materials for electronic applications. Their high charge carrier mobilities make them suitable candidates to be used for flexible field effect transistors (FETs) [2, 13]. Furthermore, optical properties of these materials can be used for solar cells, photoelectrochemical cells and photocatalytic applications [9, 17]. As the most instrumental step in the study and practical application of 2D materials, great significance is put into identifying and locating these layered materials.

There are a few techniques to identify and locate 2D materials. Raman microscopy, which can distinguish graphene monolayers, was not able to search for graphene monolayers automatically [5]. Atomic force microscopy (AFM) can determine the dimensions of nanolayers deposited on insulating substrates but is time-consuming and produces slow throughput. Scanning electron microscopy (SEM) and transmission electron microscopy (TEM) can locate the nanolayers but seriously contaminate the layer [4]. Therefore, optical imaging is predicted to be the most promising method forward among all these measures. This is because by calculating the optical contrasts, optical imaging offers the possibility of a simple, rapid and non-destructive way to characterize large-area samples. Herein, the simulations of optical imaging have been widely used to detect the 2D materials. Blake et al. suggests in the paper published in 2007 that the thin flakes of 2D materials on a certain thickness of SiO_2 are transparent enough to add to an optical path when placed on an oxidised Si wafer, creating a contrast with the case when the nano-layers are absent. This contrast is sufficient for the human brain to identify and locate the thicker flakes. Benameur et al. simulated the process by calculating the optical contrast of MoS_2, WSe_2 and $NbSe_2$ on top of SiO_2 wafer and suggested that single layers of MoS_2, WSe_2 and $NbSe_2$ could be detected on 90 and 270 nm SiO_2 using optical means. However, there has been no report on the optical contrast of $MoSe_2$ (a thorough search of the relevant literature yielded no results), neither single layer nor a few layers. $MoSe_2$ will hence be the one of the focuses of this study. This study will also investigate the optical contrast simulation of single to 5 layers (which can be regard as bulk-like materials) of MoS_2 to enhance the current existing studies on this material and make the study of MoS_2 more systematic.

2 Hypothesis

Each 2D material, characterized with thickness and refractive index, will exhibit a unique optical contrast when stacked on top of a substrate, such as SiO_2. The optimal detecting conditions of various 2D materials can be predicted from the simulations of their optical contrasts.

3   Methods and Materials

3.1   Experimental Setup

Under the microscope, the 2D material with a refractive index of n_1 and thickness of d_1 is placed on top of SiO_2 of thickness d_2 and silicon wafer. Light of different wavelengths is shone on the sample using narrow-band filters [5]. Any results involving the color of the light based on different wavelengths are in accordance with the spectrum shown in the annex. The resulting contrast can be either obtained experimentally from the microscope or calculated theoretically based on Fresnel law (Fig. 1).

3.2   Determination of Indices

As the case of normal light incident from air to the trilayer structure (nanolayer, SiO_2, Si wafer), refractive index of air (n_0) is 1, independent of all other variables.

The 2D material in test is described by a thickness of d_1 (data obtained from Blake et al. [5]; Zhang et al. [15]; Beal and Hughes [3]), and its refractive index is n_1, which is dependent on the wavelength of the light. For graphene, its refractive index was found to be well described by the refractive index of bulk graphite $n_1 = 2.6 - 1.3i$, independent of the wavelength [5, 12]. This can be attributed to the fact that the optical response of graphite with the electric field parallel to graphene planes is dominated by the in-plane electromagnetic response [5]. However, in the cases of other dichalcogenides nanolayers, a set of light wavelength dependent values of refractive indices is used, as presented in the annex. Literature has reported the effect of extinction coefficient of the 2D materials [6], however, it is not considered in this report.

SiO_2 is described by the thickness d_2 and another wavelength dependent refractive index n_2 but with a real part only: n_2 (400 nm) $= 1.47$ [14]. The Si layer is assumed

Fig. 1 Experiment setup of the substrate system for the optical imaging of 2D materials. Paths A, B, C are light paths with and without the nanolayer. The top layer is air, which is above the nanolayer, while the Si wafer is subsequently referred to as the fourth layer

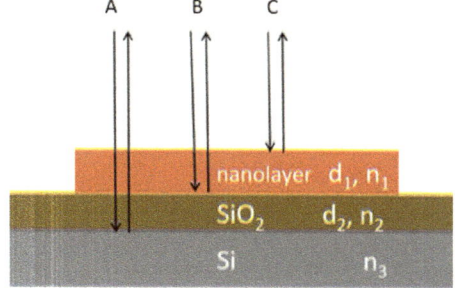

to be semi-infinite and described by a complex refractive index n_3 dependent on wavelength of light. The set of indices is found from previous literature [1, 5].

3.3 Computation

Fresnel law is used to calculate the contrast [5]. Using the described indices, the reflected light intensity (with the presence of the 2D material) can be written as:

$$P(n_1) = \left| \frac{r_1 e^{i\phi_+} + r_2 e^{-i\phi_-} + r_3 e^{-i\phi_+} + r_1 r_2 r_3 e^{i\phi_-}}{e^{i\phi_+} + r_1 r_2 e^{-i\phi_-} + r_1 r_3 e^{-i\phi_+} + r_2 r_3 e^{i\phi_-}} \right|^2,$$

where $r_1 = \frac{n_0 - n_1}{n_0 + n_1}$, $r_2 = \frac{n_1 - n_2}{n_1 + n_2}$, $r_3 = \frac{n_2 - n_3}{n_2 + n_3}$ are the relative indices of refraction between each two layer;

$\phi_1 = \frac{2\pi n_1 d_1}{\lambda}$, $\phi_2 = \frac{2\pi n_2 d_2}{\lambda}$ are the phase shifts due to changes in the optical path; $\phi_+ = \phi_1 + \phi_2$, $\phi_- = \phi_1 - \phi_2$ are for the ease of computing.

On the other hand, the reflected light intensity in the absence of the 2D material can be found using the same formula, but with $n_1 = n_0 = 1$. Similarly, $r_1 = 0$ and $\phi_1 = 0$ as there would be no refraction nor phase shifts between the first two layers. The light intensity with the absence of the 2d material can hence be written as:

$$A(n_1 = 1) = \left| \frac{r_2' e^{i\phi_2} + r_3 e^{-i\phi_2}}{e^{i\phi_2} + r_2' r_3 e^{-i\phi_2}} \right|^2,$$

where $r_2' = \frac{n_0 - n_2}{n_0 + n_2}$ is the new relative index between the second and third layer.

The contrast is defined as the relative intensity of reflected light with the presence and absence of the 2D material. It is written as:

$$C = \frac{P(n_1) - A(n_1 = 1)}{A(n_1 = 1)}.$$

Graphs are plotted to investigate the correlation between the contrast, thickness of SiO_2 and the wavelength of light to find the optimal situation to identify and locate the nanolayer. The refractive index of SiO_2 and Si is obtained from previous experimental data.

4   Results and Discussion

4.1   Benchmark

In order to test that the mathematical model derived from Fresnel law is accurate in interpreting the contrast for the experiment, a set of calculated data for the contrast of graphene on top of SiO_2 and Si wafer is compared to that obtained experimentally. Our simulated results show excellent agreement with the experimental data when the wavelength of incoming light is between 410 and 750 nm. Furthermore, the colour plot produced by the model described previously proves that graphene can be visualised on top of any thickness SiO_2 except for around 150 nm and below 30 nm, and that optimal contrast is produced with the use of 90 and 280 nm thickness of SiO_2 under green light, as shown in the annex. These conclusions are in agreement with the results in previously published literature [5]. The tests on graphene proves that the model used in this report is accurate and can give reliable result when comparing with the experiment, which suggests that the conclusions made based on this model are likely to be applicable in laboratories (Figs. 2 and 3).

4.2   Molybdenum Disulfide (MoS₂) on Top of SiO₂ and Si Wafer

This set of results was generated by simulating a MoS_2–SiO_2–Si system (Fig. 4). For a single layer of MoS_2, as shown in Fig. 4(1), the optical contrast simulation suggests that it can be detected on 70 and 240 nm SiO_2 using red light. With increasing number

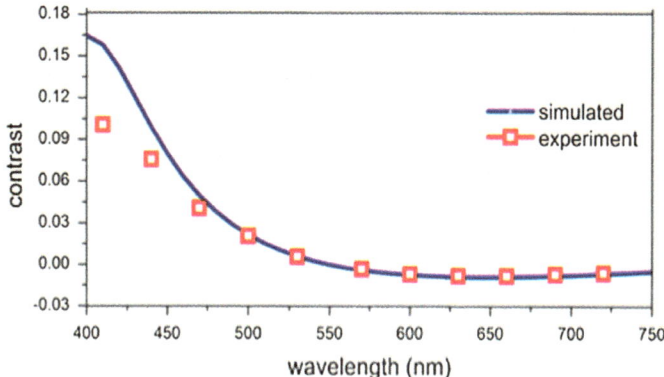

Fig. 2 Contrast as a function of wavelength for graphene–SiO_2–Si system with a 200 nm-thick SiO_2 substrate. The solid line is the simulated result and the red circles are the experimental data obtained from Blake et al. paper

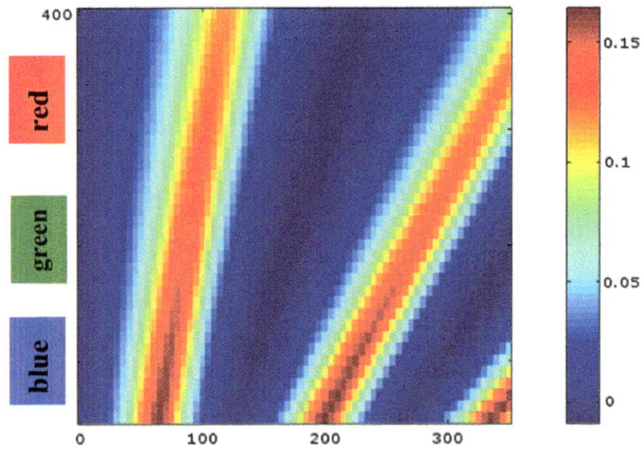

Fig. 3 Colour plot of the contrast (colour bar) as a function of wavelength (y-axis) and SiO$_2$ thickness (x-axis) for graphene–SiO$_2$–Si system. The colour bar is the expected contrast

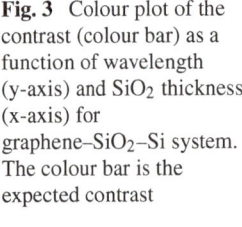

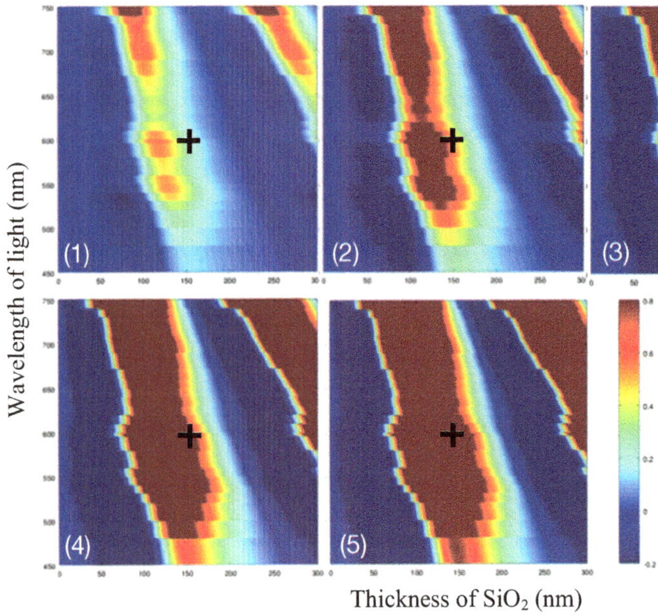

Fig. 4 Colour plots of the contrasts (colour bar) as a function of wavelength (y-axis) and SiO$_2$ thickness (x-axis) for the MoS$_2$–SiO$_2$–Si system. Graphs (1)–(5) used 1–5 layers of MoS$_2$ respectively. The crosses are marked on the (150,600) point on all the graphs

Table 1 Optimal detecting condition for different layers of MoS_2

Number of layers of MoS_2	Optimal light wavelength (nm)	Colour of the optimal light	Optimal thickness of SiO_2 (nm)
1	(a) 675–750	(a) Red	(a) 50–80
	(b) 675–750	(b) Red	(b) 220–250
2	(a) 500–750	(a) Blue–red	(a) 50–120
	(b) 580–750	(b) Yellow–red	(b) 200–300
3	(a) 480–750	(a) Blue–red	(a) 40–150
	(b) 570–750	(b) Yellow–red	(b) 190–300
4	(a) 450–750	(a) Blue–red	(a) 40–180
	(b) 560–750	(b) Green–red	(b) 190–300
5	(a) 450–750	(a) Blue–red	(a) 40–180
	(b) 550–750	(b) Green–red	(b) 180–300

of layers (from 1 up to 5 layer) of MoS_2 put on the substrate, the contrast generally increases. The appropriate condition to produce the optimal contrast in order to identify the MoS_2 is easier to achieve with more layers of MoS_2, suggesting that the thicker is the 2D material, the easier it can be detected. Also, it is found that 4 and 5 layers of MoS_2 exhibit similar optical contrast. This conclusion is agreeable with the widely accepted idea that a 2D material with thickness up to 5 layers can be regarded as a bulk material. The optimal detecting condition for different layers of MoS_2 is summarized in Table 1.

The differences in the contrast can also be used to predict the number of layers of MoS_2 in the experiment. When an appropriate point is selected from the colour plot, for example, 150 nm-thick of SiO_2 with 600 nm-wavelength of light (as indicated by the cross in Fig. 4), different number of layers of MoS_2 produces different contrast. Therefore, the experiment data can be compared with the data from the graph to determine the number of layers. This is however only applicable to one to four layers of MoS_2 because five or more layers of MoS_2 are considered as bulk materials and exhibit different properties.

4.3 Molybdenum Diselenide (MoSe₂) on Top of SiO₂ and Si Wafer

This set of results was generated by simulating a $MoSe_2$–SiO_2–Si system (Fig. 5). Similar to MoS_2, it is easier to detect more layers of $MoSe_2$ than less layers of $MoSe_2$ as the condition to produce optimal contrast is easier to be met. It also approaches bulk-like materials when there are more than 5 layers. The optimal detecting condition for different layers of $MoSe_2$ is summarised in Table 2.

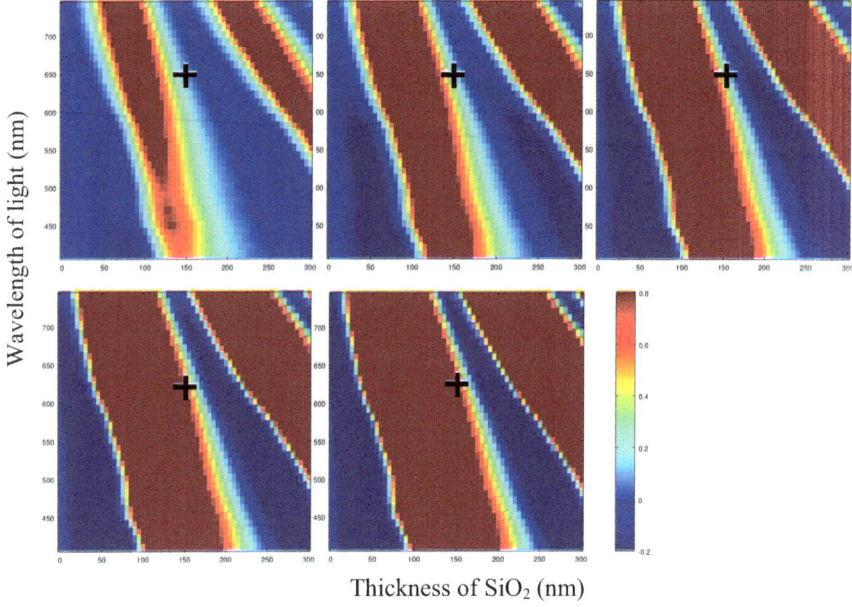

Thickness of SiO$_2$ (nm)

Fig. 5 Colour plots of the contrasts (colour bar) as a function of wavelength (y-axis) and SiO$_2$ thickness (x-axis) for the MoSe$_2$–SiO$_2$–Si system. Graphs (1)–(5) used 1–5 layers of MoSe$_2$ respectively. The crosses are marked on the (150,650) point on all the graphs

Table 2 Optimal detecting condition for different layers of MoSe$_2$

Number of layers of MoSe$_2$	Optimal light wavelength (nm)	Colour of the optimal light	Optimal thickness of SiO$_2$ (nm)
1	(a) 450–750	(a) Blue–red	(a) 50–125
	(b) 600–750	(b) Orange–red	(b) 200–300
2	(a) 450–750	(a) Blue–red	(a) 40–170
	(b) 580–750	(b) Yellow–red	(b) 190–300
3	(a) 450–750	(a) Blue–red	(a) 40–190
	(b) 570–750	(b) Yellow–red	(b) 190–300
4	(a) 450–750	(a) Blue–red	(a) 20–200
	(b) 550–750	(b) Green–red	(b) 160–300
5	(a) 450–750	(a) Blue–red	(a) 20–210
	(b) 540–750	(b) Green–red	(b) 160–300

Also, the data can be used to roughly distinguish the different number of layers of $MoSe_2$. For example, on the point when 150 nm-thick of SiO_2 and 700 nm-wavelength light are used, the contrast increases as the number of layers increases. The number of layers can be determined by comparing the experiment data with the simulated data. Similar to MoS_2, $MoSe_2$ approaches bulk-like material when there are four or more layers. Any number of layers above four are hence indistinguishable using this method.

4.4 Further Analysis of Data

In order to test the dependence of the results on the accuracy of the refractive indices used, the second graph of Fig. 6 was generated using only the real part of the MoS_2 refractive index for four layers of MoS_2 on a MoS_2–SiO_2–Si substrate system. It is observed that the two graphs show significantly different results, suggesting that the results are sensitive to refractive index value. In other word, an accurate experimental measurement of the refractive indices of the materials is essential in producing an accurate prediction on the optical imaging condition.

Furthermore, cross-referring the contrast produced by the MoS_2 and $MoSe_2$ on the same substrate system, $MoSe_2$ always produces higher contrast. Therefore, $MoSe_2$ is easier to be detected than MoS_2 using the optical imaging method.

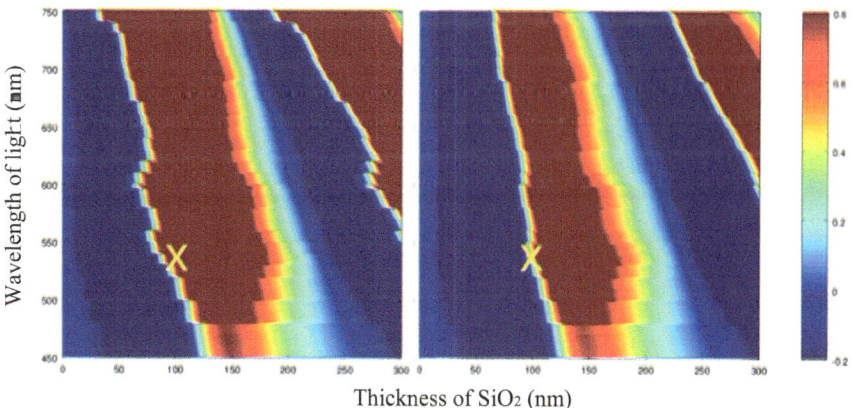

Fig. 6 Graphs using both the real part and the imaginary part of the MoS_2 refractive index (left) and using only the real part of the MoS_2 refractive index (right) for four layers of MoS_2 on a MoS_2–SiO_2–Si system. The crosses are marked on the point (100,540) on both graphs

5 Conclusion and Recommendations for Further Work

This report has simulated the optical imaging of 1–5 layers MoS_2 and $MoSe_2$ on top of SiO_2–Si substrate by calculating their optical contrasts. This simulation model has been successfully benchmarked with previous experimental data, indicating that this model is able to yield reliable results. To the best of this author's knowledge, there is no existing report on optical contrast of $MoSe_2$ and our study on MoS_2 would enhance existing studies. This report then presented the optimal conditions for detecting 1–5 layers of MoS_2 and $MoSe_2$ which are summarized in Tables 1 and 2. The results can be translated directly into laboratory works for further usage.

In the future, studies can be done to provide more accurate refractive indices of the 2D materials as the results are proven to be highly sensitive to the index values. Further development in this area can enable the computation simulation to be more accurate, hence more effectively translating into laboratory work. Moreover, studies should be done in other substrate systems, such as silicon nitride and PMMA, to find the most effective one in identifying various nano-layers, or for other 2D materials such as other transition metal dichalcogenides.

Annex

The refractive indices of Si (n_3) under different wavelength of light (λ)

λ	n_3	λ	n_3
400	5.59 + 0.30i	580	3.99 + 0.02i
410	5.31 + 0.22i	590	3.96 + 0.02i
420	5.09 + 0.17i	600	3.94 + 0.02i
430	4.93 + 0.13i	610	3.92 + 0.02i
440	4.79 + 0.11i	620	3.90 + 0.02i
450	4.68 + 0.09i	630	3.88 + 0.02i
460	4.58 + 0.08i	640	3.86 + 0.02i
470	4.49 + 0.06i	650	3.84 + 0.02i
480	4.42 + 0.06i	660	3.83 + 0.01i
490	4.35 + 0.05i	670	3.82 + 0.01i
500	4.29 + 0.05i	680	3.80 + 0.01i
510	4.24 + 0.04i	690	3.79 + 0.01i
520	4.19 + 0.04i	700	3.77 + 0.01i
530	4.15 + 0.03i	710	3.76 + 0.01i
540	4.11 + 0.03i	720	3.75 + 0.01i
550	4.08 + 0.03i	730	3.74 + 0.01i

(continued)

(continued)

λ	n_3	λ	n_3
560	4.04 + 0.03i	740	3.73 + 0.01i
570	4.02 + 0.02i	750	3.72 + 0.01i

Obtained from Aspnes and Studna [1]

The refractive indices of MoS$_2$ (n_1) under different wavelength of light (λ)

λ	n_1 (MoS$_2$)	λ	n_1 (MoS$_2$)
410	2.618907483 + 2.370021053i	580	3.815753044 + 0.700631287i
420	3.014371463 + 2.393303884i	590	3.866571732 + 0.939710275i
430	3.403675391 + 2.119403715i	600	3.982798133 + 1.152804209i
440	4.167467336 + 1.498213218i	610	4.355782042 + 1.0971244i
450	4.256552674 + 1.320743724i	620	4.432288287 + 0.873786334i
460	4.54855381 + 1.217525152i	630	4.370715779 + 0.66616927i
470	4.531845087 + 0.902397657i	640	4.280148276 + 0.769723872i
480	4.460022602 + 0.834155468i	650	4.319564865 + 0.980618163i
490	4.555527227 + 0.91077546i	660	4.798946962 + 0.753548024i
500	4.763262257 + 0.896765813i	670	4.697663505 + 0.328942748i
510	4.64436128 + 0.769289673i	680	4.35435329 + 0.26367002i
520	4.502635476 + 0.707720181i	690	4.128525765 + 0.14716099i
530	4.454385348 + 0.639439657i	700	3.905526901 + 0.108340472i
540	4.23777656 + 0.540692668i	710	3.780854838 + 0.136344553i
550	4.141289654 + 0.517426911i	720	3.697186582 + 0.166696212i
560	4.036426341 + 0.604108665i	730	3.53002181 + 0.025206426i
570	3.998970996 + 0.64626683i	740	3.043110006 + 0.034871247i

Obtained from Zhang et al. [15]

The refractive indices of MoSe$_2$ (n_1) under different wavelength of light (λ)

λ	n_1 (MoSe$_2$)	λ	n_1 (MoSe$_2$)
400	3.26 + 2.97i	580	4.93 + 1.47i
410	3.47 + 3.08i	590	4.93 + 1.47i
420	3.81 + 2.95i	600	4.90 + 1.40i
430	3.81 + 2.95i	610	4.87 + 1.34i
440	3.99 + 2.90i	620	4.83 + 1.28i
450	4.13 + 2.81i	630	4.80 + 1.25i
460	4.23 + 2.71i	640	4.77 + 1.21i
470	4.34 + 2.61i	650	4.74 + 1.18i

(continued)

(continued)

λ	n_1 ($MoSe_2$)	λ	n_1 ($MoSe_2$)
480	$4.44 + 2.52i$	660	$4.72 + 1.18i$
490	$4.53 + 2.42i$	670	$4.71 + 1.23i$
500	$4.62 + 2.32i$	680	$4.81 + 1.26i$
510	$4.62 + 2.32i$	690	$4.81 + 1.26i$
520	$4.71 + 2.22i$	700	$4.94 + 1.24i$
530	$4.78 + 2.11i$	710	$5.01 + 1.08i$
540	$4.85 + 2.00i$	720	$4.98 + 0.98i$
550	$4.91 + 1.87i$	730	$4.95 + 0.89i$
560	$4.92 + 1.70i$	740	$4.89 + 0.86i$
570	$4.94 + 1.57i$	750	$4.83 + 0.83i$

Obtained from Beal and Huges [3]

References

1. Aspnes, D. E., & Studna, A. A. (1983). Dielectric functions and optical parameters of Si, Ge, GaP, GaAs, GaSb, InP, InAs, and InSb from 1.5 to 6.0 eV. *Physical Review B, 27*(2), 985.
2. Ayari, A., Cobas, E., Ogundadegbe, O., & Fuhrer, M. S. (2007). Realization and electrical characterization of ultrathin crystals of layered transition-metal dichalcogenides. *Journal of Applied Physics, 101*(1), 014507.
3. Beal, A. R., & Hughes, H. P. (1979). Kramers-Kronig analysis of the reflectivity spectra of 2H-MoS$_2$, 2H-MoSe$_2$ and 2H-MoTe$_2$. *Journal of Physics C: Solid State Physics, 12*(5), 881.
4. Benameur, M. M., Radisavljevic, B., Heron, J. S., Sahoo, S., Berger, H., & Kis, A. (2011). Visibility of dichalcogenide nanolayers. *Nanotechnology, 22*(12), 125706.
5. Blake, P., Hill, E. W., Castro Neto, A. H., Novoselov, K. S., Jiang, D., Yang, R., et al. (2007). Making graphene visible. *Applied Physics Letters, 91*(6), 063124.
6. Castellanos-Gomez, A., Agraït, N., & Rubio-Bollinger, G. (2010). Optical identification of atomically thin dichalcogenide crystals. *Applied Physics Letters, 96*(21), 213116.
7. Geim, A. K., & Novoselov, K. S. (2007). The rise of graphene. *Nature Materials, 6*(3), 183–191.
8. Mas-Balleste, R., Gomez-Navarro, C., Gomez-Herrero, J., & Zamora, F. (2011). 2D materials: To graphene and beyond. *Nanoscale, 3*(1), 20–30.
9. Moehl, T., Kunst, M., Wuensch, F., & Tributsch, H. (2007). Consistency of photoelectrochemistry and photoelectrochemical microwave reflection demonstrated with p-and n-type layered semiconductors like MoS$_2$. *Journal of Electroanalytical Chemistry, 609*(1), 31–41.
10. Nave, C. R. (2017). *Spectral colors*. Retrieved from http://hyperphysics.phy-astr.gsu.edu/hbase/vision/specol.html.
11. Novoselov, K. S., Geim, A. K., Morozov, S. V., Jiang, D., Zhang, Y., Dubonos, S. V., et al. (2004). Electric field effect in atomically thin carbon films. *Science, 306*(5696), 666–669.
12. Palik, E. D. (Ed.). (1998). *Handbook of optical constants of solids* (Vol. 2). Academic press.
13. Podzorov, V., Gershenson, M. E., Kloc, C., Zeis, R., & Bucher, E. (2004). High-mobility field-effect transistors based on transition metal dichalcogenides. *Applied Physics Letters, 84*(17), 3301–3303.
14. Tan, C. Z. (1998). Determination of refractive index of silica glass for infrared wavelengths by IR spectroscopy. *Journal of Non-crystalline Solids, 223*(1–2), 158–163.
15. Zhang, H., Ma, Y., Wan, Y., Rong, X., Xie, Z., Wang, W., et al. (2015). Measuring the refractive index of highly crystalline monolayer MoS$_2$ with high confidence. *Scientific Reports, 5*.

16. Zhang, Y., Tan, Y. W., Stormer, H. L., & Kim, P. (2005). Experimental observation of the quantum Hall effect and Berry's phase in graphene. *Nature, 438*(7065), 201–204.
17. Zong, X., Yan, H., Wu, G., Ma, G., Wen, F., Wang, L., et al. (2008). Enhancement of photocatalytic H_2 evolution on CdS by loading MoS_2 as cocatalyst under visible light irradiation. *Journal of the American Chemical Society, 130*(23), 7176–7177.

Predicting Individual Thermal Comfort

Lim Xin Yi, Lee Jia Jia and Daren Ler

Abstract Thermal comfort is a very important factor in many people's lives; when people do not feel comfortable, their focus, productivity and performance are affected. In our research, we investigate how to help students achieve optimal comfort by predicting their comfort levels. We collected data on 5 students, and then utilised the k-nearest neighbour machine learning algorithm, in conjunction with tenfold cross-validation, to generate models of student comfort in the classroom. The features utilised are air temperature, air velocity, air relative humidity, body temperature and heart rate. In our experiments, we seek to learn if acceptable individual models may be derived, and more importantly, if combined models can help increase predictive accuracy. Our work suggests that combined models are applicable only when used to augment datasets that are applied to a subject with a more complex thermal comfort model.

Keywords k-nearest neighbor algorithm · Thermal comfort · Machine learning · Accuracy rate · Variables · Features · Relative humidity · Heart rate · Wind speed · Body temperature · Time of the day · Air temperature · Self-assessed comfort level · Comfort level · k value · Individual subjects · Merged dataset · Datasets · Classifier · Tenfold cross validation · Wrapper based technique · Instances

1 Introduction

1.1 General Problem

We are investigating on using machine learning and algorithms to produce a model that can gauge students' comfort level in their classrooms then test if it only works for specific individuals or is generalized and able to predict comfort levels of most people.

L. X. Yi (✉) · L. J. Jia · D. Ler
National Junior College, Singapore 288913, Singapore
e-mail: limxinyi02@gmail.com

© Springer Nature Singapore Pte Ltd. 2019
H. Guo et al. (eds.), *IRC-SET 2018*,
https://doi.org/10.1007/978-981-32-9828-6_29

1.2 Motivation

It is important for us to learn how to predict comfort level in classrooms as comfort level of students affect their productivity rate and in turn affect many other factors such as results and performance in school. By engineering a machine learning, we can better predict students' comfort level and the school can then access their comfort level in their classrooms and provide better learning environment for them, causing learning to be more effective. It is important to know whether such data on thermal comfort can be merged and still be used effectively as it can provide insights about creating an environment with maximum thermal comfort for everybody in the same area.

1.3 Hypothesis

We hypothesize that machine learning can accurately predict student's comfort level by using input of relative humidity level, air temperature, body temperature, wind speed and heart rate. We will be testing our machine learning by catering it for every specific individual. We will also be testing if the same machine learning can predict comfort level accurately for others, to see if it works for most people or only for an individual.

2 Background, Terminology and Problem Statement

2.1 Problem Statement

We are looking into the question of how do we know whether a person is comfortable in a certain temperature. This is important as many industries, such as air-conditioning, may have good use for such information to be used on devices customized to the person's comfort level, so as to maximize comfort levels.

2.2 Defining "Thermal Comfort"

Thermal comfort can be defined as that condition of mind which expresses satisfaction with the thermal environment and is assessed by subjective evaluation.

2.3 Specific Problem

We will find out one's comfort level on a scale of 1–5, 1 being more uncomfortable and 5 being most comfortable. We will be using 7 different features: air temperature, relative humidity, body temperature, heart rate, wind speed, time and self-assessed comfort level.

3 Experimental Setup

3.1 Plan

We are creating a model to predict comfort levels for specific individuals. Afterwards, we will try to merge the models together and see if the models work better or worse for the general. To do this, we will be collecting data from 5 students. Then, we will use the data collected to conduct machine learning and create a classifier using k-nearest neighbours and *tenfold cross validation*. The features used are: air temperature, relative humidity, wind speed, body temperature, heart rate, time of the day and self-assessed comfort level. We will first choose the best k value, that gives us the highest accuracy percentage, for each dataset. Then, we will carry out feature selection and choose the best features for each dataset. Then, we will be merging the data sets together and choose the best k value and features again. In total, we will be doing 16 experiments (Table 1).

We will then compare the accuracy rate and evaluate our results to see if the model works for most of the people or can it only work for specific individuals.

Table 1 The 16 experiments we will be working on

a	a + b	b + c	c + d	d + e	a + b + c + d + e
b	a + c	b + d	c + e		
c	a + d	b + e			
d	a + e				
e					

Let a be dataset of student 1, b be dataset of student 2, c be dataset of student 3, d be dataset of student 4 and e be dataset of student 5

3.2 Reasoning

Based on Fanger's thermal comfort model, there are 6 factors that they consider: air temperature, mean radiant temperature, air velocity, air humidity, clothing resistance and activity level.

We are including air temperature, air velocity, air relative humidity with reference to the Fanger's model; however we removed mean radiant temperature, clothing resistance and activity level due to it being not feasible due to our limited resources. We added factors that are similar or related to the factors removed. Since, it is not easy for us to measure mean radiant temperature, we included body temperature as one of the factors. Body temperature can be determined by the clothes an individual is wearing, which affects thermal comfort. We also included heart rate as there are correlations between heart rates and activity level and human metabolism rate. Whenever we have higher heart rate, it is usually because we are engaged in vigorous activities, affecting our thermal comfort. We also included time as a factor as the net temperatures changes according to time of the day, which determines thermal comfort.

We have also used a similar concept to the ASHRAE thermal sensation model as a measure for thermal comfort, which serves as our output. However, we shrunk the range from 7 to 5 as the environment that our subject matters are placed in will be constant, hence there would not be great differences in temperatures and such a large scale is unnecessary. We also changed it such that it shows comfort levels instead of level of warmth as there is a need to adjust to our local climate which is very rare for somebody to feel cold in normal room temperature. These are the adjustments we have made in order to be more practical and make our experiments more feasible.

We used kNN although it is a simple classifier because it can benefit no matter how large training data is. In addition, no assumptions has to be made during the learning process. For tenfold cross validation, all the datasets are used for training and testing, ensuring that results are more accurate and precise and all datasets will be tested. In addition, k fold cross validation provides more accurate estimates of the test error rate.

3.3 Methodology

For our experiment, we collected data of relative humidity, air temperature and wind speed in TB33 and also body temperature and heart rate of 5, aged 15, females studying in TB33. We recorded the time when readings were taken (readings were taken every hour from 8.30 am to 2.30 pm for 6 days). The students were asked to rate their comfort level from 1 to 5, 1 being least comfortable to 5 being most comfortable. The criteria for the scale of comfort levels as shown in Table 2.

After obtaining the dataset, we used kNN algorithm and tested out different k values to find the optimum value for our dataset. Then, we did feature selection and

Table 2 Guideline to rate comfort level on a scale of 1–5

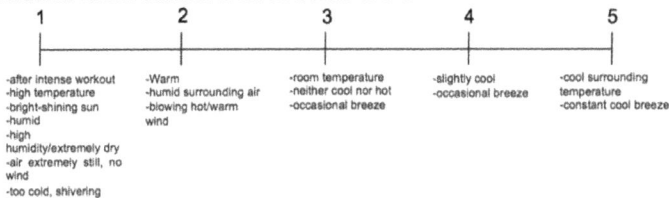

1	2	3	4	5
-after intense workout -high temperature -bright-shining sun -humid -high humidity/extremely dry -air extremely still, no wind -too cold, shivering	-Warm -humid surrounding air -blowing hot/warm wind	-room temperature -neither cool nor hot -occasional breeze	-slightly cool -occasional breeze	-cool surrounding temperature -constant cool breeze

tested for accuracy percentage for each individual feature and the accuracy for all the features. We recorded down the worst feature(s) which gave us the lowest accuracy percentage. After removing the worst features, we tested for accuracy percentage again. We then experiment with different datasets to test if the datasets only works for specific persons and is different to each person or it is generalized and applies to most of the crowd.

4 Results and Analysis

4.1 Results

See Tables 3 and 4.

For different values of k, the accuracy percentage varies. The more variable the accuracy rates over different values of k, the more susceptible that person is to comfort level changes due to external factors such as wind speed, air temperature, time of the day and relative humidity (Table 5).

Some features give low accuracy percentage while some give high accuracy percentage, showing that certain features are less important, thus it is not necessary that if all features are used, accuracy level produced will be highest. Table 6 shows the accuracy level of different individuals after removing the worst feature from its dataset (Table 7).

Table 3 Table of accuracy percentage for the best k value

	Best k value	Number of correctly classified instances	Accuracy percentage (%)	Total number of instances
Person A	22	17	41.4634	41
Person B	20	31	58.5366	41
Person C	6	22	53.6585	41
Person D	4	31	75.6098	41
Person E	18	21	51.2195	41

Table 4 Table of accuracy percentage of features

	All 7 features	Relative humidity	Wind speed	Body temperature	Air temperature	Heart rate	Time
Pereon A	41.4634%	*34.1463%*	41.4634%	41.4634%	41.4634%	43.9024%	*34.1463%*
Person B	58.5366	*53.6585%*	56.0976%	56.0976%	56.0976%	56.0976%	56.0976%
Person C	53.6585	56.0976	*34.1463*	39.0244	51.2195	43.9024	39.0244
Person D	75.6098	73.1707	73.1707	75.6098	75.6098	*70.7317*	75.6098
Person E	51.2195	51.2195	51.2195	51.2195	51.2195	*43.9024*	51.2195

Table 5 Table of accuracy percentage of best k value for pairs

Pair/group number	Individuals in the pair/group	Best k value	Number of correctly classified instances	Accuracy percentage (%)	Total number of instances
1	C & B	5	49	59.7561	82
2	C + E	14	42	51.2195	82
3	C + D	13	53	64.6341	82
4	E + D	19	52	63.4146	82
5	E + B	19	44	53.6585	82
6	D + B	13	53	64.6341	82
7	A + C	14	40	48.7805	82
8	A + E	14	49	59.7561	82
9	A + B	23	40	48.7805	82
10	A + D	14	46	56.7901	82
11	All 5	21	114	56.0976	205

4.2   Figures and Tables

See Table 8.

4.2.1   General Analysis of Individual Test Subjects' Accuracy Results

With reference to Table 3, the general best obtained accuracy of individual test subjects range from 41.4634 to 58.5366%; an anomaly in which person 4's best obtained result is far more accurate: 75.6098%. This shows that our proposed model accuracy changes according to different people but is effective to a small extent as accuracy rate is not as low as 20%—guessing rate due to 5 possible outputs: 1–5.

Table 6 Table of accuracies of 2 merged individuals' and all 5 merged individuals' datasets on different features with k = 10

Pair/group number	Description of pair	All 7 features	Relative humidity	Wind	Body temperature	Air temperature	Heart rate	Time
1	C & B	52.439	64.6341	46.3415	53.6585	54.878	53.6585	37.8049
2	C + E	51.2195	46.3415	51.2195	48.7805	48.7805	48.7805	51.2195
3	C + D	64.6341	75.6098	75.6098	63.4146	63.4146	63.4146	75.6098
4	E + D	60.9756	63.4146	63.4146	63.4146	62.1951	59.7561	59.7561
5	E + B	41.4634	51.2185	50.000	41.4634	57.3171	48.7805	47.561
6	D + B	63.4146	65.8537	64.6341	65.8537	64.6341	65.8537	60.9756
7	A + C	48.7805	40.2439	39.0244	48.7805	40.2439	43.9024	48.7805
8	A + E	59.7561	57.3171	58.5366	59.7561	58.5366	62.1951	58.5366
9	A + B	48.7805	51.2195	48.7805	48.7805	41.4634	43.9024	47.56
10	A + D	61.7284	58.0247	58.0247	59.2593	46.3415	46.3415	48.7805
11	All 5 subjects	51.7073	49.7561	55.122	55.6098	57.0732	53.6585	55.122

The boxes highlighted in italicised are the worst features of that dataset. We removed them to get the accuracy percentage for the specific datasets

Table 7 Table of accuracy percentage of different individuals

Person	Accuracy percentage (%)
A	39.0244
B	53.6585
C	46.3415
D	75.6098
E	51.2195

Table 8 Comparison of accuracy level after removing most inaccurate feature

Pair/group number	Accuracy percentage of pair (%)	Accuracy percentage of 1st individual of the pair (%)	Accuracy percentage of 2nd individual of the pair (%)	Change in accuracy percentage (%)
1	52.439	46.3415	53.6585	–
2	48.7805	46.3415	51.2195	–
3	75.6098	46.3415	75.6098	Increase
4	63.4146	51.2195	75.6098	–
5	53.6585	51.2195	53.6585	Increase
6	52.439	75.6098	53.6585	–
7	46.3415	39.0244	46.3415	Increase
8	45.122	39.0244	51.2195	–
9	53.6585	39.0244	53.6585	Increase
10	63.4146	39.0244	75.6098	–
11	54.1463	–	–	–

If the percentage of both the individuals increases, "increase" is indicated in the last column, if the percentage of one individual increases while the other decreases, a "–" is indicated

4.2.2 General Analysis of Best k Value Individually

In order to obtain the best k value, we tested the accuracy percentage for the best k value. Person A had the highest best k value of 22, 17 correctly classified instances and the lowest accuracy percentage of 41.4634%. Person D had the lowest best k value of 4, 31 correctly classified instances and the highest accuracy percentage of 75.6098%.

4.2.3 General Analysis of Best k Value in Pairs

The highest k value of 23 was obtained by pair 9 (A + B), and had the lowest accuracy of 48.7805%. However, the lowest k value of 5 was obtained by pair 1 (C + B), but did not have the highest accuracy (59.7561%). The highest accuracy was at 64.6431%, obtained by pair 3 (C + D), which had k = 13, and pair 6 (D + B), which had k = 13.

4.2.4   Analysis of Individual Accuracies in Relation to Each Feature

The results in the features are rather varied for the individual test subjects as seen in Table 6. The highlighted values are the lowest accuracy results of the individual test subjects. The features that produce 2 lowest accuracy results are relative humidity and heart rate while features that produce 1 lowest accuracy result are wind speed and time. The features that did not produce any lowest accuracy results are all of the 7 features combined, body temperature and air temperature; features that have more direct relation with thermal comfort. Through these results, it is seen that different people have different features that better determine their thermal comfort, with the exception of air and body temperature being a slightly more accurate determiner of thermal comfort.

4.2.5   General Analysis of Accuracy Percentage of Merged Test Relative to Individual Test

To test whether our proposed model will have a similar accuracy result when merged with another individual's dataset, we paired the test subjects and carried out the classification. As seen in Table 5 in comparison to Table 3, the results were rather inconsistent. When some of the individuals were paired up, accuracy rate increased drastically relative to their individual accuracy results, on the other hand, some has accuracy rate dropped relative to their individual accuracy results. We induced that this is due to the incompatibility of the two individuals that were paired up, meaning that both subjects' datasets were consistent with each other, with minimal contradictions that may cause an incorrect classification.

From Table 8 we can tell that the model can be shared between some people as the accuracy percentage increases. Accuracy percentage for Pair 3 (C + D) (75.6098%), Pair 5 (E + B) (53.6585%), Pair 7 (A + C) (46.3415%) and Pair 9 (A + B) (53.6585%) increases. Everyone has different levels of accuracy, some have better accuracies than others. From our experiment, it is shown that the model can accurately predict comfort level of certain types of people but not for all people in a crowd.

5   Conclusion and Future Work

5.1   Summarisation of Findings

For certain pairs (pair 3, 5, 7 and 9), the accuracy percentage increases but for the rest of the pairs, the accuracy percentage does not increase as the accuracy percentage of one in the pair increases while the other decreases. Thus, we can tell that the model can work on not only an individual but also some of the people of a crowd, but not all.

5.2   Link to Motivation

By creating this machine learning, we can predict others' comfort level, helping schools access comfort levels in classrooms, providing a better learning environment.

5.3   Future Work

In our data collection, we used Kestrel 4000 Pocket Weather Tracker weather tracker to measure relative humidity, air temperature and wind speed, Braun infrared thermometer and Garmin heart rate sensing watch to measure the individual's temperature and heart rate respectively. In the future, we would invest in higher quality thermometer and heart rate sensor to obtain more reliable results. In this experiment, our dataset only consist of 41 instances which is very small, causing low accuracy rate. In the future, we will increase the number of instances in the dataset. As all individuals tested were females, we would also like to collect data from males, so that we can find out if the model can produce accurate results for the crowd, which consists males.

References

1. Cover, T., & Hart, P. (1967). Nearest neighbor pattern classification. *IEEE Transactions on Information Theory, 13*(1), 21–27.
2. Fanger, P. O. (1970). *Thermal comfort. Analysis and applications in environmental engineering.*
3. Hall, M., Frank, E., Holmes, G., Pfahringer, B., Reutemann, P., & Witten, I. H. (2009). The WEKA data mining software: An update. *ACM SIGKDD Explorations Newsletter, 11*(1), 10–18.
4. Stone, M. (1974). Cross-validatory choice and assessment of statistical predictions. *Journal of the Royal Statistical Society. Series B (Methodological)*, 111–147.
5. Laftchiev, E., & Nikovoski, D. N. (2016, December). *An IoT system to estimate personal thermal comfort.* Retrieved from https://www.merl.com/publications/docs/TR2016-161.PDF.

6. Choi, J.-H., Loftness, V., & Lee, D.-W. (2011, October). *Investigation of the possibility of the use of heart rate as a human factor for thermal sensation models*. Retrieved from https://viterbipk12.usc.edu/wp-content/uploads/2017/06/2012-Building-and-Environment-Choi2c-Loftness2c-Lee-Investigation-of-the-possibility-of-the-use-of-heart-rate-as-a-human-factor-for.PDF.
7. Rabadi, N. J. (2011, August). *Developing a software to predict thermal comfort of humans at work*. Retrieved from http://jjmie.hu.edu.jo/files/v5n4/JJMIE-116-11.PDF.

Novel Design of Anode Flow Field in Proton Exchange Membrane Fuel Cell (PEMFC)

Xun Zheng Heng, Peng Cheng Wang, Hui An and Gui Qin Liu

Abstract In this paper, the authors presented a few new design flow field in hope to further optimize the fuel flow distribution within the flow field. Fuel distribution in PEMFC plays a critical role in the current density, temperature distribution and water concentration. Moreover, the effects of the geometric parameters on the cell performance are assessed.

Keywords PEMFC · CFD · Flow field design · Fuel cell simulation

1 Introduction

Fuel cell technologies are increasing getting renewed interest due to the surge in demand for clean energy. Fuel cells are mainly classified based on the electrolyte. The classifications are based on the kind of electro-chemical reactions taking place, the kind of catalyst, the temperature ranges for the fuel cell to operates, fuel required and other factors [1]. There are various types of fuel cell currently in the market as shown in Fig. 1.

Polymer Electrolyte Membrane-based Fuel Cell (PEMFC) is an energy conversion device which generates electricity through converting chemical energy between hydrogen and air [2]. It offers cleaner energy and highly efficient as compared to traditional energy conversion technologies. The chemical reaction taking place in the fuel cell are as follows:

Anode Reaction

$$\text{Anode:} \quad H_2 \rightarrow 2H^+ + 2e^- \tag{1}$$

X. Z. Heng
Engineering Cluster, Newcastle University in Singapore, Singapore Institute of Technology, Singapore, Singapore

P. C. Wang (✉) · H. An · G. Q. Liu
Engineering Cluster, Singapore Institute of Technology, Singapore, Singapore
e-mail: Victor.Wang@SingaporeTech.edu.sg

© Springer Nature Singapore Pte Ltd. 2019
H. Guo et al. (eds.), *IRC-SET 2018*,
https://doi.org/10.1007/978-981-32-9828-6_30

T →							DMFC
	DMFC	**PEMFC**	**PAFC**	**MCFC**	**SOFC**		Direct Methanol FC
Electrolyte	Proton conducting membrane	Proton conducting membrane	Phosphoric acid	Molten carbonate	Ceramic		**PEMFC**
Temperature Range (°C)	< 100	< 100	ca. 200	ca. 650	800-1000		Proton Exchange Membrane FC
Fuel	methanol	hydrogen	hydrogen	Natural gas, coal gas, biogas	Natural gas, coal gas, biogas		**PAFC** Phosphor Acid FC
Power Range	W / kW	W / kW	kW	kW / MW	kW / MW		**MCFC** Molten Carbonate FC
Application (examples)	Vehicles, portables	Vehicles, house energy, CHP	CHP	Power plants	House energy, Power plants		**SOFC** Solid Oxide FC

Fig. 1 Different types of fuel cell [1]

Cathode Reaction:

$$\text{Cathode:} \quad \tfrac{1}{2}O_2 + 2H^+ + 2e^- \rightarrow H_2O \tag{2}$$

At the Anode, the hydrogen ions are broken down into protons and electrons respectively. The electrons produce electricity current as shown in Eq. (1) to power up the circuit. While at the Cathode, the oxygen ions would react with the Hydrogen protons to produce the by-product of water reflected in Eq. (2) [2]. Figure 2 illustrated the chemical reactions reacting in PEMFC.

Although this technology has been around for more than a decade, research is still ongoing to further optimize the performance of the fuel cell—with the aim of

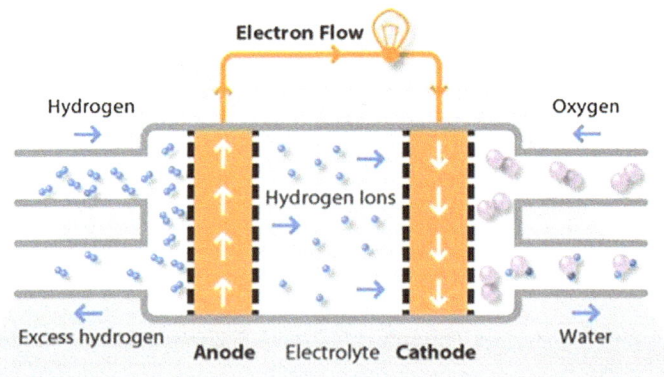

Fig. 2 Chemical reactions involved in a PEMFC

achieving a higher power output using lesser fuel (Hydrogen) whilst maintaining a reasonable heat flux generation.

Much research had been conducted to improve the performance of the fuel cell, with one key focus area on the effects of the design of the channel flow designs. The flow field is a very important and critical component that can directly affect the performance of the PEMFC. Flow field serves as the reactants distributors and current collector where an optimum design could increase up to 50% in power density [3, 4]. One crucial requirements for flow field design is to achieve uniform distribution of reactants over the entire active area of the fuel cell [5]. This is to prevent hot spots forming while keeping an optimal pressure drop to prevent water flooding within the flow field.

With the advent of computers hardware and readily available high fidelity numerical codes, computational fluid dynamics (CFD) is use to study the flow field design. CFD studies had proven to be a cheaper and faster alternative as compared to experimental prototyping. CFD studies could reduce up to 20–40% of experimental cost [6] yet achieving results accurately close to experimental data. In addition, CFD provide ease for the engineers to make amendments without incurring any additional cost.

Most research investigated on the influence of geometric parameters on the performance of the PEMFC [7–11]. Manso et al. [8] reviewed the results conducted in the recent years related to the different geometric parameters of the flow channels. They concluded that geometric parameters show great influence on the overall performance of the PEMFC. There are a few geometric parameters that should be consider in optimizing the flow field. The parameters are as follows below.

Cross sectional shapes of channels such as rectangular, semicircle, trapezoidal etc. had been widely investigated. Dewan et al. analyzed 3 different cross-sectional area namely rectangular, parallelogram and trapezoidal. They concluded that rectangular channel cross-section provides higher cell voltages while trapezoidal channel cross-section gave more uniform reactant and local current density distributions at the membrane–cathode GDL interface [12].

Two other essential geometry; Channel and Rib Widths were also investigated [12–15]. Yoon et al. [16] conducted an experiment to investigate on the effects of the different channel and rib configurations. The authors proved that narrower rib improves the overall performance of the fuel cell. However, it shows that wider rib has better water vapor retention. In general, narrower rib widths provides high current density due to higher pressure drop.

The purpose of this paper is therefore to investigate on the different anode flow field design which could lead to improving the fuel cell efficiency.

Nomenclature	
ε	Porosity of membrane
k	Permeability of membrane
τ	Fluid stress

(continued)

(continued)

Nomenclature	
ρ	Density
Y_i	Species mass fraction
$\vec{U}$	Gas velocity
Di^{eff}	Effective mass diffusivity
E	Total energy
h	Enthalpy
τ_{eff}	Effective fluid stress
S_h	Sink term
K_{eff}	Effective conductivity

2 Methodology

2.1 ANSYS Fluent 18.2

ANSYS Fluent 18.2 was used in this study to simulate the reaction of the PEMFC. ANSYS Fluent was chosen due to the fuel cell module add on in build in the software that simulate the fuel cell's reaction and behaviors through a few governing equations. The governing equations that was required for the simulation are shown below. The equations are derived based on the non-Darcy law (turbulence flow).

Mass equation:

$$\nabla\left(\varepsilon\rho\vec{U}\right) = 0 \tag{3}$$

Momentum equation:

$$\nabla\left(\varepsilon\rho\vec{U}\vec{U}\right) = -\varepsilon\nabla P + \nabla(\varepsilon\tau) + \frac{\varepsilon^2\mu\vec{U}}{k} \tag{4}$$

Species equation:

$$\nabla(\varepsilon\rho U Y_i) = \nabla Di^{eff} \cdot \nabla Yi + Si \tag{5}$$

Energy equation:

$$\nabla\left(\vec{U}(\rho E + P)\right) = \nabla\left(K_{eff}\nabla T - \sum_j h_j\vec{J}_j + \left(\tau_{eff} \cdot \vec{U}\right)\right) + S_h \tag{6}$$

Potential equation:

$$\frac{d\phi_s}{dx} = -\frac{I}{\sigma^{eff}}$$ (7)

2.2 Model Geometry

A 3-D one channel parallel flow field based reference model is developed based on the dimensions provided by Cheng et al. [7] as shown in Table 1. The model shown in Fig. 3 includes all the components in the fuel cell namely the anode current collector, anode flow channel, anode gas diffusion layer, anode catalyst layer, membrane, cathode catalyst layer, cathode gas diffusion layer, cathode flow channel and cathode current collector.

Table 1 Cell dimensions

Geometrical parameters	Dimensions (mm)
Length of gas channels	50
Height of gas channels	1
Width of gas channel	1
Width of cell	2
Thickness of catalyst layer	0.01
Thickness of gas diffusion layer	0.3
Thickness of membrane layer	0.178
Thickness of current collector	2

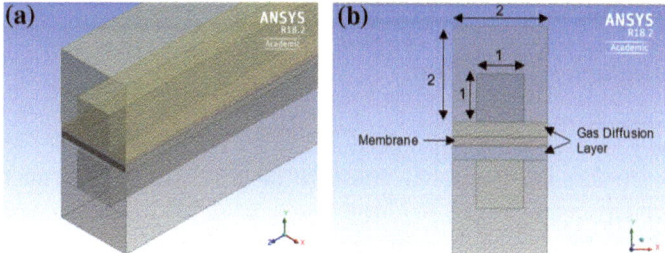

Fig. 3 a Isometric view of fuel cell model, **b** front view

Table 2 Model boundary conditions

Boundary conditions	Location		Value
Inlet mass flow rate	Inlet anode flow channel	Anode inlet mass flow rate	0.3 ms^{-1}
		Anode inlet mass fraction H$_2$	0.445
		Anode inlet mass fraction H$_2$O	0.555
	Inlet cathode flow channel	Cathode inlet mass flow rate	0.5 ms^{-1}
		Cathode inlet mass fraction O$_2$	0.212
		Cathode inlet mass fraction H$_2$O	0.079
Pressure	Operating pressure	–	101,325 Pa
	Pressure outlet anode flow channel face	Anode outlet gas pressure	0 Pa
	Pressure outlet cathode flow channel face	Cathode outlet gas pressure	0 Pa
Porosity	GDL		0.3
	Catalyst layer		0.112
Wall	The anode terminal and upper anode current collector face	Specific electric potential	0 V
		Temperature	343 K
	The cathode terminal and lower cathode current collector face	Specific electric potential	0.8–0.6 V
		Temperature	343 K
	All other cell faces	Thermal condition constant temperature	343 K

2.3 Boundaries Conditions

Anode and Cathode boundaries conditions are set as reactant species velocities at the inlet to simulate the momentum transport. Table 2 illustrated the summary of the model boundary conditions set in Fluent.

2.4 Model Validation

Validation work is conducted as a preliminary study to ensure that the computational model proposed here can accurately predict the flow field behavior within the PEMFC. The I-V graph showing both the simulated and experimental results in Fig. 4

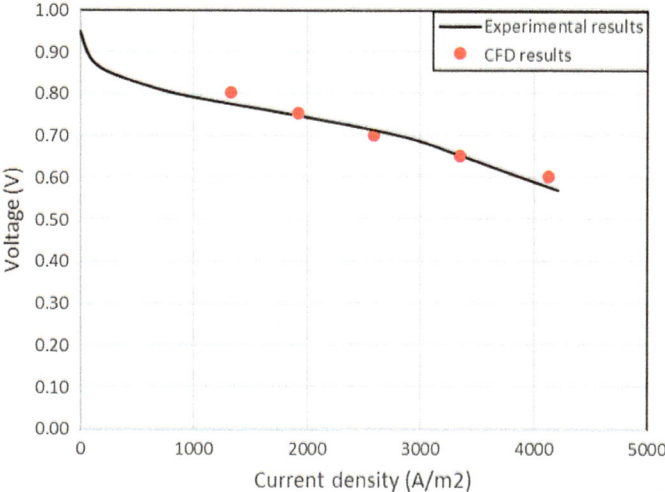

Fig. 4 Comparison of CFD model and experimental results

shows good agreement with the experimental data reported by Cheng et al [7]. There
is minimal variance between the simulated polarization curve and the experimental
data.

3 Ideation

3.1 Description of Design 1

Studies from literature reviews had shown that serpentine flow field by far are still
the best performing flow field in the PEMFC as compared to other different types of
flow field design [17, 18]. Leveraging on the advantages of serpentine flow field for
being the best performing and highest pressure drop amongst the rest, the motivation
behind this design is to increase the amount of pressure drop by putting the serpentine
design beside each other in a parallel configuration. As shown in Fig. 5, Design 1 has
two congruent serpentine flow channels next to each other while being connected by
one fuel inlet and having two outlets at the end of the channels.

3.2 Description of Design 2

Design 2 was inspired after studying the various differences in the existing flow field.
Taking into account the advantages of the serpentine and parallel flow field, Design 2

Fig. 5 Ideation of design 1

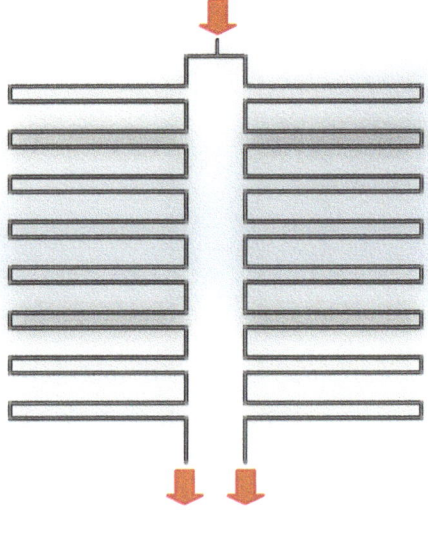

Fig. 6 Ideation of design 2

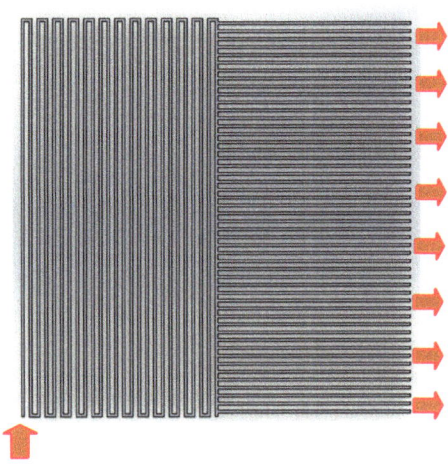

aims to strike a balance between the both. Design 2 aimed to optimize the hydrogen gas being fed through 1 inlet and evenly distributed to its various outputs channels. As shown in Fig. 6, Design 2 has 1 inlet and various outlets.

3.3 Description of Design 3

Design 3 was inspired by the air con vent which allow air flow through in a diagonal layout. As shown in Fig. 7, Design 3 was different from Design 1 and 2 as the input

Fig. 7 Ideation of design 3

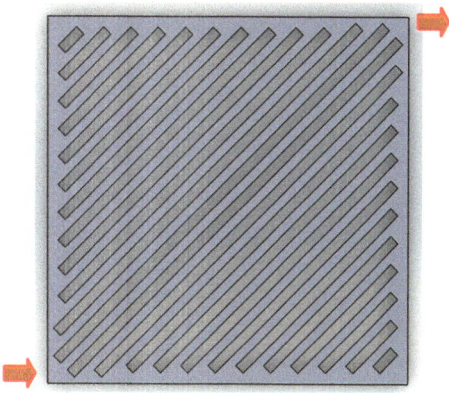

fuel is being transported freely through the diagonal channels instead of being fed through the inlet and flow within the channel design. There will be only one of each inlet and outlet respectively.

4 Results and Discussion

Before running on the fuel cell simulation, an air flow simulation was conducted on proposed designs to ensure that all the hydrogen is evenly distributed.

Prior to designing of the flow field, uniform fuel distribution and high-pressure drop are the key design criteria that are taken into considerations. Ideally, both factors should be met in order to achieve the optimal design.

Other than the key parameters required to be met, there are various design considerations that are essential in contributing to the effects of the performance of the fuel cell to be considered. Firstly, the length of channel and number of channels plays a role in affecting the performance because shorter path lengths have better reactants distribution as compared to long channels. Secondly, channel and ribs width contributed a significant effect on the performance of the fuel cell. Lastly, the cross-sectional shape which might show improvements in various aspect of the fuel cell operation.

From Fig. 8, it depicts that only the air flow in Design 1 are evenly distributed. Design 2 was observed that the last few channels in the parallel region was not utilized at all as there were almost close to zero in the velocity as seen in Fig. 7. Whereas for Design 3, only the center region is experiencing air flow.

After noticing Design 2 and 3 are experiencing some air flow difficulties in some area of the flow field, modification had been made to further optimized the air flow. The rib widths were increased and the number of channels were reduced. After which the simulation was conducted and the results are shown in Fig. 9.

(a)

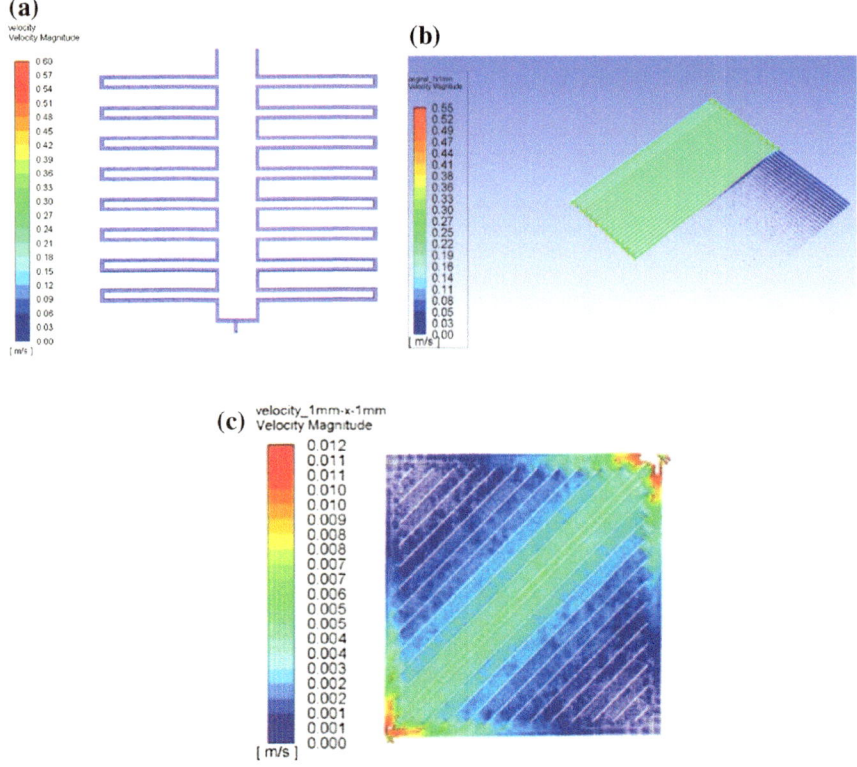

Fig. 8 Velocity magnitude of design 1 (**a**), design 2 (**b**) and design 3 (**c**)

(a)

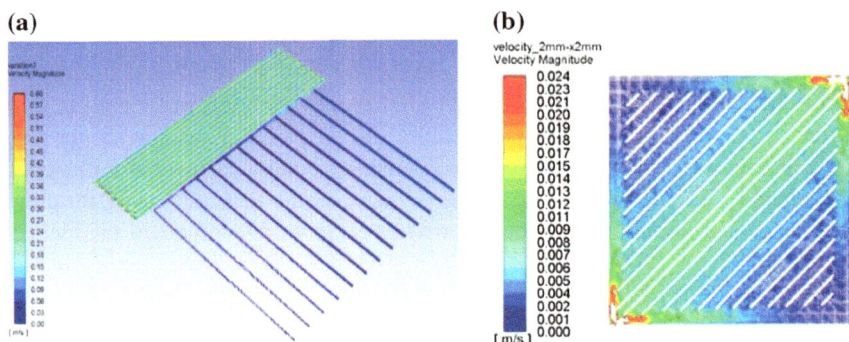

Fig. 9 Variations of geometric parameters for design 1 (**a**) and design 3 (**b**)

Table 3 Pressure drop for various geometric parameters

Cross sectional area (mm^2)	Cross sectional shape	Pressure drop (Pa)
0.25	Square	152.57
1		29.68
4		9.33
1	Trapezoidal	39.08

Table 4 Pressure drop for different cross-sectional shapes

Cross sectional area (mm^2)	Cross sectional shape	Pressure drop (Pa)
1	Square	62.59
1	Trapezoidal	65.15

It was observed that design 2 shows significant improvement in the air flow with wider rib widths and lesser channels. However, the same trend was observed. Whereas for Design 3, the same trend persists even though the modification was made. Thus, design 2 and 3 were eliminated for further analysis.

Different types of cross sectional shapes were analyzed in Design 1. For a fair comparison, the number of channels and active area are fixed. Based on Table 3, it is observed that the smaller the cross-sectional area, the larger the pressure drop. While the trapezoidal shows a slightly higher value in the pressure drop comparing with a 1×1 mm square cross-sectional area.

Moreover, Design 1 was modified by increasing the number of channels and smaller ribs width. Table 3 shows the pressure drop between the differences in cross sectional shape.

In comparison to the results shown in Tables 3 and 4, it is observed that trapezoidal cross-sectional shape shows higher pressure drop as compared to square shape. However, as the pressure drop difference between the 2 shapes are minimal, it is almost negligible in the improvement of the fuel cell performance taking into the manufacturing process and cost of trapezoidal shape.

5 Conclusion

In conclusion, geometries parameters show significant effects in the performance of the fuel cell had been studied earlier. Simulation results had shown that smaller channel width make a significant impact on the pressure drop which could let to increase of the overall current density. Although trapezoidal cross sectional shows better performance as compared to rectangular, the slight improvement is negligible taking into consideration the manufacturing cost.

Design 1 is expected to show an improvement in performance due to its higher pressure drop as achieved during the air flow simulation compared to the parallel flow field.

6 Future Work

Moving forward, the above-mentioned designs will run various CFD simulations in a fuel cell configuration to further prove its hypothesis stated by the authors.

Acknowledgements The authors would like to thank Singapore Institute of Technology for the support of this work.

References

1. Office of Energy Efficiency and Renewable Energy. (2018). *Types of fuel cells*. (Online) Available https://www.energy.gov/eere/fuelcells/types-fuel-cells. Retrieved March 26, 2018.
2. Spiegel, C. (2007). *Designing and building fuel cells*.
3. David, S. W., Kenneth, W. D., & Danny, G. E. (1991). *United States Patent, 4,988,583*.
4. David, S. W., Kenneth, W. D., & Danny, G. E. (1992). *United States Patent, 5,108,849*.
5. Arvay, A., French, J., Wang, J. C., Peng, X. H., & Kannan, A. M. (2013). Nature inspired flow field designs for proton exchange membrane fuel cell. *International Journal of Hydrogen Energy, 38*(9), 3717–3726.
6. Linfield, K. W., & Mudry, R. G. (2008). *Pros and cons of CFD and physical flow modeling*.
7. Cheng, C. H., Lin, H. H., & Lai, G. J. (2007). Design for geometric parameters of PEM fuel cell by integrating computational fluid dynamics code with optimization method. *Journal of Power Sources, 165*(2), 803–813.
8. Manso, A. P., Marzo, F. F., Barranco, J., Garikano, X., & Garmendia Mujika, M. (2012). Influence of geometric parameters of the flow fields on the performance of a PEM fuel cell. A review. *International Journal of Hydrogen Energy, 37*(20), 15256–15287.
9. Yoon, Y. G., Lee, W. Y., Park, G. G., Yang, T. H., & Kim, C. S. (2004). Effects of channel configurations of flow field plates on the performance of a PEMFC. *Electrochimica Acta, 50*(2–3 Special Issue), 709–712.
10. Wang, X., Zhang, X., Liu, T., Duan, Y.-Y., Yan, W.-M., & Lee, D.-J. (2010). Channel geometry effect for proton exchange membrane fuel cell with serpentine flow field using a three-dimensional two-phase model. *Journal of Fuel Cell Science and Technology, 7*(5), 051019/1–051019/9.
11. Jaruwasupant, N., & Khunatorn, Y. (2011). Effects of difference flow channel designs on proton exchange membrane fuel cell using 3-D Model. *Energy Procedia, 9,* 326–337.
12. Ahmed, D. H., & Sung, H. J. (2006). Effects of channel geometrical configuration and shoulder width on PEMFC performance at high current density. *Journal of Power Sources, 162*(1), 327–339.
13. Kumar, A., & Reddy, R. G. (2003). Effect of channel dimensions and shape in the flow-field distributor on the performance of polymer electrolyte membrane fuel cells. *Journal of Power Sources, 113*(1), 11–18.
14. Lee, S., Jeong, H., Ahn, B., Lim, T., & Son, Y. (2008). Parametric study of the channel design at the bipolar plate in PEMFC performances. *International Journal of Hydrogen Energy, 33*(20), 5691–5696.

15. Sun, W., Peppley, B. A., & Karan, K. (2005). Modeling the Influence of GDL and flow-field plate parameters on the reaction distribution in the PEMFC cathode catalyst layer. *Journal of Power Sources, 144*(1), 42–53.
16. Yoon, Y. G., Lee, W. Y., Park, G. G., Yang, T. H., & Kim, C. S. (2005). Effects of channel and rib widths of flow field plates on the performance of a PEMFC. *International Journal of Hydrogen Energy, 30*(12), 1363–1366.
17. Limjeerajarus, N., & Charoen-Amornkitt, P. (2015). Effect of different flow field designs and number of channels on performance of a small PEFC. *International Journal of Hydrogen Energy, 40*(22), 7144–7158.
18. Jang, J. H., Yan, W. M., Li, H. Y., & Tsai, W. C. (2008). Three-dimensional numerical study on cell performance and transport phenomena of PEM fuel cells with conventional flow fields. *International Journal of Hydrogen Energy, 33*(1), 156–164.

Simulation-Based Analysis of a Network Model for Autonomous Vehicles with Vehicle-to-Vehicle Communication

Qi Yao Yim and Kester Yew Chong Wong

Abstract Autonomous vehicle technology is an expansively researched area of transport that aims to tackle long-standing problems of traffic such as congestion, safety and efficiency. Many of these vehicles combine automated driving with communication among vehicles and infrastructure to bring about a seamless driving experience that would not have been possible with human driving. The development of computer simulations for such vehicles aims to address concerns on whether the benefits proposed by autonomous vehicle makers can be realized in various traffic environments. This paper assesses the efficiency of autonomous vehicles that are introduced on an arterial road network with features similar to Singapore's road networks. A cellular automata simulation has been developed that considers vehicle-to-vehicle communication abilities of autonomous vehicles. A traffic data collection algorithm based on web traffic services was developed to estimate real-time travel times along each stretch of road in the network simulation, from which autonomous vehicles can optimize their speed and route for a faster journey time. Based on preliminary results, the simulation was tested under multiple traffic densities and situations. The results display interesting interactions between vehicles and road elements such as lanes and traffic lights, which has allowed both autonomous and non-autonomous vehicles to travel to their designated destination faster when autonomous vehicles have been introduced.

Keywords Autonomous vehicles · Road model · Road networks · Traffic simulation · Cellular automata

Q. Y. Yim (✉)
National Junior College, Singapore, Singapore
e-mail: yimqiy@gmail.com

K. Y. C. Wong
Mathematics Department, National Junior College, Singapore, Singapore

© Springer Nature Singapore Pte Ltd. 2019
H. Guo et al. (eds.), *IRC-SET 2018*,
https://doi.org/10.1007/978-981-32-9828-6_31

389

1 Introduction

Autonomous vehicle (AV) technology, particularly when users can rely fully on the vehicle's built in driving capabilities, has been widely explored as an alternative to conventional human-driven vehicles, spurred by reasons such as convenience, safety and traffic congestion. In Singapore, the development and testing of autonomous vehicles has been picking up pace in recent years, where various vehicle types from taxis to buses have been studied. Problems addressed in the local context include land scarcity for roads and parking, mobility and shortage or limitations of drivers [1]. One useful feature of autonomous vehicles is vehicle-to-vehicle (V2V) communication, which aims to improve traffic flow in road networks. V2V communication allows vehicles to gain information about other vehicles, including location, velocity and intended route, and thereafter make decisions regarding their own movement to achieve optimal time savings while driving safely. V2V communication does not require the presence of additional infrastructure on roads, therefore making it a good starting point in implementing AV technology. Litman [2] hypothesized that while certain companies hope to make autonomous vehicles available for commercial use in a matter of years, large-scale benefits such as improving traffic flow (which cuts down on road congestion, energy costs and carbon emissions of vehicles) will only become significant when the use of the technology is common. He suggests possible obstacles in achieving the intended benefits, such as coexistence with non-autonomous vehicles and optimization between benefits.

2 Previous Work

Computer-aided traffic simulation enables developers to test the efficiency of the communication technology employed, especially when a wide range of road types and networks is considered, before implementation of vehicles and infrastructure on actual roads. Gora and Rüb [3] proposed a traffic simulation framework that incorporates communication features of autonomous vehicles. Hu et al. [4] considered a highway model with autonomous vehicles, with capabilities such as coordinated driving and lane changing. Models of network-based traffic could be applied to the scheduling of other autonomous systems such as robotic servers [5] and autonomous vehicles in ports [6].

This research aims to design a microscopic traffic simulator based on existing models to investigate the effects of introducing autonomous vehicles on a road network on both trip efficiency (measured by mean trip velocity) and road efficiency (measured by traffic flow). Parameters of this network can be modified to simulate multiple traffic conditions. These vehicles have V2V communication capabilities that enable it to coordinate its velocity with nearby vehicles as well as finding, where possible, a route taking shorter time.

3   Mathematical Model

Cellular automaton, a microscopic traffic model where vehicles' positions are represented as cells in a matrix, was used as the mathematical model. Microscopic traffic models simplify the movements of vehicles while being accurate enough to detail large-scale traffic data. The first cellular automaton model for traffic flow was devised by Nagel and Schreckenberg (referred to as the NaSch model) [7], simulating single-lane highway traffic. The model recreates traffic flow with accuracy and has become the basis for numerous studies on traffic modelling. Its stochastic (rather than deterministic) nature, brought about though a randomization process, is important when constructing a model that differentiates the behavior of autonomous and non-autonomous vehicles. In particular, spontaneous formations of jams are observed, which exemplifies the physical limitations of drivers.

3.1   Research Methodology

This research combines the NaSch model and road network models [8, 9], with features such as vehicles changing direction and lanes. For this simulation, a 2-by-2 network of two-way intersecting roads was considered (Fig. 1). Each road is of length 216 sites (including two 8×8 site junctions) and has 3 lanes per direction. The following describes the steps taken for updating a vehicle n's current velocity, v_n.

Step 1: Acceleration
$v_n \leftarrow \min(v_n + 1, v_{max})$, where $v_{max} = $ maximum velocity
Step 2: Lane Changing (see below)
Step 3: Braking due to traffic lights or vehicles in front
If both n and the next vehicle (vehicle directly in front) are autonomous vehicles:
$v_n \leftarrow \min(v_{max}, v_j + j_n - 1, s_n - 1, oldv_n + 1)$ where $v_j = $ velocity of the next vehicle, $j_n = $ headway from n to next vehicle, $s_n = $ number of sites after n's site where stopping site of a STOP signal is located, $oldv_n = v_n$ before step 1

Fig. 1 Diagram of the road network studied. The distance between any two adjacent nodes (orange) is 50 sites, and consists of a bi-directional road (blue)

Else:
$v_n \leftarrow \min(v_n, j_n - 1, s_n - 1)$
Step 4: Randomization (non-autonomous vehicles only)
$v_n \leftarrow \max(v_n - 1, 0)$ with probability p_{slow}
Step 5: Motion
n is advanced v_n sites in its respective direction, changing direction as required.

Based on the NaSch model, v_{max} was kept at 5 while p_{slow} was set to 0.5 in accordance with cellular automata studies on road network traffic. Comparing the simulation with real-world traffic systems, one site corresponds to a square of side 3.7 m (allowing each vehicle to be represented as a single cell), and one time step corresponds to 4/3 s. This gives each vehicle a maximum acceleration of one site per timestep or 2.775 ms^{-2} (10 km/h). The maximum velocity is 13.9 ms^{-1} (50 km/h), the speed limit for major roads in Singapore. Each simulation runs for 1000 timesteps and the data collected are averages over 3 simulations.

Data obtained from the simulation includes mean velocity v, mean density ρ and traffic flow q. Mean density is the mean total number of occupied sites among all time steps divided by the number of sites, and represents the average probability a vehicle can be found at a particular site and time. Traffic flow is a measure of the efficiency of the roads in the network:

$$q = \rho v \tag{1}$$

3.2 Modifications Made for a Road Network Adaptation

The road network incorporates four directions in two dimensions. Each intersection consists of a traffic light (signal) for each direction, and signals are synchronized in cycles of length $4T + 40$. Signals stop vehicles during Step 3 if the signal's status is '0' for the particular direction. When its status is '1', each signal allows traffic to pass for T steps. Traffic in the next direction then waits for 10 steps before it can pass, to allow traffic from the previous direction to clear. Vehicles are able to turn perpendicularly at intersections, or go straight on. Turning has been simplified to a single change of direction in no time when the vehicle has reached the site of its new lane [3]. In addition, turning vehicles look for vehicles along the path of their turning in both their old and new directions in Step 3. Individual lanes have been reserved for a specific turning direction, which requires lane changes to be made.

Changing lane takes a negligible amount of time, where vehicles check whether their target site is vacant and move to the site if so. An important assumption is that for road network traffic, a vehicle's main priority is to move to the correct lane for facilitating its turning desires; hence vehicles do not overtake each other. After it was observed that autonomous vehicles would be unable to change lane if travelling in parallel with another autonomous vehicle on is target lane, a rule was set such that vehicles reduce their speed conditionally by one step /site if a lane change is not possible.

3.3 Modifications Made for Autonomous Vehicles and Their V2V Capabilities

Autonomous vehicles are able to interact with those directly in front and coordinate their speeds. While this is a popular solution also known as "convoy driving" adapted for highway traffic autonomous vehicle models, it can also be applied to road network traffic especially with the presence of traffic signals that slow down groups of drivers significantly. Each autonomous vehicle communicates its intended travel speed to the autonomous vehicles behind it, until all vehicles have finalized their plans for the next time step [3]. As such, instead of the next vehicle's current position, autonomous vehicles can use the velocity of a vehicle in front and adjust its own velocity (while keeping within maximal acceleration) such that it maintains a safe distance behind.

The traffic model starts empty, and generation of new vehicles is done through Poisson-distributed inter-arrival times a, modelled using the following distribution function. λ is the rate of vehicle entry and can be adjusted independently for each origin node such that the measured ρ is indirectly affected:

$$P(a \leq x) = 1 - e^{-\lambda x} \quad \text{for} \ x \geq 0 \tag{2}$$

Each car has a pre-determined origin and destination node (which can be any node except the four intersection nodes 4, 6, 10 and 12) within the network (Fig. 1). For non-autonomous vehicles, it is assumed that drivers have planned their routes based on the shortest distance to their destination. Vehicles are generated at their origin and move through the network to their destination, after which the vehicle is removed from the system.

Autonomous vehicles can optimize their routes to save travel time based on information gathered from other vehicles in the network. Currently, certain vehicles relay real-time information on their position via in-vehicle systems or mobile devices to web services [10], such as Google Traffic, which then display the traffic condition along that road. In the simulation, each vehicle sends its location and velocity to the system with probability p_{locate} (representing the percentage of vehicles that transmit data). The collective information is then used to estimate the time taken to drive along the road. At every node where multiple routes are possible, autonomous vehicles calculate the shortest route by time instead of distance, computed with Dijkstra's shortest path algorithm. Another variable $datastep$ limits the number of time steps before the current time step from which this data is retained. For this model, $datastep$ = $4T + 40$, since flow in the network is highly dependent on traffic signals. If no data on a road is available due to a lack of vehicle information, it is assumed the road is largely clear, and the travel time can be estimated through the following formula based on the periodicity of the traffic signals:

$$\frac{\sum_{t=1}^{pt_{red}} t}{pt_{cycle}} + \frac{D}{v\text{max}} \tag{3}$$

where D = road section length, $pt_{red} = 3T + 40$ and $pt_{cycle} = 4T + 40$ for this particular implementation.

4 Results and Analysis

4.1 Impact of Signal Time, T

First, simulations for a non-autonomous vehicle network were performed. Figure 2 shows the relationship between vehicular density and traffic flow. When compared to other road network traffic studies [8], each value of T displayed a similar curve. For that model, with greater T, the stop phase for each signal also increases, thereby reducing traffic speed.

However, while the curve gradients in Fig. 2 for $T > 50$ follow the model, the curves for $T < 50$ deviate from the model and have a smaller gradient with decreasing T. Further investigation revealed that this is due to the introduction of a 10-step delay between go phases.

On the other hand, traffic flow for a fully autonomous vehicle network is observed to be directly proportional to vehicular density. This can be attributed to the communication between autonomous vehicles which enables more vehicles to pass by an intersection during each green phase. There is also a clear inverse relationship between values of T tested and traffic flow.

4.2 Impact of Autonomous Vehicle Implementation Rate

The study of traffic under varying percentage of autonomous vehicles (p_{auto}) confirms a reduction in travel time with implementation of autonomous vehicles. Increasing p_{auto} for the same entry rate value λ results in increases of mean velocity for both types of vehicles (Fig. 3a) and a decrease in measured road density shows that vehicles spend lesser time in the network (Fig. 3b). The combination of both effects leads to a stable or slightly increasing traffic flow (Fig. 3c). In both cases, the improvements become more effective with higher entry rates λ. One of the concerns of autonomous vehicle implementation is how vehicle travel and hence congestion could ultimately increase as the use of the technology increases travel convenience [2]. However, the mean velocity for various values of λ converge to a small range as the percentage of autonomous vehicles increases, indicating that such induced vehicle travel is not necessarily an issue for trip efficiency.

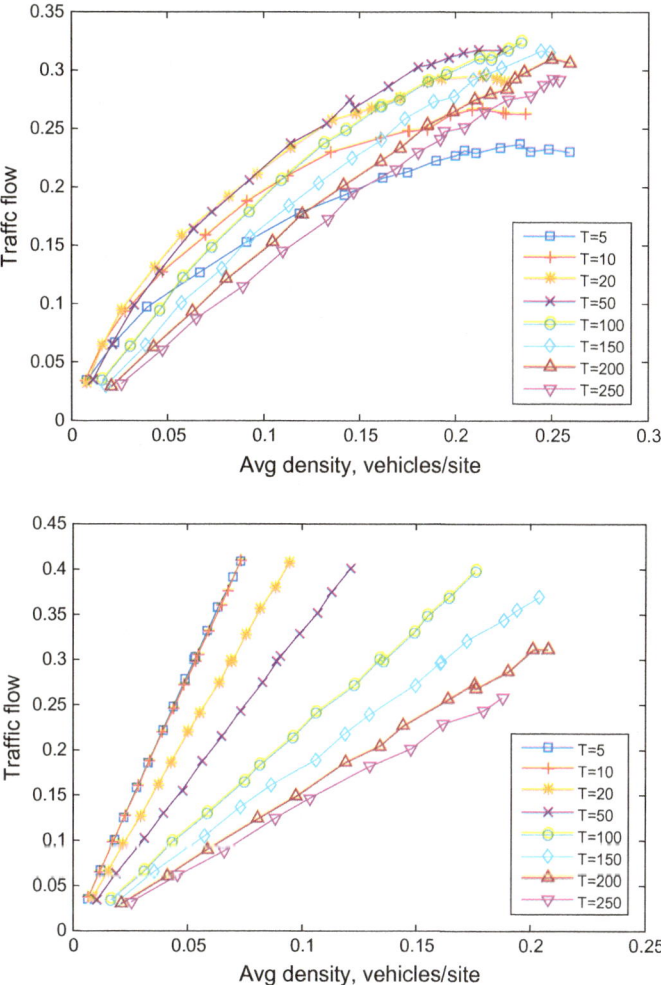

Fig. 2 Traffic flow, q against road density ρ for varying green phase duration T with $p_{auto} = 0$ (top) and $p_{auto} = 1$ (above). $p_{locate} = 1$

4.3 Impact of Shortest Path Algorithm

The shortest path algorithm was also tested for its efficiency with $p_{auto} = 0.5$ and $p_{auto} = 1$: firstly without any modifications to the road network (Fig. 4a); and secondly (Fig. 4b) with an uneven distribution of vehicles, where the rate of vehicle entry λ at node 5 was increased by 8 times, hence increasing traffic on the stretch of road 4–5–6 (see Fig. 1). Critical vehicles, defined as vehicles that have a choice of multiple paths with the same distance (e.g. journey from node 1 to node 14), were also considered.

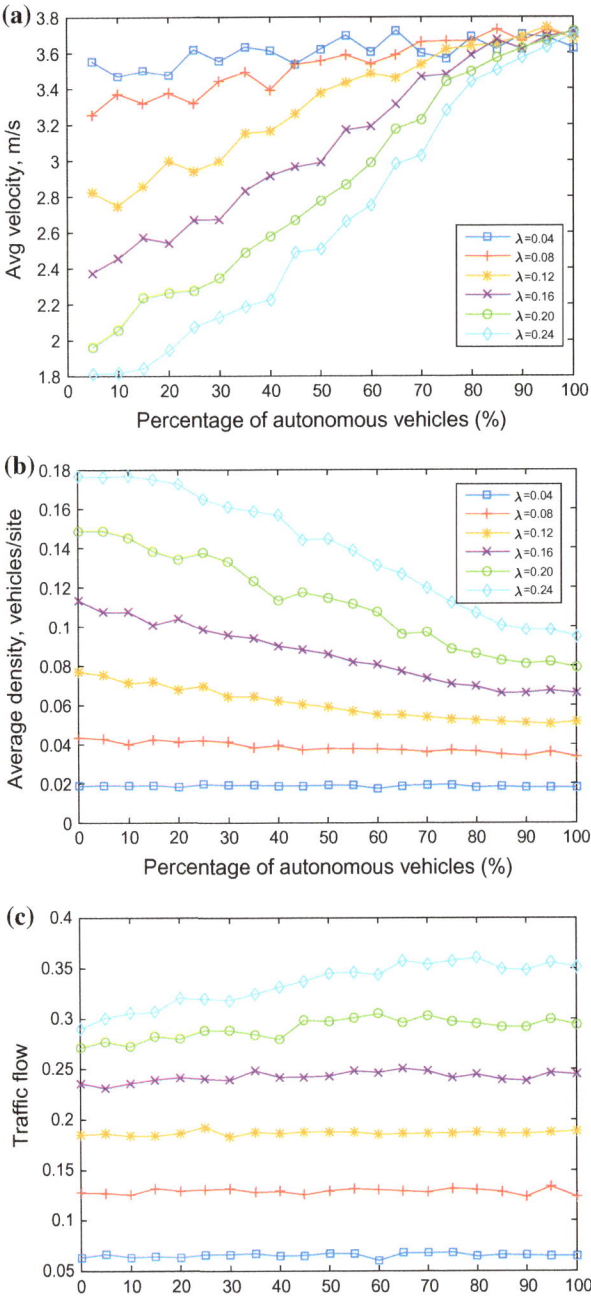

◀**Fig. 3 a** Graphs of percentage of autonomous vehicles against non-autonomous (top) and autonomous (above) mean velocity for $T = 40$, $p_{locate} = 1$ and varying λ. **b** Graphs of percentage of autonomous vehicles against actual measured density for $T = 40$, $p_{locate} = 1$ and varying λ. **c** Graphs of percentage of autonomous vehicles against traffic flow for $T = 40$, $p_{locate} = 1$ and varying λ.

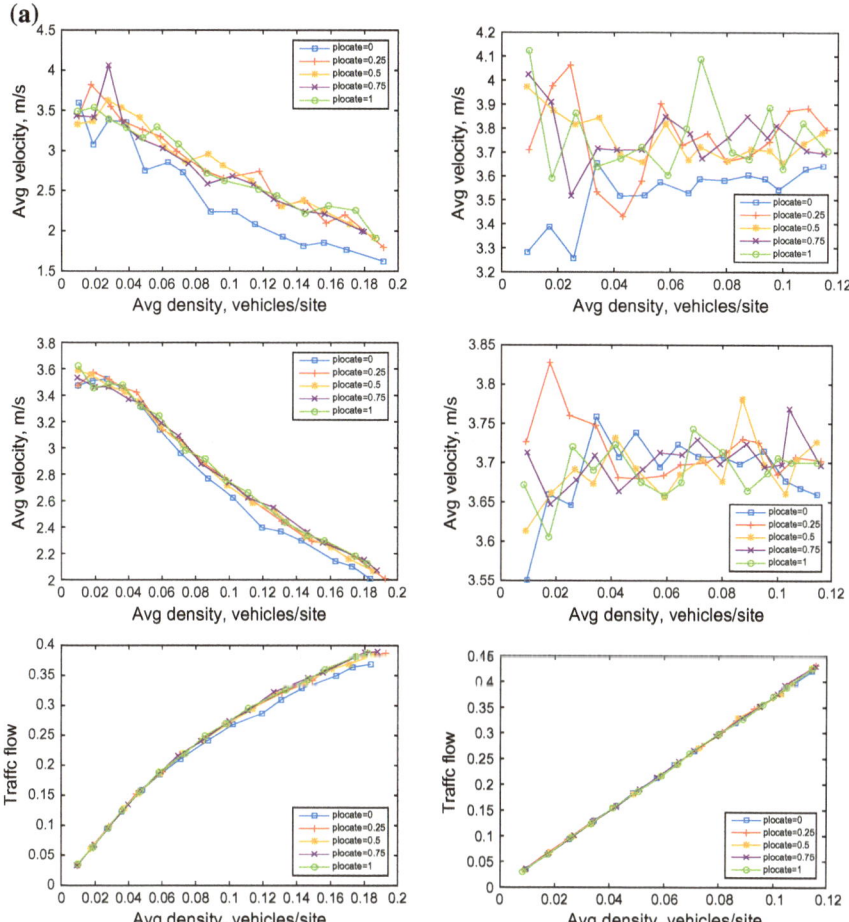

Fig. 4 a Top to bottom: graphs of average density against average velocity of critical vehicles, overall average velocity and traffic flow with $T = 40$. $p_{auto} = 0.5$ (left) and $p_{auto} = 1$ (right). **b** Top to bottom: graphs of average density against average velocity of critical vehicles, overall average velocity and traffic flow, with increased influx of vehicles at site 5 and $T = 40$. $p_{auto} = 0.5$ (left) and $p_{auto} = 1$ (right)

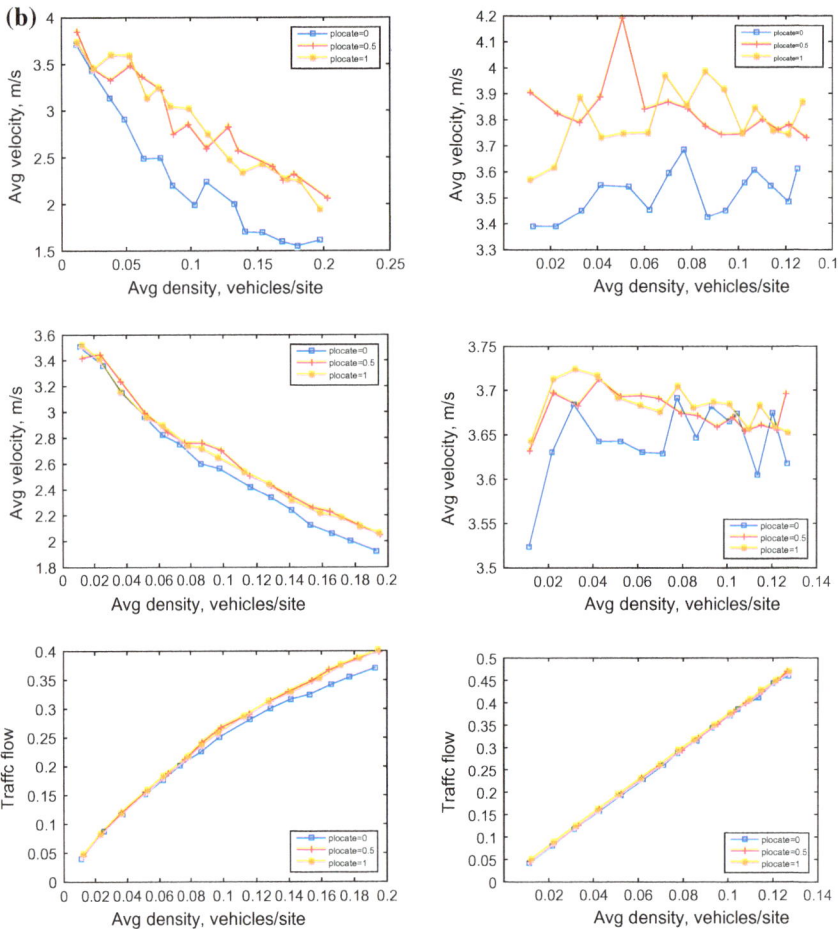

Fig. 4 (continued)

In both cases, the algorithm produced a slight improvement in overall traffic flow with high-density traffic for mixed vehicle types ($p_{auto} = 0.5$); in particular, critical vehicles displayed an improvement of 0.5 m/s for normal traffic and 1.0 m/s for the increased influx scenario. The driving behavior of the autonomous vehicles is hence able to organize the extra load at a particular node. In addition, there is no significant difference between values of $p_{locate} > 0$ tested, showing that a small amount of data collated is enough to obtain an estimate of the condition of the network as Herrera et al. [10] suggested. With a network where $p_{auto} = 1$, the algorithm showed no significant effect in optimizing the overall traffic speed and flow. This suggests that coordinated driving enables autonomous vehicles to mitigate situations which could have caused otherwise caused congestion.

5 Conclusion and Future Work

This research work has produced a model of vehicles in a small traffic network. The results show that generally, with a high density of vehicles and an ideal communication system between autonomous vehicles in place, increased efficiency and reduced congestion of road networks results. In addition, the existing traffic data collected from vehicles has proven sufficient for autonomous vehicles to find a route taking shorter time. A greater saving in time is predicted for larger road networks as well as those with larger irregularities in road density.

Further work may be conducted to improve the model's dynamics based on realistic road network traffic, as Knospe, Santen, Schadschneider and Schreckenberg [11] did with highway traffic. The study of vehicle-to-infrastructure communication (V2I) and vice versa (I2V) is also possible. For example, vehicles can get information on signal information directly rather than through vehicles located nearby. Within a small network, all AVs are ideally capable of communication with each other, so limitations of this communication have to be considered for larger road networks. It is worth noting that the durations of traffic lights in Singapore are controlled based on real-time vehicle density provided through sensors, and autonomous vehicles using these data can plan their routes better.

References

1. Tan, C. K., & Tham, K. S. (2014). Autonomous vehicles, next stop: Singapore. *Journeys*, 5–11.
2. Litman, T. (2017). *Autonomous vehicle implementation predictions* (p. 28). Victoria Transport Policy Institute.
3. Gora, P., & Rüb, I. (2016). Traffic models for self-driving connected cars *Transportation Research Procedia, 14,* 2207–2216.
4. Hu, J., Kong, L., Shu, W., & Wu, M. Y. (2012, December). Scheduling of connected autonomous vehicles on highway lanes. In: *2012 IEEE Global Communications Conference (GLOBECOM)* (pp. 5556–5561). IEEE.
5. The Rise of a Robotic Dawn in Services Industry. (2017, January 2). Retrieved from http://www.todayonline.com/singapore/rise-robotic-dawn-services-industry.
6. Singapore to Start Trials of Driverless Trucks for Port Transport. (2017, January 9). Retrieved from http://www.channelnewsasia.com/news/singapore/singapore-to-start-trials-of-driverless-trucks-for-port-transpor-7558490.
7. Nagel, K., & Schreckenberg, M. (1992). A cellular automaton model for freeway traffic. *Journal de Physique I, 2*(12), 2221–2229.
8. Schadschneider, A., Chowdhury, D., Brockfeld, E., Klauck, K., Santen, L., & Zittartz, J. (2000). A new cellular automaton model for city traffic. In *Traffic and Granular Flow '99* (pp. 437–442). Berlin, Heidelberg: Springer.
9. Gora, P. (2009). Traffic simulation framework—A cellular automaton-based tool for simulating and investigating real road network traffic. *Recent Advances in Intelligent Information Systems*, 641–653.

10. Herrera, J. C., Work, D. B., Herring, R., Ban, X. J., Jacobson, Q., & Bayen, A. M. (2010). Evaluation of traffic data obtained via GPS-enabled mobile phones: The Mobile Century field experiment. *Transportation Research Part C: Emerging Technologies, 18*(4), 568–583.
11. Knospe, W., Santen, L., Schadschneider, A., & Schreckenberg, M. (2002). A realistic two-lane traffic model for highway traffic. *Journal of Physics A: Mathematical and General, 35*(15), 3369.

Structural Biomimetic Scaffold Modifications for Bones

Xin Yi Ariel Ho, Hui Yu Cherie Lee, Jing Wen Nicole Sze and Tee Wei Teo

Abstract Over the years, biomimetic scaffolds have been commonly used in the process of tissue regeneration for treatment of bone defects. Even though current biomimetic scaffolds are easily accessible, they lack mechanical strength for ideal applications. This research aims to improve on current designs to create new three-dimensional biomimetic scaffolds for bones. Different biomimetic scaffolds were designed and 3D-printed and was tested for mechanical strength conducting a tensile strength test on the scaffolds. The tests show that the scaffold with hexagonal pores proved to be the most effective scaffold due to its geometrical properties which allows it to withstand more pressure. This was concluded according to the spread of pressure along the scaffold that is dependent on the amount of pressure exerted, and the identification of first fracture which affects the line of breakage across the entire scaffold. This research is able to better help extend the field of tissue engineering and the applicability of 3D biomimetic bone scaffolds for clinical usage.

Keywords Tissue regeneration · Bone defects · Mechanical strength · Biomimetic 3D scaffolds

1 Introduction

Bones are the second most transplanted tissues within the body, with approximately 3.5 million bone graft procedures performed each year. There have been many cases whereby a great number of bone grafts have been used to reconstruct defects of a large bone. These faults can be caused by accidents, trauma, tumors, infections, and birth defects, or in some cases where the bone is unable to regenerate itself: osteoporosis, necrosis and atrophic non-unions [1].

Being one of the most common type of treatment for such damaged bone tissues, autografting has been used to transplant tissue from one site to another in the same patient. Autografting is usually described beneficial by its high osteoconductivity, osteogenic characteristic if grafted rapidly and its minimal osteoinductivity [2]. This

X. Y. A. Ho · H. Y. C. Lee · J. W. N. Sze (✉) · T. W. Teo
National Junior College, Singapore 288913, Singapore
e-mail: nicole.sjw@gmail.com

© Springer Nature Singapore Pte Ltd. 2019
H. Guo et al. (eds.), *IRC-SET 2018*,
https://doi.org/10.1007/978-981-32-9828-6_32

method of treatment however is an expensive and painful process and is often associated with donor site morbidity caused by infection and hematoma. An alternative for treatment is by harvesting allografts, which is the transplantation of tissues from living donors to patients. This method of treatment is more efficient than autografting but carries a risk of introducing infectious agents to the patient or rejection of tissue by the patient's immune system [1, 3].

Although these two types of bone grafts are commonly used for medical treatment, they have their limitations and are not the best methods to be used. The regeneration of tissues within the patient would be more desirable in treating damaged bone tissues through tissue engineering. With the use of biomimetic scaffolds, an ideal environment is provided for regeneration of tissues, which in turn can repair bone fractures that are extremely complex and pose significant health risks to the patient [3, 4]. These biomimetic scaffolds mimic the important features of the extracellular matrix (ECM) architecture and act as templates for tissue formation, assisting functional cells physically, chemically and biologically in their growth. [3, 5] The biomimetic scaffolds are typically seeded with cells and occasional growth factors which are then cultured in vitro to synthesise tissues that will then be implanted directly into the injured body site [3].

Even though these three-dimensional biomimetic scaffolds are said to be better than other bone grafting methods, the existing three-dimensional scaffolds for tissue engineering are shown to be less than ideal for actual applications. A good scaffold should have appropriate mechanical properties to provide a suitable environment for the regeneration of tissues. Ideally, these scaffolds should be porous in structure and permeable for cells and nutrients to enter. Having an appropriate surface structure and chemistry for cell attachment would also prove its benefits in creating a more efficient environment for cell growth. Furthermore, there are many other complex requirements in the design of scaffolds for tissue engineering that are not yet fully understood. However, an important aspect of such scaffolds would be mechanical strength in which such structures must be able to withstand high pressures, a characteristic of bones. Current biomimetic scaffolds may hence be seen as advantageous because of their unlimited availability but may not be as preferred due to the lacking in mechanical strength [1, 6].

The aim of this research paper is to design an improvised version of the current mechanical structure of three-dimensional biomimetic scaffolds made of synthetic materials which are able to best support the regeneration of bones. By modifying and creating a new design of the three-dimensional scaffold, the project aims to further understand biomimetic scaffolds for the field of tissue engineering, as well as to enhance its application and efficacy for clinical use. The use of the basic and different designs in the created scaffolds would allow varied properties of these shapes to be analysed. A scaffold design with hexagonal pores is hypothesised in this research to be most effective for tissue regeneration according to its tensile strength from geometrical properties.

2   Methodology

2.1   Designing of Scaffold

In order to create an improved and efficient design for biomimetic scaffolds for bones, comparisons and contrasts are made and observed from current scaffold designs for application. Designs must consider the requirements for scaffolds which help in bone cell regeneration or consist of functions similar to bone cell structures. 3 different types of scaffolds were made with varying porosity and surface area. The volume of material of each scaffold were kept consistent at 7.00 cm^3 as an independent variable to ensure that the volume of the scaffolds does not affect the results of the strength test. A border ring of 0.5 cm (width) × 1 cm (height) was also added to each scaffold to ensure that the pores at the sides of the scaffolds do not affect the amount of pressure it is able to withstand. The scaffolds were also printed with a consistent diagonal infill pattern with a percentage of 55% and a Z resolution of 0.20 mm. A consistent circular shape of the scaffold has also been applied for closest replication of a portion of the bone.

These scaffolds are designed and 3D-printed with ABS plastic using UpBox (3D printer). ABS plastic is chosen for its high tensile strength, impact resistance, good electric insulation, moisture resistance and its high strength to weight ratio [7] (Figs. 1 and 2).

2.1.1   Circular Porous Scaffold

This scaffold design is designed to mimic the effects as that of a sponge, thus the circular porous design. By using a circular design, relatively precise designs and microstructure of the scaffolds can be incorporated. Thus the physicochemical properties of the porous scaffolds can be easily engineered to mimic the physical properties of the native ECM in target tissues, making it advantageous for making tissues that

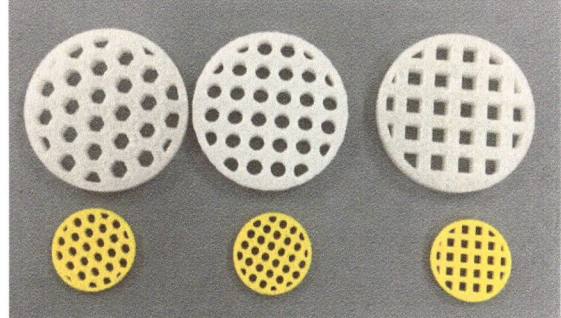

Fig. 1 Dimensional scaffolds printed using ABS plastic. From left to right: Hexagonal scaffold, circular porous scaffold, lattice scaffold

Scaffold	Details of design/mm	Volume /cm^3	Surface Area/cm^2	Mass of Material Used/g
Circular Porous	50mm X 50mm X 10mm Cylinder 5mm thick ring border 10.00mm X 10.00mm circular pores with 5.00mm space between each circular pore vertically and horizontally ABS	6.93	59.67	17.30
Hexagon	50mm X 50mm X 10mm Cylinder 5mm thick ring border 5.51mm X 5.70mm hexagonal pores with 3.86mm horizontal space and 4.00mm vertical space between ABS	6.96	59.69	17.45
Lattice	50mm X 50mm X 10mm Cylinder 5mm thick ring border 5.00mm X 4.50mm rectangular pores with 3.25mm vertical space and 3.24mm horizontal space between ABS	7.00	60.30	17.70

Fig. 2 Properties of Scaffolds

has to bear a large amount of weight which contributes to the mechanical properties desired in this project [8].

2.1.2 Hexagon Scaffold

This scaffold design was used with reference to the honeycomb structure of beehives where the sides of the hexagons are of equal length. The use of a hexagonal design would ensure a high specific stiffness and high impact absorption for the scaffolds which is a necessary property of a bone cell. The known flexibility of axial and shear tensile strength of this structural design can potentially provide very good mechanical properties of the scaffold [9].

2.1.3 Lattice Scaffold

This scaffold design was made based on current bone scaffolds that are commonly used. These lattice bone scaffolds are formed with hierarchical structures where the rods that connect to the lattice are also made of lattices and smaller rods. The hierarchy structure of the lattice allows the bone scaffolds to have their compressive strength, just like how bones do. Furthermore, this structure has a peak in surface-volume ratio when there are changes in the unit cell length which is the individual cubes and the overall size of the scaffold, this suggests that there are optimal points to be found in the design [10]. Hence, this serves as a positive control setup for comparison with other scaffolds.

2.2 Tensile Strength Test

Each designed scaffolds were placed under increasing pressure and weight across the diameter with the use of 10, 5, 2.5, 1.25 and 0.050 kg weights. The amount of pressure or weight that the scaffold could sustain before breakage was observed and the type of fractures on the scaffolds were identified.

The setup for the test shown in Fig. 3 was designed and 3D printed to hold the scaffold in place while the weights were placed above the metal rod in the middle

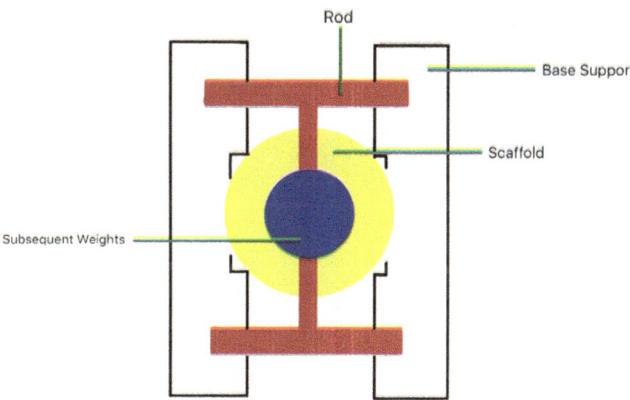

Fig. 3 Schematic diagram (top view) of strength test setup

of the scaffold. The pressure will be exerted on the center diameter of the scaffold with a metal rod held in place by supports to allow the scaffolds to break cleanly in half. This method of breakage is to determine the amount of stress each scaffold can sustain in the middle where all weakest points can be found. A high speed camera (Phantom Microlab 310) was used in order to capture the moment of fracture and breakage in the scaffold. The type of fractures in the scaffold were recorded and analysed further. This setup was conducted by a proved hypothesis in which the greater the amount of pressure and weight sustained by the scaffold results in greater strength of the scaffold.

For every test on each scaffold, two 10 kg weights were used as a starting weight since all three scaffolds could withstand the pressure exerted by the weights without any fractures occurring. Weights of smaller masses were then placed carefully above the scaffold and the maximum number of weights added would be dependent on the final breakage of the scaffold. Masses of 0.050 kg were used for more accurate and precise results for the last test.

3 Results and Discussion

3.1 Comparison of Surface Area

When the structure of the scaffold was designed and printed, dimensions were measured and recorded. The surface areas were then compared against one another to find out which had the highest exposed surface area and how the strength of the scaffold was affected by it. As shown in Fig. 2, the lattice structured scaffold had the highest amount of surface area exposed; however, the difference between the exposed surface area of the lattice scaffold as compared to other scaffold was no more than 1 cm^2. Hence, the amount of exposed surface area does not affect the results of the strength test as the difference of the surface area between each one of them is not significant (Fig. 4).

Fig. 4 Graph of average weight causing complete breakage

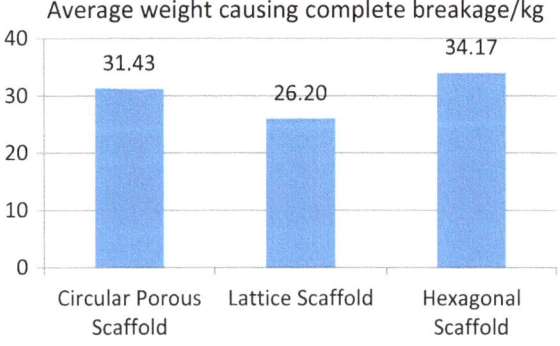

4 Strength of Scaffolds

4.1 Amount of Sustainable Pressure

A good scaffold is considered to be able to withstand a high amount of pressure, mimicking the characteristic of bones in which they are strong and flexible. When stress is applied, these scaffolds must be proven durable and effective in supporting the damaged bone of the patient. From the results shown in Fig. 5, it can be seen that the scaffold with hexagonal pores could sustain the highest amount of weight before breakage throughout all three tests on each type of scaffold. With reference to the lattice scaffold, a significant difference of 5.23 and 7.97 kg between the positive control setup with the circular porous and hexagonal scaffolds respectively was observed. This determines the effectiveness of the two designs in which they can withstand a higher exertion of stress as compared to the basic lattice structure of scaffolds commonly used for tissue engineering.

Scaffold	Weight causing complete breakage/ kg	Average weight causing complete breakage/kg	Points of fractures (where fractures first occurs)	Type of Fracture
Circular 1	31.25	31.43	Side weak points	Tensile stress
Circular 2	30.00		Side weak points	
Circular 3	33.05		Side weak points	
Lattice 1	26.50	26.20	Side weak points	
Lattice 2	25.00		Center weak points	
Lattice 3	27.10		Side weak points	
Hexagonal 1	35.00	34.17	Side weak points	
Hexagonal 2	33.75		Side weak points	
Hexagonal 3	35.75		Center weak points	

Fig. 5 Table of results taken from strength test

4.2 Analysis of Fracture

According to the test setup for finding out strength for scaffolds, tensile stress, in which equal and opposite forces were exerted on the scaffold, is identified to be applicable for all scaffolds. Over the point of weakness located in the middle, longitudinal cracks were formed due to the large amount of weight, causing a fracture and then breakage of the scaffolds [11].

According to Fig. 6, the hexagonal scaffold in Test 1 was broken into half, following the line of stress that was applied in the middle of the scaffold. The line of breakage showed the fractures of the weak points between the hexagonal pores, in which corners of the continuous shapes were split. These observations can also be seen in Test 3 where the thinner corners of connecting distance between the hexagonal pores fractured, causing the breakage. Test 2, however, showed different results in which the line of breakage was along the middle of thicker material between the hexagonal pores. Comparing the results obtained from these tests in Fig. 5, the amount of weight that the scaffold could sustain in Test 2 was lesser than the first and last.

When the fracture occurred along most of the weaker points of the scaffold, more weight could be sustained by the scaffold as compared to when fractures were seen on the structured sides. In Test 1 and 3, the pressure from the weights was exerted evenly causing the weaker points to fracture before the points with thicker material. The result in Test 3 was caused by the uneven spread of weight on the scaffold in which a higher amount of the total pressure was exerted on the thicker material. From these differing longitudinal cracks, it can be seen that the amount of pressure the scaffold can sustain would depend on the spread of pressure along the scaffold.

Based on Fig. 7, a first fracture occurred on the side of the lattice scaffold in Tests 1 and 3 when the starting weights were placed. As the sides of these scaffolds were one of the weak points along the horizontal line of breakage, the pressure exerted by the weights would be concentrated on the sides more than the center of the scaffold causing the first fracture on the side of the scaffold. On the other hand, the first fracture on Test 2 occurred in the center as a longitudinal crack along the thick material connecting the pores of this scaffold. This could be due to the application of

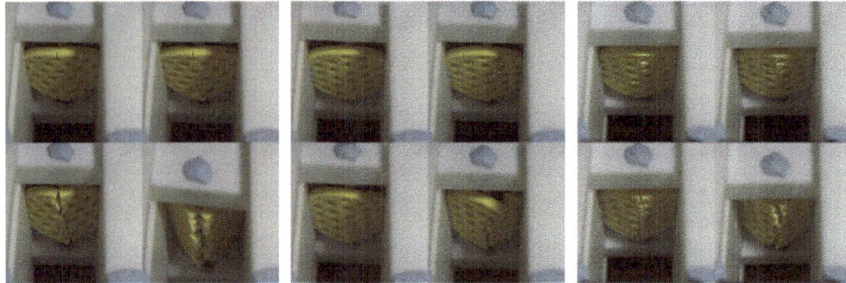

Fig. 6 Hexagonal scaffold Test 1, Test 2, Test 3

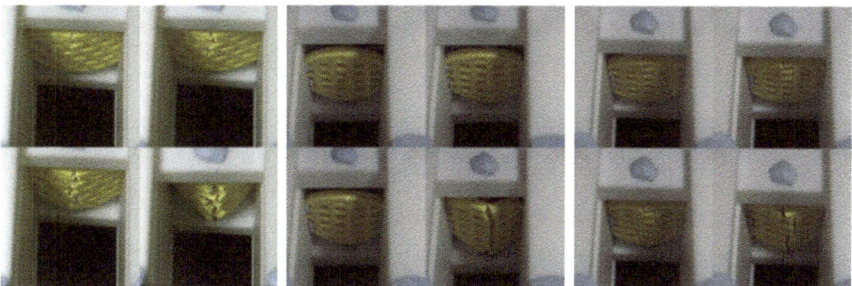

Fig. 7 Lattice scaffold Test 1, Test 2, Test 3

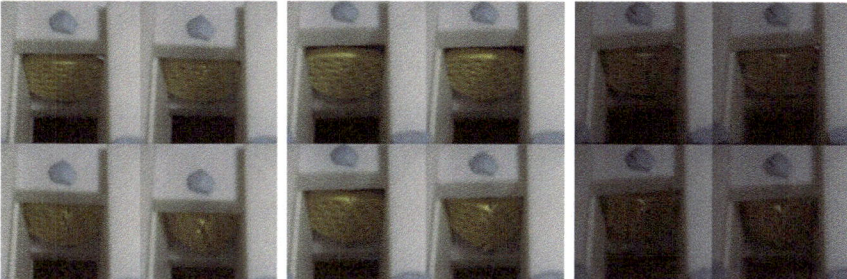

Fig. 8 Circular porous scaffold Test 1, Test 2, Test 3

the pressure on the scaffold or a limitation within the setup. It can thus be concluded that the first occurrence of fracture would usually affect the way in which the line of breakage would appear.

Throughout the 3 tests, with reference to Figs. 5 and 8, first fracture occurrences were at the side weak points. This was due to the even distribution of weight applied on the scaffolds from the geometrical properties of the circular pores. According to the shape of the pores, a circular design is able to spread the amount of pressure exerted evenly due to its lack of edges, which are the main reasons for the greater number of weak points in the other scaffolds. Since there were no specific weak points on the circular porous scaffold, the fractures occurred on random points of weaknesses found near the center of the scaffold. Hence, with a reduced number of weak points and equal amount of pressure applied on each point of the circular porous scaffold, uneven lines of breakages were formed on the scaffolds.

5 Conclusion

As discussed above, the spread of pressure along the scaffold is dependent on the amount of pressure exerted while the first occurrence of fracture affects the line of breakage across the entire scaffold. With reference to the amount of pressure a basic lattice scaffold can sustain, both the hexagonal and circular scaffolds have proven effective in having a great mechanical strength. However, although the circular porous scaffolds had fewer specific weak points, a scaffold design with hexagonal pores as able to spread the pressure evenly across its specific weak points in the design. Due to its high tensile strength from its geometrical properties, this allowed it to withstand the highest amount of weight than the rest of the scaffolds. The reason as to why the circular porous scaffold was unable to withstand a higher amount of pressure could be due to the random weak points across the scaffold, leading to uneven fractures and an easier breakage. Therefore, it is proven that a scaffold design with hexagonal pores is indeed most effective for tissue regeneration.

6 Limitations

One vital limitation of this project that cannot be overlooked is the interlayer adhesion of the 3 dimensional printed scaffolds. The layering of the infill of the three-dimensional printed items resulted in the seams between each layer and line, and the strength of adhesion between them. This caused points of weaknesses to be formed in the scaffold due to the inconsistent layering resulting in random errors in our strength test. This can be seen from Figs. 6, 7 and 8, where the points of weaknesses were typically found along the distance between the pores as they were thinner and had to resist a larger amount of weight and pressure. However, in Fig. 7, the lattice scaffold in test 2 did not show any signs of bending or lines of weaknesses on the thinner distance between the pores until it suddenly fractured into 2 clean halves along the thicker areas of the scaffold. This can be due to 3D printing defects such as weak interlayer adhesion which resulted in the bad bonding layer between the two halves. As such, the points of weaknesses in Fig. 7 lattice test 2 occurred at an unusual place where the longitudinal crack was on the thicker areas of the scaffolds instead of the thinner areas, unlike other scaffolds.

7 Future Work

Cell cultivation can be applied in future researches as these scaffolds are used in the medical field as supports for bone cell growth. Scaffolds are used to deliver growth factors to sites of fractures, speeding up recovery processes-the repair of the bones. Through cell cultivation, the efficiency of the scaffolds can be tested to find the best scaffold in terms of osteoinduction which is the capacity of attracting immature cells to the healing site and stimulating these cells to develop into bone-forming cells [12]. There are various techniques for cell cultivation, one of such is cell seeding whereby a concentrated cell suspension is passively placed on the scaffolds designed to attain a high yield, kinetic rate and spatial uniformity of cell attachment. This method can be used on the scaffolds designed in the future to take the rate of cell growth on the scaffold into consideration for the best scaffold design.

In addition, future scaffolds should be printed out to a 100% infill instead to carry out strength tests. In this research, scaffolds were printed to be 55% infill and it was unexpected that the interlayer adhesion between the fibers of the 3D-printed scaffolds would be such a serious issue that affects the accuracy of the data collected. Hence, by using a 100% infill, there will be no major defects in the scaffold such that would not affect the way in which the scaffold fractures and the points of weaknesses on the scaffold when under pressure.

Furthermore, scaffolds can also be printed using PLA plastics to make appropriate comparisons with the scaffolds printed in ABS plastics. As compared to ABS plastic, PLA plastics are stronger, more rigid and have a low melting point. From this research it can be seen that ABS plastic does not have a strong interlayer adhesion and through the benefits of PLA plastics, using PLA plastics instead may be able to reduce the 3D printing defects for data of higher accuracy to be obtained.

Acknowledgements Special thanks to Mr Teo Tee Wei from National Junior College for his patient guidance throughout the research process of this project. It is also greatly appreciated with gratitude towards National Junior College for the provision of the opportunity to engage in scientific research.

References

1. Kheirallah, M., & Almeshaly, H. (2016). Bone graft substitutes for bone defect regeneration. A collective review. *International Journal of Dentistry and Oral Science, 3*(5), 247–257.
2. Albert, A., Leenaijse, T., Druez, V., Delloye, C., & Cornu, O. (2006). Are bone autografts still necessary in 2006? A three-year retrospective study of bone grafting. *Acta Orthopuedica Belgica, 72*(6), 734.
3. O'brien, F. J. (2011). Biomaterials & scaffolds for tissue engineering. *Materials Today, 14*(3), 88–95.
4. Torkzadeh, R. A. (2010). *Accidents happen but who's going to pay the bills?: A consumers guide to the california personal injury and wrongful death system.* Cork: BookBaby.
5. Bhatnagar, R. S., & Li, S. (2004, September). Biomimetic scaffolds for tissue engineering. In: *26th Annual International Conference of the IEEE* (Vol. 2, pp. 5021–5023). *IEMBS'04.* Engineering in Medicine and Biology Society, 2004. IEEE.
6. Yang, S., Leong, K. F., Du, Z., & Chua, C. K. (2001). The design of scaffolds for use in tissue engineering. Part I. Traditional factors. *Tissue Engineering, 7*(6), 679–689.
7. 3D Printing Systems Australia. (2017). Retrieved January 31, 2017, from http://3dprintingsystems.com/products/filament/.
8. Chan, B. P., & Leong, K. W. (2008). Scaffolding in tissue engineering: General approaches and tissue-specific considerations. *European Spine Journal, 17*(4), 467–479.
9. Ju, J., Summers, J. D., Ziegert, J., & Fadel, G. (2010, January). Compliant hexagonal meso-structures having both high shear strength and high shear strain. In: *Proceedings of the ASME International Design Engineering Technical Conferences, Montreal, Quebec, Canada.*
10. Egan, P., Ferguson, S., & Shea, K. (2017). Design of hierarchical 3D printed scaffolds considering mechanical and biological factors for bone tissue engineering. *Journal of Mechanical Design, 139.* https://doi.org/10.1115/1.4036396.

11. Miranda, P., Pajares, A., Saiz, E., Tomsia, A. P., & Guiberteau, F. (2007). Fracture modes under uniaxial compression in hydroxyapatite scaffolds fabricated by robocasting. *Journal of Biomedical Materials Research, Part A, 83*(3), 646–655.
12. Tozzi, G., De Mori, A., Oliveira, A., & Roldo, M. (2016). Composite hydrogels for bone regeneration. *Materials, 9*(4), 267.

Vision-Based Navigation for Control of Micro Aerial Vehicles

Xavier WeiJie Leong and Henrik Hesse

Abstract This paper presents the use of an external vision-based positioning system for navigation and flight control of micro aerial vehicles. The motion capture system Optitrack is used to localize the vehicle which is a nano-quadcopter Crazyflie 2.0. An interface was created to pass positioning data to the Robot Operating Software (ROS). ROS acts as communication layer to bridge between Optitrack and available control nodes for Crazyflie 2.0 platforms.

Keywords Vision-based navigation · Robot operating system (ROS) · Aerial robotics · Micro aerial vehicles (MAVs)

1 Introduction

In the last decade, Micro Aerial Vehicles (MAVs) have been used extensively for research and, recently, also find applications in areas such as inspection surveillance [1–3]. To achieve autonomous operation for such applications, we require robust positioning and control performance of MAVs. To localize the MAV and estimate its state, common sensor fusion approaches typically rely on inertial sensors and GPS to obtain a position measurement. However, in GPS-denied environments the vehicle has no sense of its position [4, 5]. This lack of position information naturally hinders the control performance and is especially relevant in recent applications of MAVs for inspection of enclosed areas. For such industry applications, we typically require a reliable, inexpensive positioning system [2].

The choice of the MAV in this work is the open-sourced, open-hardware nano-quadcopter Crazyflie 2.0 as shown in Fig. 1. The Robot Operating Software (ROS) is used as the communication layer to demonstrate the flexible and modular code design for MAVs using ROS. To obtain external position measurements we use the motion capture system Optitrack.

X. W. Leong · H. Hesse (✉)
Aerospace Sciences Division, University of Glasgow, Singapore, Singapore
e-mail: henrik.hesse@glasgow.ac.uk

© Springer Nature Singapore Pte Ltd. 2019
H. Guo et al. (eds.), *IRC-SET 2018*,
https://doi.org/10.1007/978-981-32-9828-6_33

Fig. 1 The palm-sized
nanoquadcopter, Crazyflie
2.0, attached with reflective
markers

Fig. 1 The palm-sized nanoquadcopter, Crazyflie 2.0, attached with reflective markers

The main objective of this work is to implement a ROS framework for the inexpensive localization and control of MAVs using vision-based position data from Optitrack.

1.1 MAV Platform: Crazyflie 2.0

The research platform Crazyflie 2.0 is a nano-quadcopter which was developed by a team of software engineers at Bitcraze. It weighs 27 g and can be easily flown indoors with the dimensions of 92 mm × 92 mm × 29 mm.

The MAV comes equipped with a micro-processor which reads measurements from a 10-degree-of-freedom (10-DOF) inertial measurement unit (IMU). The IMU consists of a 9-DOF instrument with gyroscope, accelerometer and magnetometer and 1-DOF unit with a pressure sensor. The IMU therefore provides measurements of angular velocity, linear acceleration, orientation and lastly altitude. However, the Crazyflie 2.0 has no 3D position measurements and the pressure sensor tends to drift providing inaccurate altitude measurements due to external disturbances [6, 7].

1.2 External Positioning System: Optitrack

In the aerial robotics research domain, quadcopters are typically localized using the external positioning system VICON system [4, 8, 9]. Recent work [10], however, has demonstrated the use of the motion capturing system Optitrack as an inexpensive solution to provide 6 DOF accuracy. Optitrack is used to detect infrared light reflected from markers on the MAV. After calibration and so-called wanding [11], the system is able to stream data in real time to a 3D virtual reality graph.

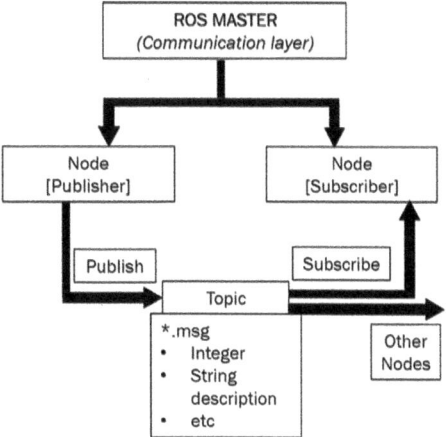

Fig. 2 ROS flowchart of nodes in publishing and subscribing

1.3 Robot Operating System: ROS

ROS is a communication interface extensively used in the robotics community to program robots. Its strength is the modular nature which allows code reuse and compatibility. It is operated on an Ubuntu operating platform and features open source libraries and tools for robotic applications. ROS nodes are bits of codes C++ or Python to perform specific tasks. In Fig. 2, ROS is illustrated as a communication layer between nodes, where nodes can publish and subscribe to topics. These topics are messages sent to ROS and which can be subscribed to by other nodes to make use of this data [11, 12].

2 Mathematical Modelling

2.1 Crazyflie Dynamics

The dynamics of the Crazyflie 2.0 quadcopter is modelled with respect to an inertial Earth frame as defined in Fig. 3. The body reference frame is also presented with the X, Y and Z axes and the Euler angles for roll, pitch and yaw denoted as φ, θ and ψ respectively [13]. Using these Euler angles, we can derive a transformation between the quadcopter body-attached frame B and the inertial frame E as,

$$H_B^E = \begin{bmatrix} C\theta\,C\psi & -C\theta\,S\psi & S\theta \\ S\varphi\,S\vartheta\,C\psi + C\varphi\,S\psi & C\varphi\,C\psi + S\varphi\,S\theta\,S\psi & -S\varphi\,C\theta \\ -C\varphi\,S\vartheta\,C\psi + S\varphi\,S\psi & S\varphi\,C\psi + C\varphi\,S\theta\,S\psi & C\varphi\,C\theta \end{bmatrix} \quad (1)$$

where C is the cosine and S the sine operator.

Fig. 3 Coordinate system
for Crazyflie 2.0

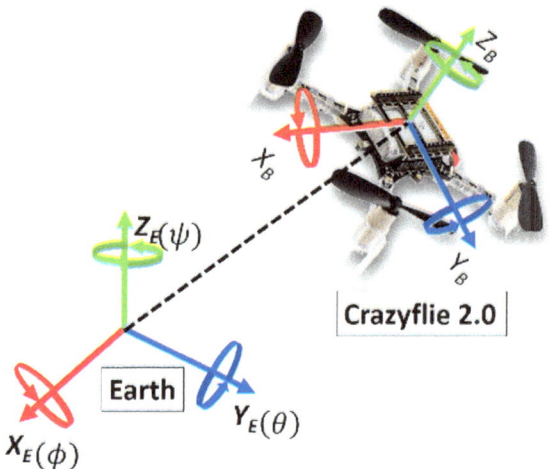

The dynamic equations are derived next under the following assumptions:

1. The quadcopter is a rigid body that cannot deform.
2. The quadcopter is symmetrical in terms of mass distribution and geometry.
3. The mass of the quadcopter is constant.

Based on Newton's 2nd Law we can find the translational equations of motion (EoM) in the body frame B as [14],

$$\sum \vec{F}_B = m\vec{a} = \left(\dot{\vec{v}}_B + \vec{\omega}_B \times \vec{v}_B \right) = H_E^B \vec{F}_g^E + \vec{F}_a^B$$

where m is mass, $\vec{a}$ is total acceleration vector, $\dot{\vec{v}}_B$ is linear acceleration vector, $\vec{\omega}_B = [p, q, r]$ is the angular velocity vector, $\vec{v}_B = [u, v, w]$ is linear velocity vector, $\vec{F}_g^E$ is the gravity vector of the Crazyflie 2.0 in the Earth from E and $\vec{F}_a^B = \left[F_x, F_y, F_z \right]$ is the vector of total motor forces in the body frame B resulting from all four propellers. The translational EoM can then be expressed in vector form as,

$$\begin{bmatrix} \dot{u} \\ \dot{v} \\ \dot{w} \end{bmatrix} = \begin{bmatrix} 0 \\ 0 \\ F_z/m \end{bmatrix} - H_E^B \begin{bmatrix} 0 \\ 0 \\ g \end{bmatrix} - \begin{bmatrix} p \\ q \\ r \end{bmatrix} \times \begin{bmatrix} u \\ v \\ w \end{bmatrix}$$

Next, the rotational dynamics are derived using the theorem of angular momentum,

$$\sum \vec{M}_B = \dot{\vec{h}} = \dot{\vec{h}}_B + \vec{\omega}_B \times \vec{h}_B$$

where $\vec{h}$ is the total angular momentum around the center of gravity, $\vec{h}_B$ the angular momentum with respect to the rotating body frame B and $\vec{M}_B$ the vector of resultant torques from the motors. With $\vec{h}_B = \mathcal{J}\vec{\omega}_B$, we can find the angular EoM as,

$$\sum \vec{M}_B = \mathcal{J}\dot{\vec{\omega}}_B + \vec{\omega}_B \times \mathcal{J}\vec{\omega}_B$$

with the inertia matrix $\mathcal{J}$ of the Crazyflie 2.0 assumed to be diagonal due to symmetry, such that

$$\mathcal{J} = \begin{bmatrix} I_{xx} & 0 & 0 \\ 0 & I_{yy} & 0 \\ 0 & 0 & I_{zz} \end{bmatrix}$$

where $I_{xx} = 1.395 \times 10^{-5}$ kg m^2, $I_{yy} = 1.436 \times 10^{-5}$ kg m^2 and $I_{zz} = 2.173 \times 10^{-5}$ kg m^2 [8].

The relationship between the translational and angular EoM is given by the transformation H_E^B defined in (1). The Euler angles are propagated through the relationship between $\vec{\omega}_B$ and Euler rates,

$$\begin{bmatrix} p \\ q \\ r \end{bmatrix} = \begin{bmatrix} 1 & 0 & -\sin\theta \\ 0 & \cos\theta & \sin\phi\cos\theta \\ 0 & -\sin\phi & \cos\phi\cos\theta \end{bmatrix} \begin{bmatrix} \dot{\phi} \\ \dot{\theta} \\ \dot{\psi} \end{bmatrix}$$

which can be re-arranged to compute the Euler rates,

$$\begin{bmatrix} \dot{\phi} \\ \dot{\theta} \\ \dot{\psi} \end{bmatrix} = \begin{bmatrix} 1 & \sin\phi\tan\theta & \cos\phi\tan\theta \\ 0 & \cos\phi & -\sin\phi \\ 0 & \sin\phi/\cos\theta & \cos\phi/\cos\theta \end{bmatrix} \begin{bmatrix} p \\ q \\ r \end{bmatrix} \quad \text{for } \theta \neq \frac{\pi}{2}$$

where the angular velocities $\vec{\omega}_B = [p, q, r]$ are now the inputs and can come from IMU measurements of the Crazyflie 2.0. To avoid gimbal lock, pitching cannot reach 90° ($\theta \neq \pi/2$). Quaternions will be introduced in Sect. 3.2 to resolve the gimbal lock.

3 Motion Tracking with the Optitrack System

Figure 3 shows the Optitrack system with six Flex 13 cameras which can detect reflected infrared light from reflective markers on the Crazyflie 2.0. With the six measurements the Optitrack System can then compute the 3D coordinates of the quadcopter with high precision and frequency. The software *Motive* is used to process the 3D position data as the virtual graph and must be calibrated through a process called *wanding* [11].

The red-dotted area in Fig. 4 is measured 3 m × 2 m and is the effective area the system can operate. The Optitrack system is validated in Motive by placing reflective balls around the borders of the effective area and calculating the distance to match

Fig. 4 Six Flex 13 cameras
detecting the reflective
markers on the Crazyflie 2.0.
The red area is the effective
field of vision

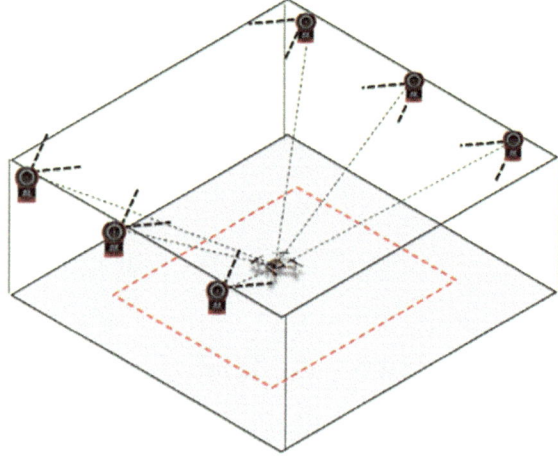

Fig. 5 Optitrack
North-Up-East (left) versus
and ROS East-North-Up
(right) coordinate frame
convention

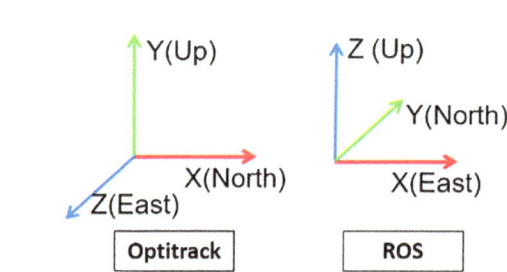

the above dimensions. The axis coordinate system for the Optitrack is right-handed
North-Up-East (NUE) as show in Fig. 5 (left).

The six Flex 13 Cameras are connected to hub (Optihub 2) which captures the
data and sends it to Motive. Motive then computes the 3D virtual graph in real time
from the measurements of the reflective markers. This process is illustrated in the
top part of the schematic in Fig. 6.

As shown in the bottom part of Fig. 6, the data is then streamed to a router
which broadcasts the position data over to ROS. Motive is Virtual Reality Peripheral
Network (VRPN) compatible which allows the data streaming to produce a report
identifiable as a movable tracker in ROS with position and orientation. However, since
ROS standards follow a different axis coordinate system, East-North-Up (ENU) as
illustrated in Fig. 5 (right), it is necessary to convert the position and orientation data.
The VRPN node therefore has additional code to transform between the Optitrack
and ROS frame conventions as explained next [11, 15].

Fig. 6 Flow of Optitrack to motive and streaming through Wi-Fi to ROS

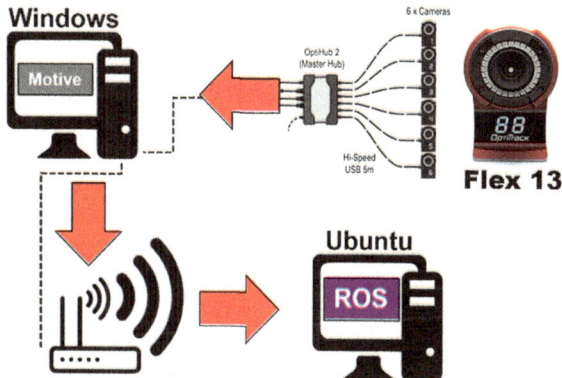

Fig. 7 Axis conversion from NUE to ENU defined in Fig. 5

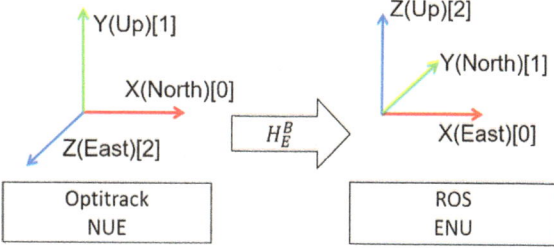

3.1 Coordinate Axis Conversion for Position

Figure 7 illustrates the transformation required to rotate between Optitrack and ROS frames [16]. This is done by applying a 90° rotation on the X axis. Using the transformation matrix in (1) and applying a roll rotation of $\varphi = \pi/2$, we can find the required transformation to be

$$H_{Opti}^{ROS} = \begin{bmatrix} 1 & 0 & 0 \\ 0 & 0 & -1 \\ 0 & 1 & 0 \end{bmatrix}$$

The transformation effectively means swapping the Y and Z coordinates and applying a negative sign to the Z coordinate.

3.2 Coordinate Axis Conversion for Orientation

To transform the orientation of the Crazyflie 2.0 from Optitrack to ROS frame conventions, we will use quaternions. Quaternions are the generalization of complex numbers to 3D and commonly used in robotics to describe rotations [13]. Following

Table 1 Example of quaternion transformation between Optitrack and ROS conventions for $\alpha = \pi/2, \vec{n}^{Opti} = [0, 1, 2]$

Orientation	w	x	y	z
Optitrack	-1	0	1	2
ROS	-1	0	-2	1

the Euler Theorem of Rotation, quaternions can be introduced as the magnitude of rotation (the angle α) and a vector as [13],

$$q^{Opti} = \left[\cos\left(\frac{\alpha}{2}\right), n_1 \sin\left(\frac{\alpha}{2}\right), n_2 \sin\left(\frac{\alpha}{2}\right), n_3 \sin\left(\frac{\alpha}{2}\right)\right] = [w, [x, y, z]]$$

where the vector

$$\vec{n}^{Opti} = [n_1, n_2, n_3] = \sin\left(\frac{\alpha}{2}\right)[x, y, z]$$

defines axis of rotation. To convert the quaternion representation in Optitrack, q^{Opti}, to ROS standards, we need to first isolate the angle part α by taking inverse cosine of w in q^{Opti}. With the angle of rotation, α, we can then compute the vector $\vec{n}^{Opti}$. Finally, to convert from Optitrack to ROS standards we simply need to apply the transformation H_{Opti}^{ROS} to $\vec{n}^{Opti}$ such that

$$\vec{n}^{ROS} = H_{Opti}^{ROS} \cdot \vec{n}^{Opti} = [N_1, N_2, N_3]$$

with $\vec{n}^{ROS}$ computed in ROS standards, we can then form the new quaternion q^{ROS} as

$$q^{ROS} = \left[\cos\left(\frac{\alpha}{2}\right), N_1 \sin\left(\frac{\alpha}{2}\right), N_2 \sin\left(\frac{\alpha}{2}\right), N_3 \sin\left(\frac{\alpha}{2}\right)\right]$$

which describes the orientation of the Crazyflie 2.0 with respect to the Earth frame E in ROS-based (ENU) convention.

Similar to the position transformation, the orientation transformation effectively swaps the Y and Z coordinates and negates the Z coordinate as illustrated in Table 1.

4 Software Integration in ROS for Path Control of the Crazyflie 2.0

Next, we will demonstrate the integration of the vision-based navigation of the Crazyflie 2.0 in ROS for a simple waypoint tracking control exercise. The software structure for the controller in ROS is illustrated in Fig. 8 where the red arrows

Fig. 8 Software structure
with ROS for communication

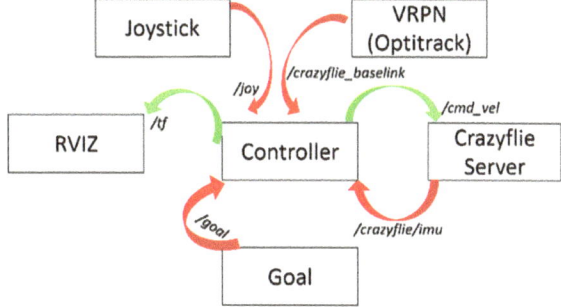

represent the topics that are subscribed by the controller and the green-colored arrows
are published by the controller. Note that the software integration for the control of
Crazyflie 2.0 has been adapted from [15].

The VRPN node accepts the position and orientation data from Optitrack and
transforms the data to ROS standards. The published data */crazyflie_baselink* has the
same structure as */tf* which is subscribed to by RVIZ [11, 15]. RVIZ is a ROS tool
that displays the 3D data on a ROS graphical user interface. The input */joy* allows
the user to command take-off, landing or emergency landing using a joystick. The
input */goal* is the desired path and coordinate point which can also be defined by
the user and is in ROS coordinates. With published IMU data */crazyflie/imu* from
the Crazyflie 2.0 and the waypoints */goal*, the controller node is able to publish the
desired velocity */cmd_vel* to the Crazyflie 2.0. These commanded velocity inputs are
subsequently processed by the onboard controller as described next.

Figure 9 shows the generic flow of the software integration from a control perspective, i.e. the position and orientation data is extracted from Optitrack which finally
lead to control inputs on the Crazyflie 2.0 itself. The boxes in green represent ROS
computing, yellow represents Motive processing and orange are operations executed
directly onboard the Crazyflie 2.0.

Based on the desired waypoints and current position data the offboard controller
in ROS produces the velocity control inputs. This input is then transmitted to the
Crazyflie 2.0 using the CrazyRadio and processed by the onboard controller [17].

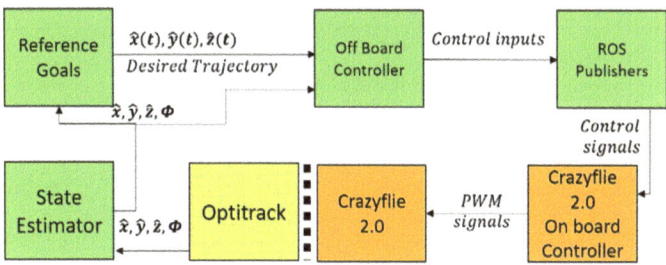

Fig. 9 Generic flow of the Crazyflie 2.0 controller

Finally, the onboard controller computes the motor inputs in the form of Pulse-Width Modulation (PWM) signals for the four motors as [9],

$$PWM_{motor_1} = Thrust - \phi/2 - \theta/2 - \psi$$

$$PWM_{motor_2} = Thrust + \phi/2 - \theta/2 + \psi$$

$$PWM_{motor_3} = Thrust - \phi/2 + \theta/2 + \psi$$

$$PWM_{motor_4} = Thrust + \phi/2 + \theta/2 - \psi$$

5 Experiments and Test Results

The following experimental results demonstrate the software integration of the Opti-track system with ROS, including the transformation of position and orientation data for the first test. The following tests show hover and waypoint tracking to demonstrate the ROS integration for autonomous flight of the Crazyflie 2.0.

5.1 Position and Orientation Axis Conversion Experiment

The first experiment is to test the conversion between Optitrack and ROS axis systems. Table 2 shows the results for a position and orientation experiment. For the position test, a location of the Crazyflie 2.0 was measured by hand. Table 2 compares the measured data to the Optitrack results and the converted ROS data.

Table 2 Experimental results to validate the position and orientation transformation from Optitrack to ROS

Position		X (m)	Y (m)	Z (m)
Measured		−0.5	0.5	−0.5
Optitrack		−0.502	0.497	−0.503
ROS		−0.502	0.503	0.497
Orientation	w	x	y	z
Optitrack	0.754	−0.0227	−0.653	0.0593
MATLAB	0.756	−0.0290	−0.0598	−0.6518
ROS	0.754	−0.0227	−0.0593	−0.653

Fig. 10 3D trajectory (blue) for hover experiment to hold 0.8 m height

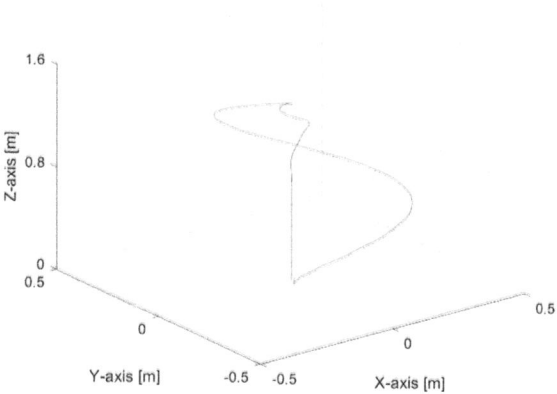

Second part of Table 2 demonstrates the transformation of the quaternion orientation extracted from Optitrack to ROS. The MATLAB test uses the corresponding Euler angles, which can also be extracted from Optitrack, converts the Euler angles to a Direct Cosine Matrix and lastly converts these to quaternions. The MATLAB exercise demonstrates the validity of the quaternion transformation proposed in Sect. 3.2.

5.2 Hover Experiment

For the hover test, the Crazyflie 2.0 is commanded to fly to a single waypoint at 0.8 m height, i.e. $[X, Y, Z] = [0, 0, 0.8$ m]. This simple experiment demonstrates the software integration and the correct interfacing between the different ROS nodes introduced in Fig. 8. The hover results in Figs. 10 and 11 further show that the Optitrack position and orientation data can be used reliably for the offboard controller to command the Crazyflie 2.0 to fly to the defined waypoint.

Figure 11 shows the corresponding time history of the Crazyflie 2.0 during hover. The measured Z data shows that the quadcopter accelerates first up to 1.2 m before leveling at 0.8 m. The X and Y position data show that the controller is able to command the Crazyflie 2.0 to the desired waypoint at $[X, Y, Z] = [0, 0, 0.8$ m]. The initial deviation from the reference is to overcome the ground effect.

5.3 Waypoint Trajectory

The final experiment demonstrates the waypoint tracking capability of the implemented vision-based control framework for the Crazyflie 2.0. Figures 12, 13 and 14 show the corresponding experimental results for waypoint tracking. Note that the

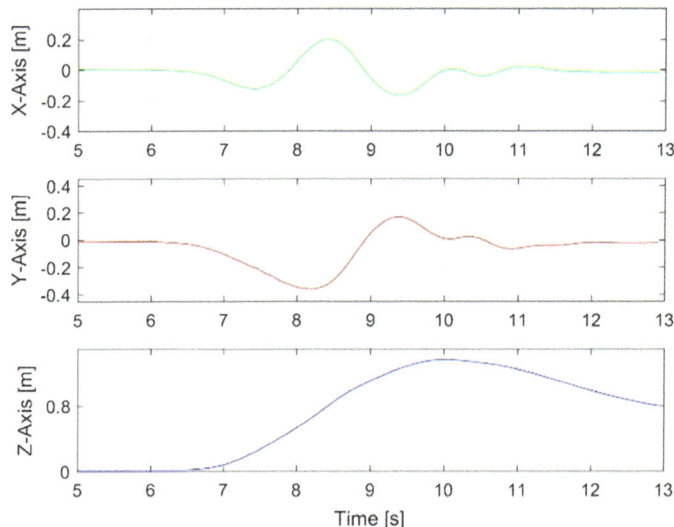

Fig. 11 Time history of 3D position components for hover experiment

Fig. 12 3D trajectory (blue) for waypoint tracking experiment with circles identifying the waypoints

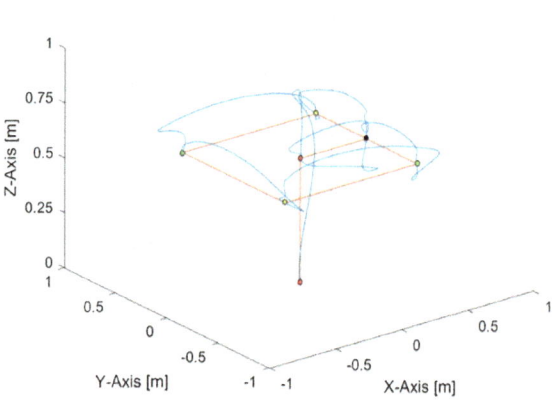

control task here is *not* to follow the trajectory indicated by the orange path in Fig. 12 but to fly to a waypoint and hold its position for 3 s at this waypoint before moving to the next waypoint. The colored circles in Fig. 12 are the waypoints defined in the code by the user.

Figure 13 shows the corresponding time history of the 3D position components for the waypoint tracking experiment. The dotted curve in Fig. 13 represents the reference position of the instantaneous waypoint that is followed at specific time

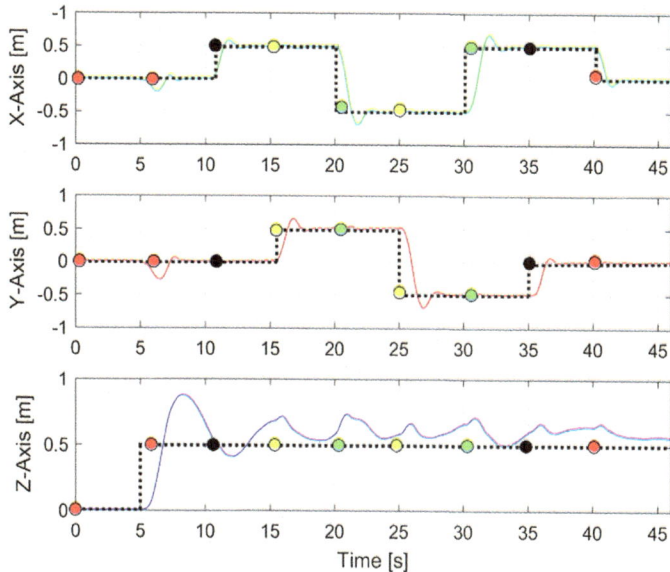

Fig. 13 Time history of 3D position components for waypoint tracking experiment. Current waypoint in circles correspond to definition in Fig. 12

Fig. 14 2D projection of the trajectory in Fig. 12

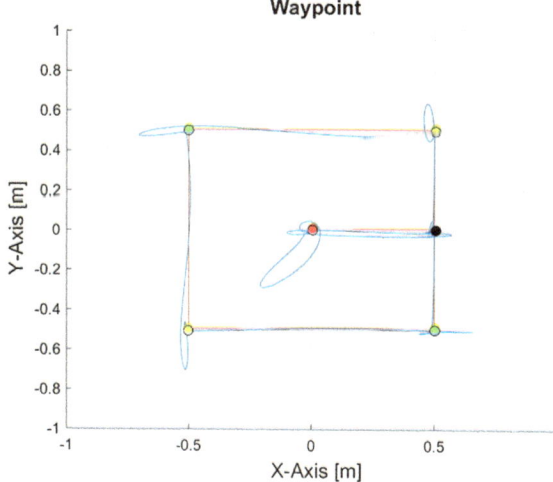

instance. It is clear that the Crazyflie 2.0 is able to reach each waypoint, hover around the waypoint for 3 s and then proceed to the next one.

The oscillations in the Z direction in Fig. 13 arise as the Crazyflie 2.0 is *not* commanded to follow a specific trajectory but to reach the next waypoint. To accelerate in the X and Y directions, the onboard controller tends to also apply excess thrust in the Z direction.

To demonstrate the waypoint tracking capabilities, Fig. 14 shows the projection of the Crazyflie 2.0 trajectory on the $X - Y$ plane only. Figure 14 highlights that the controller can successfully command the Crazyflie 2.0 in dynamic situations with only minor overshoots. The waypoint tracking can be further extended to trajectory control by discretizing the trajectory with multiple waypoints and considering the vehicle dynamics in the discretization.

6   Conclusion

The objective of this work was to use the motion capture system Optitrack as an external localization system for autonomous flight control of the nano-quadcopter Crazyflie 2.0. This paper provides a comprehensive documentation on how to integrate the Optitrack system with ROS and it especially addresses the transformation between coordinate conventions in ROS and Optitrack. This enables the use of the Optitrack localization system, as an alternative to the standard VICON positioning system, for educational purposes.

The integration of Optitrack with the communication platform ROS further makes for a versatile platform which can rely on the modular and open-source nature of ROS for a wide range of robot applications. In this work, we have demonstrated this capability by using open-source controller nodes to enable waypoint tracking of the Crazyflie 2.0 nano-copter. This simple integration further highlights the benefits of ROS in combination with Optitrack as an education platform in aerial robotics.

Acknowledgements The authors would like to thank Dr. Spot Srigrarom from the University of Glasgow Singapore to provide access to the Optitrack UAV facility. The authors would also like to acknowledge the technical advice from Paul Beuchat at ETH Zurich.

References

1. Wong, D. (2018, February 9). Plan for drones to deliver parcels takes flight. *The Straits Times*.
2. Warwick, G. (2018, April 04). Aircraft inspection drones entering service with airline MROs. *MRO Network*.
3. Kaur, K. (2018). The straits times. *Five projects to kick off unmanned aircraft system trials at Singapore's first drone estate*.
4. Bjorn, J., Kjaegaard, M., & Larsen, J. A. (2007). *Autonomous hover for a quad rotor helicopter*. Master's Thesis, Aalborg University, Aalborg, Denmark.

5. Wierema, M. (2008). *Design, implementation and flight test of indoor navigation and control system for a quadrotor UAV*. TU Delft, Master of Science Thesis, Delft, Netherlands.
6. Foerster, J. (2015). *System identification of the Crazyflie 2.0 nano quadrocopter*. Bachelor Thesis, ETH Zurich, Zurich, Switzerland.
7. Edwards, L. (2011). *Open source robotics and process control cookbook designing and building robust dependable real time system*. New York, USA: Elsevier.
8. Luis, C., & Le Ny, J. (2016). *Design of a trajectory tracking controller for a nanoquadcopter*. Technical Report, Polytechnique Montreal. Montreal, Canada.
9. Subramanian, G. P. (2015). *Nonlinear control strategies for quadrotors and CubeSats*. Master Thesis, University of Illinois, USA.
10. Hansen, C., Gibas, D., Honeine, J., Rezzoug, N., Gorce, P., & Isableu, B. (2014). An inexpensive solution for motion analysis. *Journal of Sports Engineering and Technology*.
11. Quigley, M., Gerkey, B., Conley, K., Faust, J., Foote, T., Leibs, J., Berger, E., Wheeler, R., & Ng, A. (2009). ROS: an open-source robot operating system. *ICRA Workshop on Open Source Software, Kobe, Japan*.
12. Mišeikis, J. (2017). INF3480—Introduction to robot operating system. *Lecture Slides*, University of Oslo, Norway.
13. Hanson, A. (2006). *Visualizing quaternions* (pp. 43–56). San Francisco, CA, USA: Morgan Kaufmann.
14. Mellinger, D., Lindsey, Q., Shomin, M., & Kumar, V. (2011). Design, modeling, estimation and control for aerial grasping and manipulation. In *IEEE International Conference on Intelligent Robots and Systems, San Francisco, CA, USA*.
15. Hoenig, W., Milanes, C., Scaria, L., Phan, T., Bolas, M., & Ayanian, N. (2015). Mixed reality for robotics. In *IEEE International Conference on Intelligent Robots and Systems, Hamburg, Germany*.
16. Hemingway, E.G. (2017). Perspectives on Euler angle singularities, gimbal lock, and the orthogonality of applied forces and applied moments. Article, University of California. Berkeley, CA, USA.
17. Preiss, J. A., Hoenig, W., Sukhatme, G. S., & Ayanian, N. (2017). Crazyswarm: a large nano-quadcopter swarm. In *IEEE International Conference on Robotics and Automation (ICRA), Singapore*.

Printed by Printforce, the Netherlands